PHYLOGENETICS AND ECOLOGY

LINNEAN SOCIETY SYMPOSIUM SERIES | NUMBER 17

PHYLOGENETICS AND ECOLOGY

edited by

Paul Eggleton and Richard I. Vane-Wright

An edited volume arising from papers presented at a joint symposium between the Natural History Museum and the Linnean Society of London.

Published for the Linnean Society of London by Academic Press

ACADEMIC PRESS
Harcourt Brace & Company, Publishers
London   San Diego   New York   Boston
Sydney   Tokyo   Toronto

ACADEMIC PRESS LIMITED
24/28 Oval Road
London NW1 7DX

United States edition published by
ACADEMIC PRESS INC.
San Diego, CA 92101

A catalogue record is available from the
British Library

ISBN 0-12-232990-2

This book is printed on acid-free paper

Typeset by Fakenham Photosetting Limited, Fakenham, Norfolk
Printed in Great Britain by The Bath Press, Avon

Contents

Contributors

E.N. Arnold, Zoology Department, The Natural History Museum, London SW7 5BD, U.K.

L. Belbin, CSIRO, Division of Wildlife and Ecology, PO Box 84, Lyneham, ACT 2602, AUSTRALIA.

Daniel R. Brooks, Department of Zoology, University of Toronto, Toronto, Ontario M5S 1A1, CANADA.

R.P. Brown, Department of Zoology, University of Aberdeen, Aberdeen AB9 2TN, Scotland, U.K.

James M. Carpenter, Department of Entomology, American Museum of Natural History, Central Park West at 79th Street, New York, NY 10024, U.S.A.

Jonathan A. Coddington, Department of Entomology, NHB 105, National Museum of Natural History, Smithsonian Institution, Washington, DC 20560, U.S.A.

Michael D. Crisp, Division of Botany and Zoology, Australian National University, Canberra, ACT 0200, AUSTRALIA.

M. Day, Department of Zoology, University of Aberdeen, Aberdeen AB9 2TN, Scotland, U.K.

Paul Eggleton, The Biodiversity Division, Entomology Department, The Natural History Museum, London SW7 5BD, U.K.

Paul Emerson, Department of Life Sciences, University Park, The University of Nottingham, Nottingham NG7 2RD, U.K.

Daniel P. Faith, CSIRO, Division of Wildlife and Ecology, P.O. Box 84, Lyneham, ACT 2602, AUSTRALIA.

Adrian E. Friday, Department of Zoology, University of Cambridge, Downing Street, Cambridge CB2 3EJ, U.K.

Francis Gilbert, Department of Life Sciences, University Park, The University of Nottingham, Nottingham NG7 2RD, U.K.

John L. Gittleman, Department of Zoology, University of Tennessee, Knoxville, Tennessee 37996–0810, U.S.A

Paul H. Harvey, Department of Zoology, University of Oxford, South Parks Road, Oxford OX1 3PS, U.K.

Hang-Kwang Luh, Department of Zoology, University of Tennessee, Knoxville, Tennessee 37996–0810, U.S.A.

D.P. McGregor, Department of Molecular and Cell Biology, University of Aberdeen, Aberdeen AB9 2TN, Scotland, U.K.

Deborah A. McLennan, Department of Zoology, University of Toronto, Toronto, Ontario M5S 1A1, CANADA.

A. Malhotra, Department of Zoology, University of Aberdeen, Aberdeen AB9 2TN, Scotland, U.K.

Arne Ø. Mooers, Department of Zoology, University of Oxford, South Parks Road, Oxford OX1 3PS, U.K.

Sean Nee, Agriculture and Food Research Council Unit of Ecology and Behaviour, Department of Zoology, University of Oxford, South Parks Road, Oxford OX1 3PS, U.K.

Sören Nylin, Department of Zoology, Stockholm University, S-106 91 Stockholm, SWEDEN.

Mark D. Pagel, Department of Zoology, University of Oxford, South Parks Road, Oxford OX1 3PS, U.K.

Graham Rotheray, The Royal Museum of Scotland, Chambers Street, Edinburgh EH1 1JF, Scotland, U.K.

Birgitta Sillén-Tullberg, Department of Zoology, Stockholm University, S-106 91 Stockholm, SWEDEN.

Hans Temrin, Department of Zoology, Stockholm University, S-106 91 Stockholm, SWEDEN.

R.S. Thorpe, School of Biological Sciences, University College of North Wales, Bangor, Gwynedd LL57 2UW, Wales, U.K.

R.I. Vane-Wright, The Biodiversity Division, Entomology Department, The Natural History Museum, London SW7 5BD, U.K.

Nina Wedell, Department of Zoology, Stockholm University, S-106 91 Stockholm, SWEDEN.

John W. Wenzel, Ohio State University, Department of Entomology, 103 Botany and Zoology Building, 1735 Neil Avenue, Columbus, OH 43210–1220, U.S.A.

W. Wüster, Department of Zoology, University of Aberdeen, Aberdeen AB9 2TN, Scotland, U.K.

Rehana Zafar, Department of Life Sciences, University Park, The University of Nottingham, Nottingham NG7 2RD, U.K.

Introduction

This book is concerned with the impact of modern phylogenetic methods on ecology, especially those branches of ecology that use comparative methods. Its scope is broad, but all the chapters emerge from the re-invigorated relationship between systematics and ecology. Advances in phylogenetic research have allowed explicit phylogenetic hypotheses to be constructed for a range of different groups of organisms. Alongside this, ecologists are more aware than ever that traits of organisms are influenced by the interaction of the past and present.

These new approaches fit neatly under the old umbrella of 'comparative biology' as they all deal with evolutionarily meaningful comparisons between traits and taxa. There are two general and opposed approaches, however, within the field: 'systematic' (emphasizing homology) and 'ecological' (emphasizing convergence). This is strikingly obvious within this book, which has strongly argued cases made for and against both. In choosing contributors to this book we have attempted to balance as far as possible these two approaches (although as systematists ourselves we must declare an interest). Unfortunately, there is still a way to go before the two convergent methodologies find extensive common ground. Perhaps one role of this book, at least, will be to clarify the areas of disagreement.

This book covers a wide range of topics. Chapter 1 introduces the whole area of historical ecology and discusses its wide potential remit. Chapters 2, 3 and 4 argue the theoretical cases for the homology and convergence approaches. Chapters 5 and 6 discuss the application of the two approaches, while chapter 7 proposes a somewhat different (although essentially convergence) multivariate approach. To this point the chapters have all dealt with taxa at the species level and above. In contrast, chapter 8 explores intraspecific microevolutionary studies. Chapter 9 deals with tree construction and the way that adaptation itself might be employed to help reconstruct phylogeny.

Chapters 10 and 11 use an existing phylogenetic hypothesis (the DNA hybridization phylogenetic tree of Sibley & Ahlquist) to explore hypotheses of ecological change over evolutionary time. These approaches are of special interest as they also explore the validity of the used phylogenetic tree itself, which has been controversial. Chapters 12 to 15 move on to explore particular cases where comparative approaches have been employed, and chapter 12, especially, emphasizes the complexity and difficulties of the whole subject area. In the final chapter the 'givens' of any comparative study, phylogenetic trees, are discussed – how they are obtained and how much faith we should have in any given tree. Readers who are unfamiliar with the aims and methods of phylogenetics might find it helpful to read this chapter first.

Although it is difficult to bring together the strands of such a disparate subject area, we still feel that some overall conclusions have emerged within the book. The first is that a bridge needs to be found between those evolutionary biologists who employ a logic-based (broadly Popperian) methods (for example a systematic approach to homology, essentially discontinuous in nature), and those who employ probability-based statistical models (for example an evolutionary ecological approach using convergence, essentially continuous in nature). Maybe no such bridge exists, but up to now it has seemed as if neither side has been talking in the same language. As Rieppel (1988)[1] makes clear, these two concerns represent "different 'ways of seeing' this world around us, each one with its merits and faults, each incomplete in itself, but complementary to each other". If this is true, then the bridge should be concerned less with a unifying theory, and more with developing a critical understanding of the strengths, weaknesses and complementary relations of these two world views.

The second general conclusion is that we should all be careful about the kinds of trees that are used in comparative studies. We believe that conventional classifications, consensus trees, or trees with very high levels of homoplasy should be avoided. Workers should state what degree of confidence they have in the trees they are using. Indeed, until a reasonable supply of phylogenetic trees is available the general scope and power of comparative approaches will be limited. The way ahead for tree construction itself is, however, unclear. Workers have not reached firm conclusions neither about the relative values of character types (e.g. molecular *versus* morphological) nor about tree-building methods (e.g. parsimony *versus* strong evolutionary models). Comparative biology may run aground if it cannot reach a consensus on how a tree is constructed and interpreted.

Third, systematists might attempt to analyse ecologically important groups. There is a paradox here, however, in that groups with trees should become of interest to comparative biologists, simply because such trees have become available. But gone are the days when a systematic study was undertaken due on the whim of a systematist – systematic research needs to be as directed as any other research. It is increasingly obvious that 'comparative ecologists' and systematists need to sit down together and devise a coherent strategy for the rejuvenation of comparative biology. The present volume may represent a rather argumentative start.

P. Eggleton
R.I. Vane-Wright

---

[1] RIEPPEL, O.C., 1988. *Fundamentals of Comparative Biology.* Basel: Birkhäuser.

Acknowledgements

This book arose from a joint symposium between the Natural History Museum and the Linnean Society of London. Many people at both institutions helped in various ways, especially John Marsden, Macretia Baird, Chris Humphries, Neil Chalmers, Peter Forey, Robert Belshaw, Campbell Smith, Sasha Barrow, David Shuker, Sam Bhattacharyya and Derek Jones.

We also wish to thank the referees: Nick Arnold, David Baum, Jim Carpenter, Jonathan Coddington, Henry Disney, Ove Eriksson, Kevin Gaston, Francis Gilbert, Paul Harvey, Chris Humphries, Ian Kitching, Chris Lyal, David Mindell, E. Oakes, Robert O'Hara, Mark Pagel, Joseph Slowinski, Dave Williams and John Wenzel who took much time and trouble reviewing the manuscripts. Academic Press, Andy Richford, and our long suffering authors have been very patient while we slowly gathered the manuscripts together. To all of these friends and colleagues we offer our most grateful thanks.

1

Historical ecology as a research programme: scope, limitations and the future

DANIEL R. BROOKS & DEBORAH A. McLENNAN

CONTENTS

Keywords: Historical ecology – phylogeny – speciation – adaptation – adaptive radiation – community evolution – biodiversity.

Abstract

Historical ecology is a research programme that allows communication between ecology and systematics to be re-established. By integrating ecological, behavioural and phylogenetic information to produce a more robust picture of evolution, historical ecology encompasses macroevolutionary studies of two general evolutionary processes, speciation and adaptation. Historical ecologists ask four basic questions: how did a given species arise; how did a given species acquire its repertoire of behavioural/ecological characters; how did a set of co-occurring species come to be associated; and how did the traits characterizing those inter-actions come to be? Answering these fundamental questions provides the database for a more encompassing picture of evolution, and a wider range of options for policy decisions concerning the preservation of biological diversity. For historical ecology to flourish, there must be active collaboration based on mutual respect among systematists, ecologists, ethologists and other evolutionary biologists. Ecolo-gists must support the training and hiring of more phylogeneticists, provide more support for museum collections and museum systematists, and become better versed about phylogenetic methodology. Systematists must provide more databases, encourage students to work on groups that are classically of interest to ecologists and ethologists, and develop better ways to explain their ideas.

INTRODUCTION

The Darwinian revolution was founded on the concept that biological diversity evolved through a combination of genealogical and environmental processes. Although in theory the majority of biologists still adhere to this proposition, in practice phylogenetic and ecological studies are often conducted quite independently. This independence was at first a necessary component of a rapidly expanding and increasingly complex discipline. Unfortunately, ecology and systematics followed very different evolutionary trajectories during this severance of information flow. Ecology experienced a theoretical explosion during the 1950s–1970s, followed by the painstak-ing collection of empirical data and re-evaluation of general models that has absorbed the last 20 years. During the same time period, systematists passed through two theoretical revolutions, spending the last two decades developing increasingly sophis-ticated methods of phylogenetic reconstruction. The end result of this separation has been the fragmentation of evolutionary biology into 'macroevolutionary' and 'microevolutionary' components, notwithstanding hopes for a newer, extended, or unified evolutionary synthesis (Wicken, 1987; Endler & McLellan, 1988; Brooks *et al.*, 1989).

Given that both disciplines appear to be surviving in their allopatric states, is there any reason to agitate for reunification? There are two answers to this question. The first answer is pragmatic: although ecology and systematics are surviving, they are not flourishing. The second answer is more esoteric: 'evolution' encompasses the interactions among mechanisms of origin, maintenance and diversification. Without information from all of these components, we can never fully understand the evolution of characters, species and biotas. And without an understanding of evolution, we must continue to base our sociological and environmental policies about the preservation and use of biodiversity resources on haphazard guesses, rather than on general biological principles.

So now, 135 years after Darwin's original intuition, we find ourselves in the unenviable position of recognizing the importance of reunification, while realizing that we no longer understand one another's language or evolutionary perspective. How can communication between ecology and systematics be re-established? Answering this question requires the development of a research programme that will allow us to integrate ecological, behavioural and phylogenetic information to produce a more robust picture of evolution. We believe that 'historical ecology' is just such a research programme.

Historical ecology encompasses macroevolutionary studies of two general evolutionary processes, speciation and adaptation. Historical ecologists explore the macroevolutionary effects of these processes by asking two basic questions. First, how did a given species arise? In order to answer this, we must explore a variety of ways in which descendant sister-species are produced from an ancestral species (speciation). Second, how did a given species acquire its repertoire of behavioural/ecological characters? This question moves us into the more familiar realm of the relationships between an organism and its environment (adaptation). In this case, however, these relationships are examined within the context of phylogeny. Having investigated speciation and adaptation processes within a group of organisms, historical ecologists then expand their temporal and spatial scale to include the effects of these processes on interactions between groups. At this level, the two basic questions can be restated as: 'how did a set of co-occurring species come to be associated?' and, 'how did the traits characterizing those interactions come to be?'. Answering these fundamental questions provides the database for investigating the macroevolutionary components of biological diversity.

SCOPE OF HISTORICAL ECOLOGY

Historical ecology and speciation

Mayr (1963) recognized three general categories of speciation processes: reductive speciation, in which two existing species fuse to form a third; phyletic speciation, in which a progression of forms within a single lineage is broken into different species at different points in time; and additive speciation, characterized by a net increase in the number of species in a clade. Over the 30 years following this suggestion, reductive speciation has been recognized more as a theoretical possibility than a notable evolutionary outcome. Phyletic speciation has suffered a somewhat worse fate, being regarded as an artefact of biologists' predisposition to break the ontogeny of a species (its anagenetic history) into arbitrary subunits, rather than as a mode of speciation. The majority of speciation models, although based on several different mechanisms, are models of additive speciation.

To historical ecologists, the most important aspect of additive speciation is that because it increases the number of species, so evolutionary explanations cannot be formulated based on analysis of a single species; sister-species and clades must be taken into account. This recognition plays an important role in our discussions about the factors that allow descendant species to maintain their identity as separate species in the presence of each other. Ever since the pioneering work of Theodosius Dobzhansky (Dobzhansky, 1937), this aspect of speciation has been synonymous with the study of 'isolating mechanisms' (see Mayr, 1963 for an extensive description). Studies of isolating mechanisms have often been based on comparison of two sympatric congeners. Within the context of historical ecology, however, we must begin with the question: 'are these sympatric congeners sister-species?' because only sister-species are linked by a common speciation event. Mechanisms that prevent hybridization of sympatric congeners that are not each others' sister-species are not attributes of the speciation process, and interspecific hybridization among sympatric congeners that are not sister-species does not indicate 'incomplete' speciation. Only when such comparisons are made between sympatric sister-species can we fully assess the relative importance of various types of isolating mechanisms.

Although speciation is a complex process, most phylogeneticists have turned their attention to the conditions under which speciation is initiated. At the coarsest level, there are two modes of initiation of speciation: those involving the physical disruption of gene flow by geographical isolation (allopatric modes), and those which do not require isolation for speciation to occur (non-allopatric modes). The allopatric category can be further subdivided depending on whether disruption of gene flow is accomplished through geological alteration (passive allopatric or vicariant speciation) or through movements of members of the ancestral species that eventually result in their geographical isolation (active allopatric speciation). In a different (but complementary) vein, adaptive changes within populations play different roles in each of these three general classes of speciation processes. Adaptive changes are not required to initiate passive allopatric speciation, although they may accompany such speciation events; conversely, adaptive changes are often postulated to facilitate active allopatric speciation and are a necessary component in initiating non-allopatric speciation.

We have suggested (Brooks & McLennan, 1991) that passive allopatric speciation or vicariant allopatric speciation be considered the null hypothesis for speciation studies because, being independent of any particular underlying biological properties, it is a mode of speciation that could occur in any group of organisms. All that is required is for an ancestral species to 'get separate and get different'. Since most models were developed to explain the breakdown of a single ancestral species into descendants, an entire clade is not necessarily expected to be the product of a single speciation process. From a conceptual standpoint then, uncovering incidents of active allopatric and non-allopatric speciation is just cause for celebration because these modes represent departures from the historical background of vicariance, and give us insights into the possible roles of a variety of environmental and behavioural processes in speciation.

An example: South American horned frogs

One of the most diverse and widespread of all frog groups is the family Leptodactylidae. Among South American leptodactylids, the subfamily Ceratophryinae comprises two genera, *Lepidobatrachus* with three species and *Ceratophrys*, with six species. *Ceratophrys* species are boldly coloured, voracious predators that are well-known to aquarists and tropical hobbyists as 'horned frogs'. They dwell in a variety of different habitats, ranging from neotropical rainforests (*C. aurita* and *C. cornuta*) through grasslands (*C. ornata*) to semi-xeric (*C. calcarata*) and xeric regions (*C. stolzmanni* and *C. cranwelli*). Lynch (1982) presented a phylogenetic analysis of the six species of *Ceratophrys*, using the monophyletic sister-group of *Ceratophrys*, the genus *Lepidobatrachus*, as the outgroup (Fig. 1).

The genus is comprised of two clades of three species each, one occurring from just north of the central Amazon northwards (*C. stolzmanni*, *C. calcarata*, *C. cornuta*) and the other associated with the Parana River system and coastal areas of south-eastern Brazil, Uruguay and Argentina (*C. cranwelli*, *C. ornata*, *C. aurita*). The three species in the 'northern' clade are all allopatric. At first glance, their

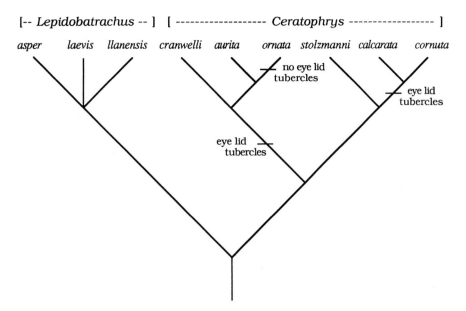

Figure 1 Speciation of frogs. Phylogenetic tree for the frog genus *Ceratophrys* based on 18 adult morphological characters, 1 larval morphological character and 1 karyotypic character (redrawn and modified from Brooks & McLennan, 1991).

distributional pattern conforms to a classical peripheral isolates scenario of a widespread central species, *Ceratophrys cornuta* (filled circles), with two smaller species located on the periphery of its range. However, Lynch's analysis uncovered the one phylogenetic pattern which specifically refutes this hypothesis. The large central species cannot be considered ancestral because *C. cornuta* is not the sister-group of the other two members of the clade, conflicting with the phylogenetic pattern predicted if speciation were due to a repeated cycle of sequential dispersal, isolation and speciation. Nor does *C. cornuta* occur in a polytomy with *C. stolzmanni* and *C. calcarata*, conflicting with the phylogenetic pattern predicted if speciation were due to random settlements of individuals around the margins of the ancestral species' range, or to a series of microvicariance events. In addition, it is *C. cornuta*, and not the peripheral species, that is the most divergent member of the clade (i.e. it exhibits the largest number of autapomorphies). This pattern, however, does support the hypothesis that evolutionary diversification in this clade has been associated with two vicariance events. The sundering of the first ancestral species resulted in the appearance of *C. stolzmanni* and the ancestor of the *C. calcarata* + *C. cornuta* clade, while the second vicariance event fragmented that ancestor, resulting in the emergence of *C. calcarata* and *C. cornuta*.

The situation is complex for the species comprising the 'southern clade'. The ranges of these species are relatively equal in size and overlap in two areas. Although *Ceratophrys ornata* and *C. cranwelli* are parapatric, they are not sister-species, so their zone of contact might initially be regarded as unimportant to speciation studies. However, this interpretation changes somewhat after a more detailed examination of the relationships depicted in Fig. 1. The sister-species of *C. cranwelli* was the ancestor (x in Fig. 2) of the clade *C. ornata* + *C. aurita*. Both of these descendent species display autapomorphies; therefore, neither of them can immediately be identified as a persistent ancestor. However, the autapomorphy for *C. ornata* is the postulated secondary loss of eyelid tubercles and there are three equally parsimonious interpretations of the transformation series for this character (Fig. 2).

Since the transformation shown in Fig. 2a eliminates the only autapomorphy postulated for *C. ornata*, we cannot eliminate the possibility that this species might, in reality, be ancestor x, based upon the information we have to date. If this possibility is realized, then the overlap between *Ceratophrys ornata* and *C. cranwelli* represents a parapatric speciation event in which both species are still extant, and thus within the scope of ongoing research. However, if future investigations uncover a convincing autapomorphy for *C. ornata*, then the origins of *C. cranwelli* and the ancestor of the *C. aurita* + *C. ornata* clade are embedded deeper within the cladogram and no longer subject to experimental investigation. *Ceratophrys cranwelli* and *C. ornata* are ecologically (xeric versus grasslands) and karyotypically distinct (both *C. ornata* and *C. aurita* are octoploid, while *C. cranwelli* is a diploid). It is possible, therefore, that either the ecological and/or the chromosomal change may have been associated with the parapatric speciation of *C. cranwelli* and ancestor x.

Significant for phylogenetic studies of speciation is that the ranges of the sister-species *Ceratophrys ornata* and *C. aurita* overlap in one locality. Both these species are ecologically isolated (Lynch, 1982), *C. ornata* in the grasslands and *C. aurita* in the rainforests. The plesiomorphic habitat preference for the genus is hypothesized to be a xeric, non-forest environment. Thus, in both the northern and southern subgenera, there has been a movement, correlated with speciation, towards the rainforests. In the northern clade this change in habitat preference is associated with two vicariance events, and thus is not the driving force behind the initiation of speciation. In the southern clade, the change in habitat preference may have been associated with two parapatric speciation events. If ecological segregation has been the motivating force behind the *C. ornata* and *C. aurita* (and *C. cranwelli*/ancestor x) differentiation, then we would expect to find evidence of ecological interactions between overlapping populations of these species. Total ecological and behavioural segregation will not provide support for the parapatric model, nor will it refute it; this type of 'absence of data' can only fail to refute the hypothesis, leaving us still partially in the dark.

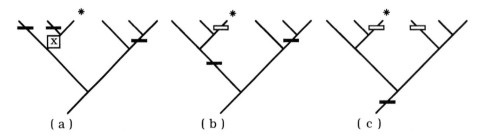

Figure 2 Three equally parsimonious transformations for the presence of eyelid tubercles mapped onto the cladogram for the frog genus *Ceratophrys*. White bars = secondary loss of eyelid tubercles; black bars = presence of tubercles. * = *C. ornata*; x = the ancestor of *C. cornuta* and *C. aurita* (redrawn and modified from Brooks & Mclennan, 1991).

There appears to be a relationship between the mode of initiating speciation, the rate of divergence and the factors that maintain species as cohesive evolutionary units. For example, in sympatric modes of speciation the initiation conditions are also the completion conditions, and in many forms of this mode, speciation is essentially instantaneous. Vicariant speciation is at the opposite end of the scale – the initiation conditions need have nothing to do with the completion conditions, and divergence rates for vicariance may be the slowest of all modes of speciation. Given these observations, it is tempting to suggest that there is a positive correlation between the strength of the connection between initiating and completing factors and speciation rates.

It is therefore interesting to note that the only extensive study undertaken to date addressing the frequency of speciation modes within a phylogenetic context suggests that there is a negative correlation between rapid modes of speciation and relative rates of their occurrence. Lynch (1989) discovered that vicariance (passive allopatric speciation), the most plausible speciation mode on theoretical grounds, also seems to have been the most prevalent on empirical grounds (see also Wiley, 1981; Wiley & Mayden, 1985). Speciation modes, such as parapatric and sympatric speciation, that require adaptive changes to initiate and/or complete the process are relatively unlikely on theoretical grounds and very few putative examples of these modes have been documented. This implies that speciation and adaptation need not always be tightly coupled evolutionarily. If that is true, the relationship between speciation, adaptation and diversity may not be as straightforward as previously thought.

Historical ecology and adaptation

Adaptation complements speciation in evolutionary biology by focusing attention on particular attributes of species rather than on the production of species *per se*. Contemporary adaptive explanations refer to an individual's response to some problem set by nature. Adaptation is thus related to notions of a functional 'fit' between an organism and its environment, implying that adaptive characters can be identified by seeking correlations between particular traits, or combination of traits, and the relevant environmental variable (Dunbar, 1982). Traits that are 'the same' in two species are postulated to have arisen as a common adaptive response to similar selection pressures. This leads to the assumption that if the two species do not live in similar environments today, then they must have lived in similar environments in the past. Traits that differ between two species are postulated to have arisen as differential adaptive responses to different selection pressures. This leads to another assumption, namely that if the species live in similar environments today, then they either lived in different environments in the past, or the current environments have not been partitioned on a fine enough scale to determine that they really are different in some evolutionarily significant way.

One way to characterize this traditional point of view is to contrast the traits ('similar' or 'not similar') with the species possessing them ('related' or 'not related': Table 1). The classical explanation for 'unrelated' taxa bearing the same trait is convergent adaptation (II), whereas the explanation for related taxa displaying different traits is divergent adaptation (III). This leaves two possibilities unexplored. In the case of unrelated taxa bearing different traits (IV) there is nothing of interest

Table 1 Pre-historical ecology adaptation scenarios. Only type II and type III patterns are considered to be of interest to students of adaptation (Tinbergen, 1964).

Traits	Taxa	
	'Related'	'Not Related'
'Similar'	not interesting I	convergent adaptation II
'Not Similar'	divergent adaptation III	not relevant IV

to study because there is no reason to believe the taxa have anything in common evolutionarily, either ecologically or genealogically. In the case of related taxa displaying the same trait (I), we are dealing with a phylogenetic scale that is uninformative to traditional evolutionary ecology, i.e. there is nothing to study because the trait has not changed evolutionarily. Stabilizing selection is often invoked to explain the absence of change in this sort of situation, based on the assumption that the two related species occur in similar environments (or the environments have changed so recently that there has not yet been 'time' for an adaptive response). Note that in this approach to explanations, there is no direct comparison between the species and its environment; the functional fit is assumed, and adaptation is assumed to be the explanation for the fit. Historical ecology allows us to expand our evolutionary perspective to encompass macroevolutionary patterns (patterns that appear on protracted temporal scales) and thus dispense with the assumptions of adaptation theory listed above without losing the robustness of adaptive explanations. When the four classes of patterns depicted in Table 1 are examined within two environmental contexts, eight possible outcomes are produced from the interactions between phylogenetic and environmental information (Table 2).

Under this view, there are two manifestations of phylogenetic constraints. Patterns of type I may provide weak evidence of adaptation if a researcher wishes to invoke stabilizing selection to explain the absence of evolutionary change in a trait displayed by related species inhabiting similar environments (x_0 in Fig. 3a). This is a weak test of adaptation because there is no macroevolutionary evidence of a functional change involved with an environmental change. We need to increase the level of the phylogenetic analysis until the origin of x_0 is uncovered. If there is a macroevolutionary association between a shift in the environment and a change in x_0, then we have strong support for the adaptational hypothesis for that trait. The type II pattern does not corroborate hypotheses of adaptive evolution because x_0 does not change even though speciation has been accompanied by movement into new environments (Fig. 3b). Of the eight patterns depicted in Table 2, this is the strongest 'falsifier' of adaptation scenarios. Types I and II are the 'absence of evolutionary change' patterns and serve as 'null hypotheses' in the discussions of adaptational studies.

Patterns of type III provide strong support for convergent adaptation. This can be viewed as a form of environmentally driven change (Fig. 4a). Pattern IV, on the other hand, postulates the appearance of convergent traits in different environments

Table 2 Historical ecological adaptation scenarios based upon the interaction between phylogenetic and environmental information.

Traits	Environment	
	'Similar'	'Different'
'Similar' (homologous)	phylogenetic constraints (stabilizing selection?) I	phylogenetic constraints II
'Similar' (homoplasious)	convergent adaptation III	convergence not due to environment IV
'Not Similar' (homologous transformation series)	divergence not due to environment V	divergent adaptation VI
'Not Similar' (non-homologous)	not relevant VII	not relevant VIII

and thus does not support an hypothesis of convergent adaptation (Fig. 4b). We suspect that this pattern usually represents instances of developmentally driven (or developmentally allowed) evolutionary change. Similar homoplasious characters keep arising regardless of the environment due, perhaps, to the presence of a developmental programme which is manifested in only a limited number of ways and which can be 'pushed' down the different pathways quite easily. For example, viviparity is a character which shows up repeatedly within several frog families whose members live in different habitats (Duellman, 1985).

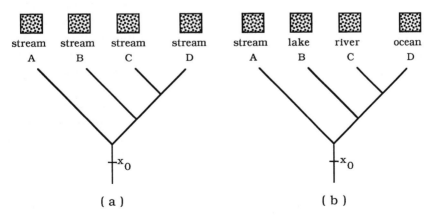

Figure 3 Patterns depicting two types of phylogenetic constraints. (a) Homologous trait and similar environments, postulate that stabilizing selection reinforces the maintenance of trait x_0 in all the members of the monophyletic group A+B+C+D. (b) Homologous trait and different environments; environment changes, trait remains the same. This does not corroborate an hypothesis of a 'functional fit' between trait x_0 and the environment (from Brooks & McLennan, 1991).

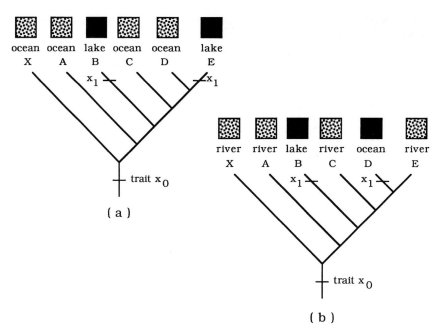

Figure 4 Patterns depicting two types of convergent evolution. (a) Homoplasious traits and similar environments, postulate that selection drives the modification of trait x_0 to x_1 in the similar lake environment. (b) Homoplasious traits and different environments; both environment and trait change but there is no evolutionary association between the changes. This does not corroborate a hypothesis of a convergent 'functional fit' between trait x_1 and the environment (from Brooks & McLennan, 1991).

Patterns of type VI, providing support for divergent adaptation, can also be viewed as a form of environmentally driven change (Fig. 5a). Pattern V, on the other hand, postulates the appearance of divergent homologous traits in similar environments and thus does not support an hypothesis of divergent adaptation (Fig. 5b). This scenario depicts the classical case of passive allopatric speciation; i.e. random developmental changes, which are 'allowed' by the developmental programme, arise and are then incorporated into the isolated population following cessation of gene flow. Like pattern type IV, then, this incorporates a developmentally driven component into explanations of evolutionary change within groups of organisms.

There are two manifestations of evolutionary changes that are not related to each other either phylogenetically or ecologically. These situations are not pertinent to studies of adaptive change (VII and VIII). As with all complex systems, explanations about adaptational changes within a given clade often contain combinations of the above eight patterns. For example, the traditional story about the adaptive radiation of Galapagos finches invokes among-islands convergent adaptation (III) and within-island divergence in similar environments (V) (Coddington, 1988).

Historical ecological studies of adaptation encompass six main areas: (1) formulating hypotheses of adaptation within a phylogenetic framework; (2) examining the temporal sequence of adaptive changes; (3) examining the evolutionary association of traits ('co-adapted traits'); (4) examples of convergent adaptation; (5) examples of divergent adaptation; and (6) examining phylogenetic constraints on ecological diversification of characters and species (adaptive radiations). Reference to a phylo-

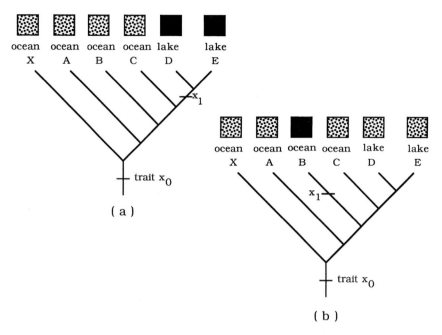

Figure 5 Patterns depicting two types of divergent evolution. (a) Homologous traits and different environments, postulate that selection drives the modification of trait x_0 to x_1 in association with the shift from an ocean to a lake environment. (b) Homologous traits and similar environments; environment stays the same, trait changes. This does not corroborate an hypothesis of a divergent 'functional fit' between trait x_1 and the environment (from Brooks & McLennan, 1991).

genetic context will allow us to ask questions relevant to studies of adaptation for single traits, such as: (1) is the trait unique to one species?; (2) if the trait is unique to one species, is it an ancestral remnant, or is it recently evolved?; and (3) if the trait is found in more than one species, how many times has the trait evolved? If we are concerned with possible evolutionary correlations among two or more traits, we may use phylogenetic analysis to identify cases in which the traits do not co-occur and did not co-originate in any members of the study group, other cases in which the traits co-occur but did not co-originate in some members of the study group, and yet other cases in which the traits both co-occur and co-originated in members of the study group.

Combining speciation and adaptation I: adaptive radiations

The disproportionate representation of one taxon within an assemblage is widespread throughout biological classifications. Some have argued that this is a taxonomic artifact (Mayr, 1969; Raikow, 1986, 1988). Others have argued that the patterns reflect biological reality but can be explained by stochastic speciation and extinction processes (Raup *et al.*, 1973; Raup & Gould, 1974; Gould *et al.*, 1977; Raup, 1984). Still others (e.g., Flessa & Levinton, 1975; cited in Dial & Marzluff, 1989) conclude that 'deterministic processes so pervade the evolution of diversity that stochastic processes must be regarded as secondary in importance' (see also Stanley *et al.*, 1981; Ehrlich & Wilson, 1991). Dial & Marzluff (1989) compared patterns of species–richness within 85 clades to the patterns predicted by five null models (Poisson: Raup *et al.*'s (1973) simulation; Anderson & Anderson's (1975) simulation; simultaneous broken-stick; and canonical lognormal distribution). They discovered that nearly all of the groups were dominated to a significantly greater extent by one taxon than predicted by any of the null models, supporting the hypothesis that biological mechanisms are responsible for differences in species number among clades (Fitzpatrick, 1988; Vermeij, 1988; Dial & Marzluff, 1989). This begs the question of what those mechanisms could be.

The answer has traditionally centred around the concept of 'adaptive radiation'; differences among clades result from high speciation rates in the more species-rich group rather than high extinction rates in the species-poor groups. Some authors have suggested that there should be an adaptive explanation for all speciation events (Stanley, 1979; Stanley *et al.*, 1981). Simpson (1953) thought that adaptive radiations resulted from diversification accelerated by ecological opportunity, such as dispersal into new territory, extinction of competitors or adoption of a new way of life (i.e. an adaptive change in ecology or behaviour). Other factors, including the adoption of a specialist foraging mode (Eldredge, 1976; Eldredge & Cracraft, 1980; Vrba, 1980, 1984a,b; Cracraft, 1984; Novacek, 1984; Mitter, Farrel & Wiegemann, 1988), sexual selection and population structure (Spieth, 1974; Wilson *et al.*, 1975; Carson & Kaneshiro, 1976; Ringo, 1977; Templeton, 1979; Gilinsky, 1981; West-Eberhard, 1983; Barton & Charlesworth, 1984; Carson & Templeton, 1984) or the origin of key ecological innovations in an ancestral species (Cracraft, 1982; Mishler & Churchill, 1984; Brooks, O'Grady & Glen, 1985; Duellman, 1985), have also been postulated to have a positive effect on speciation rates. The consensus view of adaptive radiations today remains one with emphasis on 'adaptive' (Futuyma, 1986: 32):

> . . . a lineage may enter an adaptive zone and proliferate either because it was pre-adapted for niches that became available, or because it evolves 'key innovations' enabling it to use resources from which it was previously barred.

Because of this focus on adaptation, there has been a tendency to equate species-richness with 'evolutionary success' and this, in turn, has led to the implicit assumption that speciation = adaptive radiation. If, as suggested previously, speciation and adaptation are distinct, albeit historically correlated, evolutionary processes, we cannot assume a priori that all radiations are adaptive, and must search for criteria by which we can distinguish 'adaptive radiations' from 'radiations' (or 'non-extinctions') (Brooks & McLennan, 1993a).

Studies of adaptive radiations must occur within the larger framework of phylogeny because it is necessary to identify 'evolutionarily equivalent' groups for comparisons of species-richness. Groups of species can be considered evolutionarily equivalent if they are all monophyletic and the same age. There have been a number of criteria proposed by which the relative ages of clades can be assessed, such as stratigraphic, biogeographic or molecular clock comparisons. Comparison of sister-groups is the most objective means possible for making certain we are dealing with evolutionarily equivalent units because sister-groups are, by definition, of equal age (Mayden, 1986). The search for correlates of species-richness can be made by comparing either species-rich groups with each other, or by comparing species-rich groups with their species-poor relatives. We believe that the latter comparison provides the best context for determining if the species-rich group is characterized by evolutionarily derived traits associated with species-richness. Comparisons among species-rich groups alone will not determine if any common characters are also common to some of their species-poor sister-groups, and will determine if any characters unique to a particular species-rich clade might be the key to explaining species-richness in that group. In other words, we should begin phylogenetic studies of adaptive radiations by assuming that adaptive radiations, like adaptations, are clade specific. Once a significant database has been assembled, we may ask if there are any common characteristics of species-rich clades.

Merely comparing the species-rich group with its species-poor sister group, however, will not tell us which of the two groups might require special explanation. We can only assume that the species-rich groups require explanation if we always assume that extinction rates are equivalent across taxa. For example, consider the hypothetical family of fish depicted in Fig. 6. An ichthyologist comparing sister genera D and E might conclude that E was species-rich (Fig. 6a). In light of the larger phylogenetic picture, that conclusion might be premature. The pattern depicted in Fig. 6b corroborates the conclusion; however, the pattern depicted in

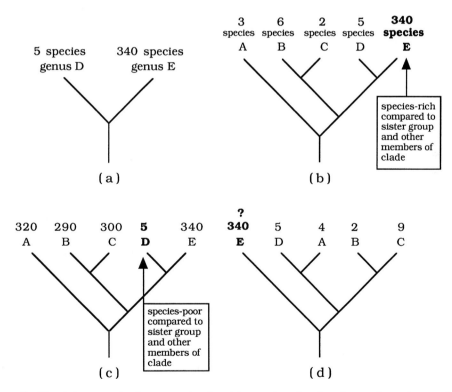

Figure 6 Heuristic example demonstrating how the explanation for observed species number is dependent upon phylogenetic patterns and the level of the analysis. (a) A two taxon statement provides no information about origins of species-richness. (b) Species-richness is derived within the family. (c) Species-richness is plesiomorphic (ancestral) within the family. (d) Pattern that might confound a statistical analysis of diversity because, without expanding the scope of the analysis to include information from outgroups, it is impossible to tell whether species-richness is derived or plesiomorphic within the family (from Brooks & McLennan, 1993a).

Fig. 6c indicates that E is not species-rich compared with the other members of the family except its sister-group. Within the context of species distributions across genera of the family, genus E is not species-rich; rather, its sister-group, genus D, is species-poor. The situation is further complicated when the pattern depicted in Fig. 6d is uncovered. Plotting the number of species/genus against number of genera would produce the typical U-shaped curve (Dial & Marzluff, 1989), in which the dominance of genus E would be postulated to represent a rare event. However, because the species-rich group is also the oldest member of a clade composed of relatively depauperate groups, it is impossible to determine whether 'species-richness' is plesiomorphic (ancestral) or apomorphic (derived) for the family. To do this, we need to expand the phylogenetic framework to include information about species number in the closest relatives of the ABCDEidae. If the sister-group of the ABCDEidae is species-poor, then group E might represent a rare event in the family. If the sister-group is species-rich, then we need to search for explanations about the species-poor clades A, B, C and D. This is an important distinction because, in the absence of the additional phylogenetic information, a statistical analysis might give an incorrect picture of the evolutionary trends in the family. Having documented pairs of species-poor and species-rich sister-groups within a phylogenetic context in which species-richness is apomorphic, we can proceed to ask the question of primary importance in studies of adaptive radiations: what, if anything, is adaptive about the radiation?

Some researchers have sought the answer to this question in the concept of 'key innovations'. A key innovation was originally defined as any novel feature that characterized a clade (i.e. any synapomorphy) and was therefore correlated with its adaptive radiation (Mayr, 1960; Liem, 1973). Liem (1973), for example, suggested that the extensive diversification of cichlid fishes in the African Rift lakes was due to the origin of a lower pharyngeal jaw suspended in a muscular sling in their common ancestor. Although intuitively pleasing, the concept as stated falters upon closer examination (Lauder, 1981; Liem & Wake, 1985; Stiassny & Jensen, 1987; Lauder & Liem, 1989). First, simply asserting that every apomorphy on a phylogenetic tree is an adaptation, and therefore potentially responsible for an observed 'adaptive' radiation, is of limited value at best and circular at worst. The question, then, is how to define the term adaptive in a manner that allows an hypothesis of adaptation to be falsified. Since the diversification of fixed characters can only be identified from a macroevolutionary perspective, the concept of adaptation must incorporate phylogenetic information, as well as information about the functional superiority of the putative adaptive character (Arnold, 1983; Greene, 1986; Coddington, 1988; Baum & Larson, 1991).

Second, if the species-rich clade is characterized basally by more than one synapomorphy, there is no a priori way to determine which of those traits might be the key innovation (perhaps even a combination of traits could be 'the innovation'). This problem is made more complex because there is no theoretical reason to expect that the innovation originated with the ancestor of the species-rich clade. It may have appeared in one of the sister-species of the original ancestor, or the clade may be comprised of several species-rich subunits, each of which may be characterized by a different 'key innovation'.

The third problem with the concept of a key innovation is that there is rarely strong evidence about the manner in which the innovative trait affects speciation rates in the clade. Key innovations can influence speciation rates directly if they increase the likelihood that species within the clade will participate in relatively rapid, 'adaptive' modes of speciation, such as peripheral isolates allopatric, parapatric and sympatric speciation. Vicariance biogeographic methods are a useful way to study this interaction of adaptation and speciation because episodes of 'adaptive' speciation are highlighted against a background of vicariance (Brooks & McLennan, 1991).

Key innovations can also influence speciation rates indirectly by decreasing the chance that species will go extinct before they have a chance to be affected by vicariant speciation (Cracraft, 1982). Larson *et al.* (1981), for example, proposed a species selection argument in which key innovations gave the descendent species in the clade an advantage over competitors. Such lineages would be expected to survive longer and extend over a larger geographic range than other lineages, thus increasing the likelihood of vicariant speciation, which would result in increased species-richness for a clade. This raises the question of whether there need be anything adaptive about simply surviving long enough to participate in many episodes of vicariant speciation. In general, the probability of speciation should increase with increased species longevity; however, the key innovation need not play a role in initiating those bifurcations. For example, long-lived species residing in areas subject to repeated vicariant episodes (hot spots: Cracraft, 1982) may show more bifurcations than equally long-lived species inhabiting more stable environments. The problem arises out of the assumption that adaptation = speciation. If the two processes are decoupled, then we must incorporate both processes into our definition of a key adaptation. Given this, we believe that a key innovation should be defined as any apomorphic character that can be demonstrated to have adaptive value (fitness advantage: see Coddington, 1988) relative to its plesiomorphic antecedent, and can be demonstrated to play a direct role in initiating speciation (and thus increasing speciation rates). Characters which simply promote the longevity of a species so that its chances of participating in vicariant speciation are increased are not key innovations.

Combining speciation and adaptation II: community evolution

Historical ecological studies of community structure do not begin with 'how does this complex system work?' or 'what are the macroscopic properties (e.g. stability, resilience) of this complex system?' but rather 'how was this system assembled historically and how did the parts evolve?'. In order to answer these questions we need information about speciation and adaptation on a large temporal scale involving comparisons among, rather than within, clades (Brooks & McLennan, 1993b).

One of the fundamental advances of the past 30 years in evolutionary biology has been the formal articulation of two different perspectives on the general manner by which species achieve their geographical distributions. These perspectives may be categorized loosely as island biogeography (MacArthur & Wilson, 1967), which calls attention to the propensity for organisms to move about, and vicariance biogeography (Croizat, Nelson & Rosen, 1974; Rosen, 1975, 1985; Platnick & Nelson, 1978; Nelson & Platnick, 1981; Humphries & Parenti, 1986), which reminds us that those movements may not be unconstrained. Both research programmes have contributed valuable insights into the problem of species co-occurrence, leading historical ecologists to this perspective: two or more species may be ecologically associated today either because their ancestors were associated with each other in the past (association by descent), or because at least one of the species dispersed into the area (association by colonization) (Brooks, 1979; Mitter & Brooks, 1983; Brooks & Mitter, 1984). In the first instance, the contemporaneous relationship is a persistent ancestral component of the biotic structure within which the interacting species reside, so the species' history must be congruent with the history of the area. In the second instance, the dispersal may have occurred a long time ago or relatively recently; in any event, there is no reason to expect the species' history to be congruent with the history of the area into which it dispersed. Both association by descent and association by colonization may be manifested in geographical patterns or in purely ecological patterns of species co-occurrence. Of course, any species that originally appears as a colonizing influence in an ecological association can become phylogenetically associated in the subsequent evolutionary history of that association. It is likely that many, if not most, ecological associations contain both vicariant and dispersalist elements. Co-speciation analyses that include both historical and non-historical influences on geographic distributions and ecological associations will thus be richer than analyses that concentrate on either one component or the other.

Co-adaptation studies are designed to uncover the adaptive components within the macroevolutionary patterns of association between and among clades. Although the microevolutionary and macroevolutionary levels of co-evolution are closely linked by the process of co-adaptation, surprisingly few phylogenetic studies have examined this problem. Studies of co-adaptation must begin with a co-speciation analysis to provide the phylogenetic background against which episodes of mutual modification can be highlighted. Without this analysis, it is impossible to objectively differentiate scenarios based on the assumption that current associations reflect historical associations from other scenarios which presuppose little, or no, history of interaction.

Ecological associations do not evolve in the same way that species evolve; rather, they are 'assembled' evolutionarily and some aspects of the assembly may be phylogenetically based. In theory, the outcome of the evolutionary interchange between co-speciation and co-adaptation reflects the importance of four influences on community structure:

1. Phylogenetic history (Table 3: type I). The conservative, homeostatic portion of any community is composed of species that evolved *in situ* through the persistence of an ancestral association. Such species display the plesiomorphic condition for characters involved in interactions with other community members and with the environment. Since this section of the community is characterized by a stable relationship across evolutionary time, it may act as a stabilizing selection force on other members of the community by resisting the colonization of competing species.
2. Stochastic (non-equilibrium) effects (Table 3: type II). If there are unused types of resources in a community over extended periods of time, stochastic evolutionary changes, operating on resident species, may result in the use of some of that previously unoccupied space. In this scenario, evolutionary changes in ecological

characters occur within a co-speciation framework. Species contributing to this portion of the community structure can be recognized by their historical congruence with other community members, coupled with the presence of apomorphic traits characterizing their interactions with other species and with the environment. Since these changes do not affect the evolution of other community members, they may represent a type of stochastic wandering through modifications allowed by the existing community structure. Such species appear to diverge in ecological traits for no apparent reason, although care must be taken to rule out the effects of previous competition. The longer a community exists below equilibrium numbers of species, through any of the processes described under colonization by pre-adapted species, the greater the possibility that resident species will experience these sorts of evolutionary changes.

3. Colonization by 'pre-adapted' species: (Table 3: type III). This portion of the community contains species that have been added by colonization. Such species can be recognized in part because their phylogenetic history is incongruent with the histories of other community members. In this context, the term 'pre-adapted' implies only that these individuals are able to colonize the area because they already possess traits which do not conflict with the existing community structure. This scenario postulates that, initially at least, there is no competition between colonizing individuals and established (resident) members of the community. If the appearance of these species reduces the possibilities for the subsequent addition of species into the community, then this type of macroevolutionary pattern corresponds to the asymptotic equilibrium model of MacArthur & Wilson (1967). If the rates of colonization are low enough the community may persist below expected equilibrium numbers. Finally, if the colonizers are so specialized ecologically that they do not affect other members of the community or pre-existing potential niche space, community diversity may increase without approaching an apparent equilibrium.

4. Colonization by competing species (Table 3: type IV). All species that colonize a community will exhibit incongruence in a co-speciation analysis. However, unlike the conservative situation depicted for pre-adapted species, varying patterns of character evolution will be traced upon this phylogenetic framework if colonizing individuals compete with resident species. In this situation, at least

Table 3 Heuristic depiction of four classes of species contributing to community structure. There are two components to ecological associations: (1) species composition; the occurrence of each species in an association is either ancestral or derived and (2) species interactions; the characters involved in interactions among members of the association are either ancestral or derived (from Brooks & McLennan, 1993b).

Species occurrence	Species interactions	
	Ancestral	Derived
Ancestral (resident)	I historically constrained residents	II residents change stochastically
Derived (colonizer)	III colonization by pre-adapted species	IV colonization by competing species

one of three things must happen in order for the colonizer to become established: the colonizing species will change, the resident species competing with the colonizer will change, or both the resident and the colonizer will change. This will produce a pattern in which the colonizer, the resident, or both, exhibit an apomorphic condition of the traits relevant to the competitive interaction. Replacement of the resident by the colonizer would be indicated on a cospeciation analysis if the extinction event is coupled with the colonization event and if other members of the extinct species' clade have similar resource requirements to the colonizer. These macroevolutionary patterns correspond most closely with the asymptotic equilibrium model of MacArthur & Wilson (1967).

LIMITATIONS OF HISTORICAL ECOLOGY

Historical ecology, as well as all other areas of contemporary comparative evolutionary biology, are victims of their own success. The excitement about, and demand for, phylogenetic information has outstripped the spread of information about the fundamentals of phylogenetic reconstruction. This situation reflects, in part, the development of that methodology in isolation from much of evolutionary biology, and the apparent inability or unwillingness of many systematists to communicate these developments outside of the discipline. At the other end of the comparative biology continuum, many ecologists and ethologists have resisted the idea of wading through 20 years of primary literature on phylogenetic methodology, succumbing instead to the seduction of cheap and user-friendly computer programs. These programs are designed for use by systematists familiar with the assumptions of phylogenetics, yet, like the widespread abuse of statistics for exactly the same reason, can be accessed by anyone able to input basic data. Unfortunately, this has led to a profusion of incorrectly interpreted, weakly supported analyses masquerading as phylogenetic trees in almost every scientific journal displaying an interest in comparative biology. There is only one answer to this problem: increased communication between systematists and ecologists who recognize that the interaction can be nothing but mutually beneficial to all concerned. Until this occurs, students of the comparative method will continue to be plagued by at least two problems.

A shortage of robust phylogenies

The availability of robust and explicit estimates of phylogeny is the primary limiting resource in comparative evolutionary studies. This situation is further exacerbated by the fact that systematists and ecologists have traditionally focused their attention on different groups of organisms, thus increasing the probability that any given taxon will have either an abundance of either systematic or ecological information, but not both. For example, many behavioural ecologists have turned to passerine birds for studies of mating system evolution and to poeciliid fishes for studies of sexual selection. At the moment, there are no robust estimates of phylogeny at the species level for either of these groups. However, there is room for optimism here. Phylogenetic trees are being produced at a more rapid rate, there is an increased breadth of taxa being investigated, including those of general conceptual interest to evolutionary biologists, and we are finally beginning to build a large enough database to compare the outcomes of analyses based upon morphological, behavioural and molecular data. These results are encouraging because, when the data are subjected to rigorous phylogenetic analysis, they generally tend to produce congruent trees.

Not enough familiarity with phylogenetic systematic methods

Phylogenetic trees are the starting point of any historical ecological study. Researchers must therefore be familiar enough with the methodology responsible for the construction of that tree in order to assess its potential strengths and weaknesses. There are number of pitfalls awaiting the unwary historical ecologist in this area. For example: (1) some published diagrams are not the phylogenetic trees that are best supported by the data; (2) some information cannot be interpreted phylogenetically (e.g. genetic distances, immunological distances, DNA-DNA hybridization); and (3) some published trees, including many based on nucleotide sequence data, represent only one of many equally parsimonious (or statistically indistinguishable, in the case of distances) representations of the data. There is a simple solution to this problem. All published trees must be accompanied by descriptions of goodness-of-fit statistics, optimization assumptions used, the number of equally parsimonious solutions and, if there is more than one tree, why the tree presented is the preferred one. Anything less makes it difficult for researchers to evaluate the robustness of the trees that form the basis of their evolutionary explorations.

The fourth pitfall may be termed 'Beware Imitations'. Evolutionary taxonomists, pheneticists and phylogenetic systematists all produce branching diagrams. Evolutionary taxonomic diagrams are based upon intuition, and phenetic diagrams are based upon shared overall similarity, be it plesiomorphic, derived or convergent. Only phylogenetic systematic diagrams are based upon shared, derived characters. It is therefore inappropriate to use the results of a non-phylogenetic systematic analysis as a phylogenetic tree in a comparative study. One of the major problems with such diagrams is that they portray paraphyletic (or polyphyletic) taxa as monophyletic groups. This has two outcomes. First, it will bias the mean in quantitative comparative studies because plesiomorphies are counted more than once (Clutton-Brock & Harvey, 1978; Wiley, 1981; Brooks & McLennan, 1991; Harvey & Pagel, 1991). Second, it may overestimate the importance of adaptive plasticity because synapomorphies may be counted more than once. This gives the impression that these traits are actually examples of parallel or convergent evolution and such homoplasy, in turn, is often considered strong evidence of adaptive evolution. Consider the following example. Figure 7a depicts the phylogenetic tree for a group of hypothetical taxa. Although the presence of five synapomorphic traits (characters 6–10) distinguishes species D and E from species A, B, and C, these taxa are all members of a monophyletic group characterized by the presence of traits 1–5. Now, suppose we have a classification scheme that places species D and E in one taxon because they are so distinct, and places species A, B and C in another taxon. Reconstruction of phylogenetic relationships based on this classification will produce the tree depicted in Fig. 7b. This arrangement forces us to postulate that characters 2–5 evolved twice, overestimating the amount of adaptive evolution. Since most commonly accepted classifications include paraphyletic groups, they cannot serve as independent templates for estimating the origins, elaborations and

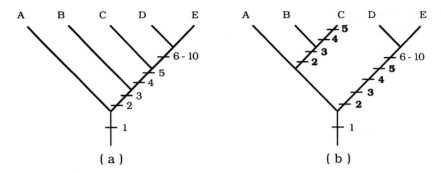

Figure 7 Example demonstrating why it is incorrect to equate 'classification (or taxonomy)' with 'phylogeny'. (a) Phylogenetic tree reconstructed for species A–E based upon phylogenetic systematic methods. (b) Tree reconstructed from a taxonomic classification scheme which includes the paraphyletic group A+B+C. This forces us to postulate that characters 2, 3, 4 and 5 evolved twice (from Brooks & McLennan, 1991).

associations of ecological characters through evolutionary time. Unfortunately, given the current dearth of available phylogenetic trees, many researchers have been forced to utilize such classifications in their preliminary analyses of ecological/ behavioural evolution.

The fifth pitfall involves the misconception that outgroup comparison is an exercise in circular reasoning, and can therefore be dispensed with in a phylogenetic analysis. Outgroup comparison lies at the heart of phylogenetic methodology because this is the only way to construct hypotheses of character polarizations independently of any hypothesis of relationships among the taxa bearing the characters (for an extensive discussion see Wiley *et al.*, 1991). It is true that some types of data appear to be less amenable than others to phylogenetic systematic analysis because they are composed of a restricted number of alternatives that changes consistently in ingroups and outgroups, and which re-occur (this is particularly true of much nucleotide sequence data). These data do not always permit the determination of consensus outgroup states and thus of robust estimates of plesiomorphic conditions. This is a problem with the type of data, not the method of analysis.

The sixth and final pitfall stems from the desire of historical ecologists not to bias their analyses by using the ecological information they want to study in building their phylogenetic trees (Brooks & McLennan, 1991). Many have assumed that using the characters of interest as part of a phylogenetic analysis and then deriving explanations about their evolution constitutes such bias and is an exercise in circular reasoning. This would be true only if the method of analysis pre-determines explanations of homology and homoplasy. However, since phylogenetic analysis is based on the a priori assumption of homology (Hennig's Auxiliary Principle: Hennig, 1966; Wiley *et al.*, 1991) and since characters are argued independently against outgroups, it is the overall weight of evidence that determines the ultimate interpretation of character evolution, not the a priori assumption (homology) of the investigator. Although we can dispense once again with the spectre of circularity, this question does raise an equally important issue: the strength of the conclusions that can be drawn by using data in this manner is dependent upon the robustness of the initial phylogenetic tree. For example, if removing the characters of interest from the phylogenetic tree changes the topology of the tree, then you do not have a robust phylogenetic hypothesis. You can still do the comparative study, and it will not be circular, but it will not be strongly supported by the data. In general: (1) if you are studying convergent adaptation, the characters of interest will show homoplasy so you will need additional characters to get a phylogeny in the first place; (2) if you are studying continuous variable characters, such as many life history traits, you will need other characters to formulate a phylogenetic hypothesis because phylogenetics has not yet discovered a way to handle such data in a manner that meets with broad consensus approval; and (3) if you are studying divergent adaptation, you will be studying homologous traits, in which case optimizing them on a tree is the same as having used them as characters in the tree's construction. Thus, this is a question of confidence in the original phylogenetic analysis, and such confidence is based upon how well supported the tree is by the data set. In closing this section we can only reiterate our earlier (Brooks & McLennan, 1991) assertion that there is only one response to a weakly supported tree: gather more data and produce a more robust phylogenetic hypothesis. Phylogenetic systematics is an open-ended process.

HISTORICAL ECOLOGY IN THE 21st CENTURY

Historical ecology, with its focus on the interaction between speciation and adaptation in the evolution of characters, species and biotas, has much to offer studies of biodiversity and conservation. For example, very little is known about the ecology and behaviour of most species, making the job of protecting them frustrating and potentially disastrous (Kleiman, 1980). However, if we have access to a phylogeny, we can optimize ecological characters from well-studied members of the group on to the tree and predict the probable characteristics of the poorly studied species. Consider the elusive, and endangered, boulder darter, *Etheostoma wapiti*, which is a member of a clade comprising six species, five of whigh display the derived character of depositing their eggs beneath or behind large slab rocks (Etnier & Williams, 1989; Fig. 8). *Etheostoma wapiti* is found in habitats containing slab rocks. Based on its phylogenetic relationships and habitat preference, Brooks, Mayden & McLennan (1992a) predicted that the boulder darter should have the same breeding strategy as other members of its clade. They proposed that this was an example in which both a rare species and its widely distributed phylogenetically close relatives shared a common ecology involving resources that are severely limited in the areas inhabited by the rare species. A recent study by Burkhead & Williams (1992) confirmed both their prediction and their proposal (Brooks, Mayden & McLennan, 1992b).

If more information about the behaviour and ecology of the boulder darter is required to formulate conservation policies, studies could be conducted without disturbing the already endangered populations of *E. wapiti* by using its widespread and relatively common sister-species, *E. vulneratum*. This method is not foolproof, but it does provide a starting point based on information other than inspired guesswork. It is also the way most managers work intuitively; phylogenetic systematics brings more rigorous methodology to bear on the problem because it allows us to identify the evolutionary sequence of sister-species production, as well as the origin and evolutionary stasis/divergence of characters.

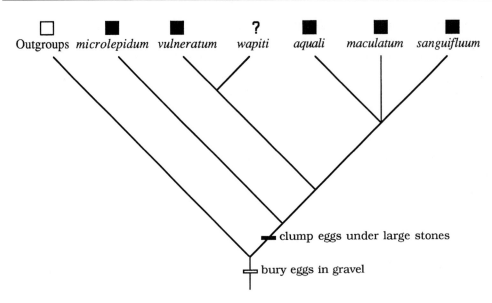

Figure 8 Phylogenetic tree for six members of the maculatum species-group of darters in the subgenus *Nothonotus* (genus *Etheostoma*) showing distribution of egg deposition behaviour. Outgroups contain the two remaining species within maculatum species-group and the eight additional species in the subgenus Nothonotus. Open squares indicate that eggs are buried in gravel; solid squares indicate that eggs are deposited on the underside of slab rocks; ? = egg deposition unknown. *Etheostoma wapiti* is an endangered species (from Brooks, Mayden & McLennan, 1992b).

Examining speciation and adaption within a phylogenetic framework can also provide information to complement current conservation/management practices used on theories about the relationship between species number and area size or number. Should areas with the greatest number of species be given priority in conservation decisions? In order to answer this, we need information about both species number and the evolutionary history of those species. For example, Hybrid areas, encompassing regions of overlap between two or more biotas, may appear to have higher biodiversity than either of the 'parent' areas. These areas, however, may represent marginal habitats for members of the overlapping biotas that have limited their expansion. Additionally, it has been postulated that the more closely related the species, the more likely they are to utilize the same resources (MacArthur, 1958, 1972; Root, 1967). In undisturbed comunities, congeneric species are found together less often than would be expected by a process combining taxa randomly from a pool of all possible species with access to the communities (Bowers & Brown, 1982). Since congeners rarely occur together often enough to allow adaptations for competitive exclusion to evolve (Maurer, 1989), artificially confining closely-related species from separate communities that evolved allopatrically may increase the likelihood of competitive interactions, leading to a rapidly cascading series of extinctions. Overall, large reserves comprising small chunks of many different communities may be as unsuccessful in preserving biodiversity as many small reserves. In both situations we would be trading evolutionary potential for current diversity, and possibly harming the current diversity in the process. We can circumvent this problem to some degree by giving priority to areas of endemism, rather than to areas of highest diversity. From a historical ecological perspective, species that evolve *in situ* (residents) are endemics. A concentration of endemic species in one location forms an area of endemism. Areas of endemism are important because they have been the focus of biodiversity production in the past, and thus may be 'hot spots' of evolutionary potential for the future. They do not always encompass the greatest number of extant species, nor are they always extremely large or centrally located. As a consequence, if a choice must be made about areas requiring protection, historical ecological data can provide information about regions that have been very important in the evolution of biological diversity.

We believe that this historical ecological perspective sheds a different light on the age-old problem of documenting biodiversity. The 'collect, classify and store everything found in a given area' approach has limited evolutionary utility because it is environmentally, rather than phylogenetically, based. It is important to document the 'elements of biodiversity', but we can only formulate a strategy for preserving that biodiversity when these elements are viewed, together with information about their interactions, within an evolutionary framework. This innovative step in management and policy formulation is critical, for as species disappear from this planet we are endangering our own ability to survive. More subtly, but equally disturbing, we are destroying our ability to study evolution, and thus to understand ourselves in relation to our world.

CONCLUSION

Historical ecology is a general term for comparative research that involves both documenting the macroevolutionary patterns of biological diversification and searching for the mechanisms underlying those patterns. In order to do this, we must generate evolutionary patterns in a manner that does not presuppose any particular evolutionary mechanism, and we must treat those patterns as the beginning point for studies of mechanism (i.e. experimental studies). There are two levels of patterns in evolutionary biology, phylogenetic trees and statistical comparisons. Although study of phylogenies and functional traits, and of phylogenies coupled with statistical analysis of variable traits, are rarely adequate tests of mechanisms by themselves, they are often critically important components of the search for mechanism, because they tell us how and where to look.

While the successes of the comparative approach are reason for celebration in themselves, they also highlight some harsh realities: (1) although their number is growing, well supported phylogenies are rare, so most are still the first and only phylogenetic hypotheses for a group; (2) the groups that have attracted phylogeneticists are rarely the groups that have attracted ecologists and ethologists (although this is changing); and (3) the number of active systematists is decreasing yearly. The solutions to these problems is straightforward, albeit logistically difficult without a groundswell of support from biologists in general. Ecologists must (1) support the training and hiring of more phylogeneticists; (2) provide more support for museum collections and museum systematists; and (3) become better versed about phylogenetic methodology. Systematists, for their part, must (1) provide more databases; (2) encourage students to work on groups that are classically of interest to ecologists and ethologists; and (3) develop better ways to explain their ideas to a naive, but enthusiastic, audience. In other words, there must be active collaboration based on mutual respect among systematists, ecologists, ethologists and other evolutionary biologists.

Historical ecology is a young discipline, and as such is not committed to any particular theoretical perspective, beyond the belief that evolution unifies all living organisms. So this is truly the most exciting time in the development of any scientific discipline; the stage of discovery. This journey of discovery will lead us down many pathways and open many previously inaccessible doors, leading to a richer explanatory framework for, and a truly unified theory of that most important of biological processes, evolution.

REFERENCES

ANDERSON, S. & ANDERSON, C.S., 1975. Three Monte Carlo models of faunal evolution. *American Museum Novitates*, (2563): 1–6.

ARNOLD, S.J., 1983. Morphology, performance and fitness. *American Naturalist, 23*: 347–361.

BARTON, N.H. & CHARLESWORTH, B., 1984. Genetic revolutions, founder effects, and speciation. *Annual Review of Ecology and Systematics, 15*: 133–164.

BAUM, D.A. & LARSON, A., 1991. Adaptation reviewed: a phylogenetic methodology for studying character macroevolution. *Systematic Zoology, 40*: 1–18.

BOWERS, M.A. & BROWN, J.H., 1982. Body size and coexistence in desert rodents: chance or community structure? *Ecology, 63*: 391–400.

BROOKS, D.R., 1979. Testing the context and extent of host–parasite coevolution. *Systematic Zoology, 28*: 299–307.

BROOKS, D.R. & McLENNAN, D.A., 1991. *Phylogeny, Ecology and Behavior: A Research Program in Comparative Biology*. Chicago: University of Chicago Press.

BROOKS, D.R. & McLENNAN, D.A., 1993a. Comparative studies of adaptive radiations with an example using parasitic flatworms (Platyhelminthes: Cercomeria). *American Naturalist*.

BROOKS, D.R. & McLENNAN, D.A., 1993b. Historical ecology: examining phylogenetic components of community evolution. In R.E. Ricklefs & D. Schluter (Eds), *Species Diversity in Ecological Communties*, 267–280. Chicago: University of Chicago Press.

BROOKS, D.R. & MITTER, C., 1984. Analytical basis of coevolution. In Q. Wheeler & M. Blackwell (Eds), *Fungus-Insect Relationships: Perspectives in Ecology & Evolution*, 42–53. New York: Columbia University Press.

BROOKS, D.R., O'GRADY, R.T. & GLEN, D.R., 1985. The phylogeny of the *Cercomeria* Brooks, 1982 (Platyhelminthes). *Proceedings of the Helminthological Society of Washington, 52*: 1–20.

BROOKS, D.R., COLLIER, J., MAURER, B.A., SMITH, J.D.H. & WILEY, E.O., 1989. Entropy and information in evolving biological systems. *Biology and Philosophy, 4*: 407–432.

BROOKS, D.R., MAYDEN, R.L. & McLENNAN, D.A., 1992a. Phylogeny and biodiversity: conserving our evolutionary legacy. *Trends in Ecology and Evolution, 7*: 55–59.

BROOKS, D.R., MAYDEN, R.L. & McLENNAN, D.A., 1992b. Phylogenetics and conservation. *Trends in Ecology and Evolution, 7*: 353.

BURKHEAD, N.M. & WILLIAMS, J.D., 1992. *Proceedings of the American Society for Ichthyology and Herpetology, 1992*: 76.

CARSON, H.L. & KANESHIRO, K.Y., 1976. *Drosophila* of Hawaii: systematics and evolutionary genetics. *Annual Review of Ecology and Systematics, 7*: 311–346.

CARSON, H.L. & TEMPLETON, A.R., 1984. Genetic revolutions in relation to speciation phenomena: the founding of new populations. *Annual Review of Ecology and Systematics, 15*: 97–131.

CLUTTON-BROCK, T.H. & HARVEY, P.H., 1978. Comparative approaches to investigating adaptation. In J.R. Krebs & N.B. Davies (Eds), *Behavioural Ecology: An Evolutionary Approach*, 7–29. Sunderland, Mass: Sinauer.

CODDINGTON, J.A., 1988. Cladistic tests of adaptational hypotheses. *Cladistics, 4*: 3–22.

CRACRAFT, J., 1982. A nonequilibrium theory for the rate-control of speciation and extinction and the origin of macroevolutionary patterns. *Systematic Zoology, 31*: 348–365.

CRACRAFT, J., 1984. Conceptual and methodological aspects of the study of evolutionary rates, with some comments on bradytely in birds. In N. Eldredge & S.M. Stanley (Eds), *Living Fossils*, 95–104. New York: Springer Verlag.

CROIZAT, L., NELSON, G. & ROSEN, D.E., 1974. Centers of origin and related concepts. *Systematic Zoology, 23*: 265–287.

DIAL, K.P. & MARZLUFF, J.M., 1989. Nonrandom diversification within taxonomic assemblages. *Systematic Zoology, 38*: 26–37.

DOBZHANSKY, T., 1937. *Genetics and the Origin of Species.* New York: Columbia University Press.

DUELLMAN, W.E., 1985. Reproductive modes in anuran amphibians: phylogenetic significance of adaptive strategies. *South African Journal of Science, 81*: 174–178.

DUNBAR, R.I.M., 1982. Adaptation, fitness and the evolutionary tautology. In King's College Sociobiology Group (Eds), *Current Problems in Sociobiology*, 9–28. Cambridge: Cambridge University Press.

EHRLICH, P.R. & WILSON, E.O., 1991. Biodiversity studies: science and policy. *Science, 253*: 758–762.

ELDREDGE, N., 1976. Differential evolutionary rates. *Paleobiology, 2*: 174–177.

ELDREDGE, N. & CRACRAFT, J., 1980. *Phylogenetic Patterns and the Evolutionary Process.* New York: Columbia University Press.

ENDLER, J.A. & McLELLAN, T., 1988. The processes of evolution: toward a newer synthesis. *Annual Review of Ecology and Systematics, 19*: 395–421.

ETNIER, D.A. & WILLIAMS, J.D., 1989. *Etheostoma (Nothonotus) wapiti* (Osteichthyes: Percidae), a new darter from the southern bend of the Tennessee River system in Alabama and Tennessee. *Proceedings of the Biological Society of Washington, 102*: 987–1000.

FITZPATRICK, J.W., 1988. Why so many passerine birds? A response to Raikow. *Systematic Zoology, 37*: 71–76.

FLESSA, K.W. & LEVINTON, J.S., 1975. Phanerozoic diversity patterns: tests for randomness. *Journal of Geology, 83*: 239–248.

FUTUYMA, D.J., 1986. *Evolutionary Biology*, 2nd edition. Sunderland, Mass: Sinauer.

GILINSKY, N.L., 1981. Stabilizing selection in the Archaeogastropoda. *Paleobiology, 7*: 316–331.

GOULD, S.J., RAUP, D.M., SEPKOSKI, J.J., SCHOPF, T.J.M. & SIMBERLOFF, D.S., 1977. The shape of evolution: a comparison of real and random clades. *Paleobiology, 3*: 23–40.

GREENE, H.W., 1986. Diet and arboreality in the emerald monitor, Varanus prasinus, with comments on the study of adaptation. *Fieldiana Zoology (N.S.), 31*: 1–12.

HARVEY, P.H. & PAGEL, M., 1991. *The Comparative Method in Evolutionary Biology.* Oxford: Oxford University Press.

HENNIG, W., 1966. *Phylogenetic Systematics.* Urbana: University of Illinois Press.

HUMPHRIES, C.J. & PARENTI, L., 1986. *Cladistic Biogeography.* Oxford: Oxford University Press.

KLEIMAN, D.G., 1980. The sociobiology of captive propagation. In M.E. Soule & B.A. Wilcox (Eds), *Conservation Biology*, 243–261. Sunderland, Mass: Sinauer.

LARSON, A., WAKE, D.B., MAXSON, L.R. & HIGHTON, R., 1981. A molecular phylogenetic perspective on the origins of morphological novelties in the salamanders of the tribe Plethodontini (Amphibia, Plethodontidae). *Evolution, 35*: 405–422.

LAUDER, G.V., 1981. Form and function: structural analysis in evolutionary biology. *Paleobiology, 7*: 430–442.

LAUDER, G.V. & LIEM, K.F., 1989. The role of historical factors in the evolution of complex organismal functions. In D.B. Wake & G. Roth (Eds), *Complex Organismal Functions: Integration and Evolution in Vertebrates*, 63–78. Dahlem Conferenzen: S. Bernhard.

LIEM, K.F., 1973. Evolutionary strategies and morphological innovations: cichlid pharyngeal jaws. *Systematic Zoology, 22*: 424–441.

LIEM, K.F. & WAKE, D.B., 1985. Morphology: current approaches and concepts. In M. Hildebrand, D.M. Bramble, K.F. Liem & D.B. Wake (Eds), *Functional Vertebrate Morphology*, 366–377. Cambridge, Mass: Harvard University Press.

LYNCH, J.D., 1982. Relationships of the frogs of the genus *Ceratophrys* (Leptodactylidae) and their bearing on hypotheses of Pleistocene forest refugia in South America and punctuated equilibria. *Systematic Zoology, 31*: 166–179.

LYNCH, J.D., 1989. The gauge of speciation: on the frequencies of modes of speciation. In D. Otte & J. Endler (Eds), *Speciation and Its Consequences*, 527–553. Sunderland, Mass: Sinauer.

MacARTHUR, R.H., 1958. Population ecology of some warblers of northeastern coniferous forests. *Ecology, 39*: 599–619.

MacARTHUR, R.H., 1972. *Geographical Ecology*. New York: Harper & Row.

MacARTHUR, R.H. & WILSON, E.O., 1967. *The Theory of Island Biogeography*. Princeton: Princeton University Press.

MAURER, B.A., 1985. On the ecological and evolutionary roles of interspecific competition. *Oikos, 45*: 300–302.

MAYDEN, R.L., 1986. Speciose and depauperate phylads and tests of punctuated and gradual evolution: fact or artifact? *Systematic Zoology, 35*: 591–602.

MAYR, E., 1960. The emergence of evolutionary novelties. In *Evolution After Darwin*, ed. S. Tax, pp 349–380. Chicago: Univ. Chicago Press.

MAYR, E., 1963. *Animal Species and Evolution*. Cambridge, Mass: Belknap Press, Harvard University.

MAYR, E., 1969. *Principles of Systematic Zoology*. McGraw-Hill: New York.

MISHLER, B.D. & CHURCHILL, S.P., 1984. A cladistic approach to the phylogeny of the 'Bryophytes'. *Brittonia, 36*: 406–424.

MITTER, C. & BROOKS, D.R., 1983. Phylogenetic aspects of coevolution. In D.J. Futuyma & M. Slatkin (Eds), *Coevolution*, 65–98. Sunderland, Mass: Sinauer.

MITTER, C., FARREL, B. & WIEGEMANN, B., 1988. The phylogenetic study of adaptive zones: has phytophagy promoted insect diversification? *American Naturalist, 132*: 107–128.

NELSON, G. & PLATNICK, N., 1981. *Systematics and Biogeography: Cladistics and Vicariance*. New York: Columbia University Press.

NOVACEK, M.J., 1984. Evolutionary stasis in the elephant-shrew, *Rhynchocyon*. In N. Eldredge & S.M. Stanley (Eds), *Living Fossils*, 4–22. New York: Springer Verlag.

PLATNICK, N.I. & NELSON, G., 1978. A method of analysis for historical biogeography. *Systematic Zoology, 27*: 1–16.

RAIKOW, R.J., 1986. Why are there so many kinds of passerine birds? *Systematic Zoology, 35*: 255–259.

RAIKOW, R.J., 1988. An analysis of evolutionary success. *Systematic Zoology, 37*: 76–79.

RAUP, D.M., 1984. Mathematical models of cladogenesis. *Paleobiology, 11*: 42–52.

RAUP, D.M. & GOULD, S.J., 1974. Stochastic simulation and evolution of morphology – towards a nomothetic paleontology. *Systematic Zoology, 23*: 305–322.

RAUP, D.M., GOULD, S.J., SCHOPF, T.J.M. & SIMBERLOFF, D.S., 1973. Stochastic models of phylogeny and the evolution of diversity. *Journal of Geology, 8*: 525–542.

RINGO, J.M., 1977. Why 300 species of Hawaiian *Drosophila? Evolution, 31*: 695–754.

ROOT, R.B., 1967. The niche exploitation theory of the blue-grey gnatcatcher. *Ecological Monographs, 37*: 317–350.

ROSEN, D.E., 1975. A vicariance model of Caribbean biogeography. *Systematic Zoology, 24*: 431–464.

ROSEN, D.E., 1985. Geological hierarchies and biogeographic congruence in the Caribbean. *Annals of the Missouri Botanic Garden, 72*: 636–659.

SIMPSON, G.G., 1953. *The Major Features of Evolution.* New York: Columbia University Press.

SPIETH, H.T., 1974. Mating behavior and evolution of the Hawaiian Drosophila. In M.J.D. White (Ed.), *Genetic Mechanisms of Speciation in Insects*, 94–101. Sydney: Australia and New Zealand Book Co.

STANLEY, S.M., 1979. *Macroevolution: Pattern and Process.* San Francisco: Freeman.

STANLEY, S.M., SIGNOR, P.W., LIDGARD, S. & KARR, A.F., 1981. Natural clades differ from 'random' clades: simulations and analysis. *Paleobiology, 7*: 115–127.

STIASSNY, M.L.J. & JENSEN, J., 1987. Labroid interrelationships revisited: morphological complexity, key innovations, and the study of comparative diversity. *Bulletin of the Museum of Comparative Zoology, 151*: 269–319.

TEMPLETON, A.R., 1979. Once again, why 300 species of Hawaiian *Drosophila? Evolution, 33*: 513–517.

TINBERGEN, N., 1964. On aims and methods of ethology. *Zeitschrift für Tierpsychologie 20*: 410–433.

VERMEIJ, G.J., 1988. The evolutionary success of passerines: a question of semantics? *Systematic Zoology, 37*: 69–71.

VRBA, E.S., 1980. Evolution, species and fossils: how does life evolve? *South African Journal of Science, 76*: 61–84.

VRBA, E.S., 1984a. What is species selection? *Systematic Zoology, 33*: 318–328.

VRBA, E.S., 1984b. Evolutionary pattern and process in the sister-group Alcelaphini-Aepycerotini (Mammalia: Bovidae). In N. Eldredge & S.M. Stanley (Eds), *Living Fossils*, 62–79. New York: Springer Verlag.

WEST-EBEHARD, M.J., 1983. Sexual selection, social competition and speciation. *Quarterly Review of Biology, 58*: 155–183.

WICKEN, J.S., 1987. *Evolution, Thermodynamics and Information: Extending the Darwinian Paradigm.* Oxford: Oxford University Press.

WILEY, E.O., 1981. *Phylogenetics: The Theory and Practice of Phylogenetic Systematics.* New York: Wiley-Interscience.

WILEY, E.O. & MAYDEN, R.L., 1985. Species and speciation in phylogenetic systematics, with examples from the North American fish fauna. *Annals of the Missouri Botanic Garden, 72*: 596–635.

WILEY, E.O., SIEGEL-CAUSEY, D.J., BROOKS, D.R. & FUNK, V.A., 1991. *The Compleat Cladist: A Primer of Phylogenetic Procedures.* Lawrence, Kansas: Special Publication of the Museum of Natural History, University of Kansas.

WILSON, A.C., BUSH, G.L. CASE, S.M. & KING, M.C., 1975. Social structuring of mammalian populations and rate of chromosomal evolution. *Proceedings of the National Academy of Science, USA, 72*: 5061–5065.

CHAPTER

2

The adaptationist wager

MARK D. PAGEL

CONTENTS

Keywords: Homology – convergence – adaptation – comparative methods – statistics – evolution.

Abstract

Two different branches of comparative biology espouse different criteria and methods for drawing inferences about adaptation. The homology approach is specialized for the study of single, often unique, evolutionary events, the convergence approach is designed to detect evolutionary patterns or correlations across a number of evolutionary events. Despite having complementary purposes, the methods share an interest in attempting to understand the origin and maintenance of complex traits. However, convergence approaches are often criticized for being unable to distinguish the current utility of a trait from the function for which the trait originally evolved. In this chapter I examine this and many of the other criticisms of the convergence approaches and argue that, if traits are properly defined, the recurrent fit between a trait and some environmental feature provides some of the best evidence available that the trait is an adaptation to its current function. Indeed, I suggest that to believe otherwise requires a peculiar view of evolution.

Phylogenetics and Ecology
ISBN 0-12-232990-2

INTRODUCTION

Blaise Pascal's second best-known contribution to philosophy and mathematics is an argument for the existence of God that came to be known as 'Pascal's wager'. Uncommonly for a philosopher or a mathematician, Pascal adopted the pragmatic view that there was little to be lost by believing in God, but potentially much to be gained should Heaven or Hell turn out to be real. Nietzsche would later hold in contempt what he considered was Pascal's intellectual decadence. But most people cannot think of even one of Nietzsche's contributions to philosophy, let alone two. Contempt alone is not a sufficient adversary to pragmatism. Could this be a homily for cladists and statistically minded comparative biologists?

I begin this chapter with the premise that pragmatism is a virtue in science, and that strict adherence to epistemological criteria, although laudable in principle, can often hinder rather than promote the understanding of empirical phenomena. Philosophy of science, for all its success at clarifying what scientists do, has had little impact on the course of science. This may be a good thing. Neils Bohr never saw the atom for which he constructed his greatly influential model. Bohr and his model might have been ignored had scientists strictly adhered to the demands of the logical positivists who followed him. Models, theories and ideas often reach well beyond the available data. Purely data-driven science will form few generalities. Closer to the concerns of this volume, comparative biologists seek to explain complex traits, usually with a view towards understanding whether and how they are adaptations. The problem confronting comparative biologists is that direct evidence will seldom be available to support the claim that a trait evolved as an adaptation to its current function.

It is useful in discussions of adaptations to be very clear about one's meanings. One can in principle separate the current function that a trait serves from the function for which it arose. For purposes of this discussion, I shall take a trait to be an adaptation to its current function if the trait arose or originated in response to natural selection for the purpose it currently serves (Coddington, 1988, Chapter 3; Harvey and Pagel, 1991). Some traits, although adaptive in the sense of promoting the fitness of their bearers, may not have arisen in response to the function they currently serve. Rhinoceros' horns are undoubtedly adaptive, and probably arose as a means of defence against predators. They are very likely still used for that purpose (even if somewhat ineffective against bullets). If, however, horns are important in repelling predators, and horn size or shape is heritable, then males and females may also make use of them when choosing a mate: possession of an effective horn is a trait that is desirable to have in one's offspring. But upon discovering this fact, one would not necessarily want to say that rhinoceros' horns evolved to attract the opposite sex. Thus, the rhinoceros' horn may be 'adaptive' in the sense of performing the useful function of being attractive, without being an adaptation for that function.

The distinction between 'adaptation' and 'adaptive' is useful, therefore, to separate factors responsible for the origin from those responsible for the maintenance of a trait. But how widespread are examples such as that given for the rhinoceros, and how should comparative biologists go about studying current function and adaptation? Two different branches of comparative biology espouse different criteria and methods for drawing inferences about adaptation. One, which I shall refer to as the cladistic or homology approach, emphasizes the study of single evolutionary events within a clade or lineage. A derived feature (apomorphy) such as a rhinoceros' horn is compared to what is thought to have been the ancestral condition. Evidence to test the hypothesis of adaptation is obtained from detailed investigations of the historical and ecological context surrounding the evolution of the trait, specifically investigations of the function of the derived trait compared to the ancestral condition. This approach then is a search for the factors responsible for the origin of a trait. The other approach to comparative studies emphasizes evolutionary convergences or the repeated evolution of analogous traits in each of several independent lineages. It is broadly the search for the 'adaptive significance' of traits as evidenced by their repeated association with some environmental feature taken to be the common selective force. I shall refer to this method as the convergence, functional or homoplasy approach.

These differences between the homology and convergence approaches have led some commentators to suggest that the convergence approach does not produce strong evidence for the inference of adaptation (Coddington, 1988, 1992; Carpenter, 1992). Indeed, despite grounds for peaceful co-existence between convergence and homology approaches, criticisms – sometimes pugnacious – have been levelled, especially at convergence methods (e.g. Carpenter, 1992). In the following sections I shall investigate the strengths and weaknesses of the homology and convergence approaches for investigating and drawing inferences about adaptations. I shall argue that direct evidence for 'adaptation' is rarely available to either approach. None the less, my view, to anticipate the course of this essay, is that convergence carefully defined and studied provides the strongest evidence available for both the current function and the origin of traits that have evolved more than once. Further, convergence actually strengthens our inferences about particular cases of homology, and in a way not otherwise available to the homology approach.

The 'adaptationist wager' of my title is the belief that convergence makes acceptable the inference of adaptation because no other single explanation of organismic diversity can plausibly explain repeated and independent convergence of derived traits. I do not take, and nor should readers, the previous statements to be criticisms of the homology approach. Indeed, comparing the two methods is somewhat artificial because, as I have mentioned (see also Coddington, Chapter 3) they have different purposes. Moreover, investigations of single homologous traits are often conducted with greater attention to the definition and measurement of traits than are studies of convergence (however, this is a sociological and not a logical statement). The result is often a story that will be persuasive to all but the most unreconstructed sceptic.

In the next section I shall discuss further what I see to be the philosophical difference between the homology and convergence approaches. In later sections I will take up some of the criticisms of convergence methods including issues of avoiding phylogenetic bias, of distinguishing function from adaptation, of scientific corroboration, of the inference of causality in comparative studies, and finally of controlling for phylogenetic effects.

ANALOGOUS TRAITS OR UNIQUE HOMOLOGIES

The homology approach studies what it regards as unique historical events. It investigates whether a derived trait is an adaptation in the species that share it as a homology. Figure 1 makes the homology approach more clear (see also Codding-ton, 1988, and Chapter 3; Harvey & Pagel, 1991, fig. 1.6). The derived trait D is apomorphic and shared by species 1, 2 and 3. The test of adaptation is to ask whether trait D originated in the lineage between species 3 and 4 in response to natural selection for derived function F_D. The test proceeds by comparing the function of trait D in species 3 to the function of the ancestral character A in species 4 (species 1 and 2 and species 5 can be included but are not logically necessary). The trait A in species 4 is taken to be an indicator of the ancestral condition that existed when trait D evolved. If the function of the derived trait in species 3 is absent in species 4, then that function, F_D, is a candidate explanation for what trait D is an adaptation for. This methodology for studying adaptations makes the inference of adaptation rest on evidence that a trait arose for the function it currently serves.

Homologists do not generally seek convergent instances of the evolution of a trait such as D in Fig. 1. Two different derived features, even if performing the same ostensive function, are not considered the same adaptation, having evolved from materially different ancestral conditions. In this sense all adaptations are unique, even though several species can share a homologous derived trait (a shared derived feature). These considerations may make unnatural to the cladist the clumping of species into groups of say, lekking species and non-lekking species. Species themselves become the centre-piece of evolution in this distinctly Aristotelian view of Nature – that every substance (species) has unique properties that explain its behaviour (traits) – but commendably lacking Aristotle's teleology. It is a view of Nature, that not only does not seek reduction to a smaller number of convergent processes, it is at odds with such a view. This is why the homology approach to comparative studies includes roughly only the clade including those species that

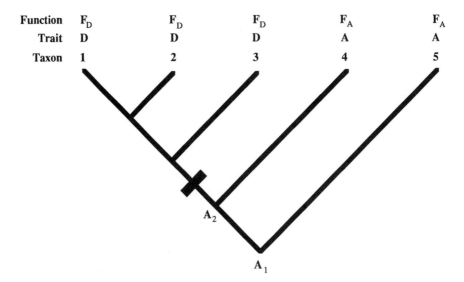

Figure 1 Five species and their cladogram showing the evolution of a derived feature D along the branch between species 3 and 4. Trait A is taken to represent the ancestral condition and thus D is the apomorphy. The functions F_D and F_A refer respectively to the function of the derived character and the function of the ancestral character.

share the derived feature or features under investigation, and an outgroup to establish the ancestral or primitive condition (Coddington, 1988, Carpenter, 1992).

To a functionalist (that is, to an investigator who studies repeated associations between traits and functions) there is something awkward about relying exclusively on single homologous traits to investigate adaptations. Evidence for convergence seems to be everywhere. Admittedly, convergence is at one level a construct of our imaginations. However, convergence happens, and can be expected *a priori* owing to fundamental asymmetries in Nature, and to fundamental similarities among organisms (e.g. all mammals produce milk) which reduce the number of likely ways that they may be expected to respond to selective forces. The first organism was not a predator, but the second one might have been. Whenever a niche exists for some organism, a predatory niche opens up. Predation as a way of life should have evolved independently many different times. We can make this prediction for worlds we have not seen, it is not dependent upon our observations of this world. If predators are common, then, most animals must evade predators in one way or another and hence crypsis, armament, warnings, speed and agility as mechanisms to avoid predation are also expected to have evolved many times.

Anisogamous sexual reproduction makes likely differential parental investment in offspring, and thus prepares the ground for sexual selection and for potentially differential investment in males and females. Thus, males develop showy ornaments and loud calls to attract females. They also develop big bodies, long teeth, loud calls, garish tails, big horns or antlers, the better to compete over females, or large testes and trenchantly designed penises, the better to compete in sperm competition. Females for their part may evolve preferences for certain kinds of males, chambered reproductive tracts to control sperm, and may invest more in one sex depending upon its expected reproductive success. Convergence is real. The success of theories of sexual selection, life history variation and sex-ratios attests to this.

Convergence methods initially seek conformity of function among analogous traits. They seek many independent instances of the evolution of a trait such as D in Fig. 1. This is a search for the 'adaptive significance' of traits. Thus, it is the function of traits that is of principal interest in a study of convergence, because the current utility of a trait can explain its diversity and maintenance. Whether in a particular instance a trait is an adaptation in the sense of having evolved for that purpose is of secondary interest. This is not to say that convergence methods fail to identify adaptations, a point I shall return to in the section 'Current utility and adaptation'. The study of convergence, in sharp contrast to the approach using homology, is a Newtonian search for a small number of shared processes. It depends upon the validity of reduction. Accordingly, I suspect that what motivates the convergence view is very likely a feeling that it is not so much species *per se* that are important for understanding function and adaptation as broad associations between two or more characters or between characters and environments. It is these associations that reveal shared patterns of selective forces (either those giving rise to or maintaining traits) and hence suggest and test general theory: do animals with intense conflict among offspring produce biased sex-ratios, are polygynous species more sexually dimorphic than non-polygynous species, is size of the litter a response to levels of juvenile mortality? This interest in generalizations may explain why comparative biologists who study convergence have (incorrectly) been slow to embrace phylogenies as an essential part of their methodology. However, even this has changed, and now the methods of cladistic inference are seen as a subset of the collection of techniques one might use to investigate evolution (Brooks & McLennan, 1991; Harvey & Pagel, 1991).

These philosophical differences between homologists and functionalists produce differences in practice, and thereby create the potential for disagreement. The homologist's view of evolutionary events as unique dictates that traits be studied as single historical events. The functionalist's view is that groups of organisms possessing analogous features are what reveal nomothetic evolutionary patterns. The homologist may shudder at the functionalists' apparently cavalier lumping. The functionalist will want to know if a single homologous trait is part of a general pattern, possibly suggesting a broader theory. This view dictates that where possible it is groups of organisms that should be studied, not single events. But what groups and which analogous traits?

THE PROBLEM OF PHYLOGENETIC BIAS

An oft repeated objection to the convergence approach is that several apparently convergent traits in fact may not be functionally similar, and therefore to include them in a single analysis is meaningless. Eyes have evolved a number of times and probably for seeing, but have they evolved for the same kind of seeing? Wings have evolved in birds, bats, flies and possibly some fish, but for what specific functions? This uncertainty about function has two implications. One is the narrow question about how to define function in any particular case. The other is the broader question of how to define and justify criteria for including or excluding species from study (Maynard Smith & Holliday, 1979; Coddington, 1988, 1992, Chapter 3). This is the question of phylogenetic bias. It is an important question because the status of an evolutionary 'law' should not depend upon the whims of the investigator in choosing the species to be included for study.

The narrow question is the less troublesome of the two. The exercise of defining analogous functions is subjective, but scientists can state clearly in any particular case what they mean by the function of a trait. Given clear definitions, it is up to others to agree or disagree. But it is the operational definition that matters. If I think that the crypsis of some ungulates is functionally equivalent to the crypsis of some rodents, it is up to me to state clearly my criteria. More pragmatically, sloppy definitions of convergence are probably more likely to lead investigators to miss relationships than to find them – to make statistical Type II rather than Type I errors. If the definition of convergence is sufficiently sloppy so as to include traits that are not really convergent, then one does not expect to find statistically improbable associations between this mixed bag of traits and some environmental force.

This answer, then, to the narrow question about defining analogous function is that it is self-correcting in some sense. Hoglund's (1989) study of lekking and size dimorphism in birds illustrates some of the issues of developing definitions of analogous functions. Hoglund's interest was in whether the intense sexual selection in lek breeding systems would favour larger body size in males. Lekking is used as a surrogate for sexual selection. Hoglund did not find that lekking was associated with greater size dimorphism when he measured the relationship across the 114 bird species for which he had data. However, he did find evidence for increased size dimorphism associated with lekking in the grouse family Tetraonidae. The Tetraonidae hold their leks on the ground, whereas some other bird species have arboreal leks. Hoglund speculated that possibly only in ground-based leks does sexual selection favour increased male size. The relationship across all birds mixes different kinds of selective pressures, and so it is not surprising that different functional responses have occurred.

The initially broad hypothesis has been replaced by one that specifies more clearly the nature of the selective force. The process of testing hypotheses and seeking relationships has found the level of generality appropriate to the hypothesis: ground-based leks rather than leks in general. The worry of course is that Hoglund's hypothesis about ground-based leks was entirely *ad hoc*. This should only be a concern, however, if the idea cannot be tested independently in other groups with and without ground-based leks.

A direct attack on the problem of inclusion criteria must give rational prescriptions for what species to include and what species not to include, and these prescriptions must be general. Inclusion criteria for the homology approach are entailed by the view of apomorphies as unique, and by the logic of adaptation, and so avoid the question of sampling. The logic of testing adaptation (Fig. 1) requires (minimally) that only the species that share the evolutionary novelty, plus sufficient outgroup species to define the ancestral condition, be included in the study. It is an atheoretical approach in that the homology itself delineates which species to include and therefore the criteria for inclusion need not make any direct reference either to an evolutionary hypothesis of function or of convergence (a cladistic hypothesis, that is, a proposed phylogeny of several species is a hypothesis about the pattern of speciation, but is not a hypothesis about the function of a trait).

The inclusion criteria for the homology approach cannot be justified on functional grounds for the simple reason that possession of the apomorphic character does not ensure that the character currently performs the function for which it evolved, much less that it performs the same function in the several species that may share it. I do not mean by these comments that the investigator lacks a hypothesis about function, only that one cannot use shared function to justify homology as one's inclusion criterion. Thus, Coddington's (1988) view that a cladogram showing the evolution of a derived feature provides logical information about how to test for the function of the trait, including what groups to use as controls, is correct, but does not concern us in this context. Rather, the inclusion criteria for the homology approach follow strictly from the view that all apomorphies are unique, whatever their current function.

In contrast to the homology view, the criteria for inclusion in a study of convergence are inseparable from the evolutionary idea one wants to test. That idea will make a claim (the hypothesis) about the similarity of the functions performed by several different convergent traits. All of the species that possess a trait that performs the hypothesized function are eligible, along with their immediate outgroups, for inclusion. However, logic and practice may not be equally clear. Coddington (1992) expresses concern with reference to Hoglund's (1989) lekking study on grounds that the hypothesis linking lekking to sexual dimorphism does not clearly spell out which species should be included and which should not. One implication of Coddington's remarks is that if Hoglund had only studied grouse, he might have reached a different conclusion about lekking and dimorphism. Coddington also wonders why flies, frogs and other species that lek were not included. These are important questions.

The first issue, that of confining the study to grouse, does not raise logical difficulties if one's conclusions are also confined to the organisms one has studied. Intuition and experience may tempt us to generalize, for example, that bird and frog leks will be similar. But to be valid, any empirical evolutionary generalization to emerge from a study of grouse would have to carry the qualifier 'in the grouse' to the extent that what was being studied was specific to grouse. Thus, if as a first step Hoglund had only studied grouse, no scientific problems would arise: indeed in the limit, such a strategy converges on the homology approach. Regarding the difference between the results Hoglund obtained from the full data set versus the restricted sample, if he had sampled in a biased way from within the grouse, or from within the rest of the birds, one would have reason to question his conclusions. On the assumption that he did use all data, or sampled from it at random, Hoglund has apparently discovered something unusual in the birds about grouse. Further investigation might reveal why they differ from the others. This is an interesting and potentially useful result.

The second issue – why confine the study only to birds – is more subtle. The hypothesis in a study of convergence must specify a relationship between the presence of some selective force, and the possession of a trait thought to be a response to that selective force. The precise specification of selective forces logically defines which species to include in the study. In practice inclusion criteria will often be defined on the basis of monophyletic groups, such as the birds. This is probably just convenience on the part of investigators, but the convenience can be shown to have a deeper logic. In the case of birds, one has greater reason to believe that they share many other factors of anatomy and development, ecology, lifestyle and so on, that may be relevant to the form of the selective force and to the number of different ways that the convergences are likely to be expressed. The similarities among birds probably mean that bird leks approximate each other to a greater degree than they do frog leks. Similarities among birds probably also mean that they can be expected to respond in similar ways to the sexual selection imposed by a lek. Note that the logic of this argument is emphatically not dependent upon homologies defining the birds, but rather rests on the belief that similar organisms will tend to live in similar environments, and be subject to similar selective pressures.

Hoglund, therefore, should have included flies and frogs in his study if his ideas about the nature and functions of leks applied to them (given differences in development, mating, life history and ecology) and if he wanted to be able to generalize to them. However, there is no necessity to include these other groups, because there may be strong reasons to believe that the nature of the convergences and selective forces differ greatly among flies, frogs and birds. A study by Read & Nee (1991) of 'Haldane's Rule', however, provides an example of a comparative hypothesis that logically demands to be tested across a very wide range of taxa. Haldane's Rule refers to the eponymous scientist's observation that in hybrid sexual crosses among animals, the heterogametic offspring are frequently absent, rare or sterile. Thus, in most mammal crosses, it is the male offspring that are typically sterile, whereas in birds it is the females that are affected. Haldane's rule, as stated, applies to any animal with sexual reproduction and in which the sex of offspring is determined by sex-chromosomes. Accordingly, Read & Nee searched right across the animal kingdom – eventually including data from mammals, birds, lizards, butter-flies, drosophila, mosquitoes, tsetse flies and even reduviid bugs – for evidence bearing upon Haldane's rule. Surprisingly, even this wide taxonomic search failed to turn up a sufficient number of independent instances of the evolution of hetero-gamety to test the hypothesis powerfully, leaving Read & Nee (1991) to conclude that there is not at present statistical support for Haldane's rule.

A different form of phylogenetic bias is that introduced by extinct species for which data are not available. Depending upon their character states, extinct species may change the inferred ancestral character states and thus what one takes to be the derived trait. This form of bias plagues both the convergence and homology approaches, but may be especially troublesome where conclusions are drawn on the basis of a single clade.

CURRENT UTILITY AND ADAPTATION

As discussed at the outset to this essay, the current utility of a trait may not be the same as the function for which the trait arose. Therefore, the observation of an association across a number of species between a set of convergent traits and an environmental feature does not ensure that those traits are adaptations to the problem posed by the environmental feature: one would need to know that each of those traits arose in response to the selective force represented by the environmental factor. This point is often used to criticize the convergence approaches, especially when simple correlations are calculated between two traits or between a trait and an environmental feature across a number of species. However, both homology and convergence approaches are confronted by the problem of distinguishing current utility from adaptation.

If an environmental feature thought to be the selective force behind a homology is present in those species with the trait, but is absent in the outgroup, then one has evidence consistent with the trait being an adaptation to that environmental condi-tion. Such evidence does not prove that the trait is an adaptation to this environ-mental condition: the trait may function for the reasons ascribed to it, but it may have arisen for some other purpose. Thus, in Fig. 1 the fact that species 1, 2 and 3 use the derived feature for derived function F_D, which in turn is missing in the outgroup species, does not prove that F_D was the reason why D evolved. The only significant advantage that this evidence has over the correlation is that the trait has been established to be derived, and is thus a candidate for being explained as an adaptation to the environmental feature. However, I will describe later in this section how this problem of the correlation approach can be easily overcome in con-vergence studies. My view is that overall, the problem of current utility is potentially more troublesome for the homology approaches.

For purposes of this section we will assume that natural selection was the original cause of the trait (as opposed to, say, drift), but that we cannot be sure whether the function the trait currently performs is the one for which it originally evolved. It will seldom be possible to provide direct evidence that a trait is an adaptation to its current function. Comparing the species with the derived feature to a sister-species which lacks the derived feature will not suffice. This is because, strictly speaking, the ancestral condition no longer exists: extant species lacking the derived feature may have evolved some other means of responding to the original selective force posed by the environment. Thus, we cannot be sure that trait A in Fig. 1 is equivalent to traits A_1 or A_2, even though most techniques for reconstructing ancestral values would make the nodes represented by A_1 and A_2 equal to A. One might, for example, have an interest in whether insect warning coloration evolved to advertise unpalatability, as a way to reduce predation. However, there is no particular reason to believe that predation rates will be lower in extant warningly coloured versus non-warningly coloured insects: the character state represented by the extant non-warningly coloured insects may not be the ancestral condition. Selection experiments on organisms that can be studied for several generations can demonstrate adaptations but will seldom be useful for studying the complex features of interest to many comparative biologists. Similarly, experiments that remove or alter the derived feature demonstrate its current function but not necessarily the function for which the feature arose. In short, the strict definition of adaptation although central to evolution by natural selection is something of a Platonic ideal for empiricists: they can partake in it but never achieve it for interesting cases. I suspect that Calvinist-like adherence to strictures preventing the inference of adaptation save for cases where adaptation has been demonstrated directly will do much to inhibit the flow of ideas and the understanding of Nature. I do not know of a single study of a complex trait that meets the strict criteria to prove 'adaptation' in the strong sense defined at the outset of this essay.

Given that adaptation is virtually impossible to show except in limited laboratory situations, how can empiricists strengthen their cases? The homology approach is to study intensively a single instance of the evolution of a derived feature and compare it to the putative ancestral condition in an attempt to build a detailed and complex web of interrelated facts that point to the original function of the trait. I will have more to say on this in the penultimate section. Here it is sufficient to say that such stories are often very persuasive, although I think they gain their greatest plausibility from the fact that we simply lack any other explanation for complex traits than natural selection. The convergence approach to the current utility versus adaptation problem is to seek repeated instances of analogous derived traits performing similar functions. Repeated convergence does not of course prove that the traits are adaptations to their current function, but it weakens many of the objections to the inference of adaptation that might apply in cases of a single homologous trait. The sheer improbability that each of many independently derived traits arose for some other purpose, each of which traits then came to be used for the same purpose, lends considerable weight to the inference that they are adaptations to their current function. Indeed, to hold any other view requires a peculiar view of evolution: that traits are labile and evolve for various functions until the organism chances upon using the trait for its current function, at which time there is no further modification of the trait. Traits evolved for some other purpose must, on this view, just by chance be optimally suited to their new function, or mysteriously lose the ability to be modified to their new function.

I have glossed over an important point in my description of the convergence methods: how to define characters. Studies of convergence must identify derived traits, and the several instances of convergence must be independent. For example, if one's idea is that leks cause great competition among males thereby favouring the evolution of larger male body size, one must show that male body size evolves to become larger in species that adopt lekking. Demonstrating that there is sexual size dimorphism is not sufficient: perhaps the females got smaller. The new comparative methods for studying convergence (reviewed in Harvey & Pagel, 1991), especially those designed for the study of continuously varying traits identify independent instances of convergence. These methods compute what are, for technical reasons, called 'contrasts' but which are more colloquially often little more than the pairwise difference between two sister-species on the values of their trait. Figure 2 shows a simple idealized phylogeny that can be used to illustrate the notion of a contrast or difference, and why it is a measure of a derived state. The two traits Y and X in Fig. 2 can represent any quantitative trait such as body size, generation time, home range area, foraging time, and so on. The difference between species 1 and 2 on the Y variable measures a derived state. It is the amount of evolutionary divergence in the trait between the two species since the two species split from their common ancestor. A similar contrast can be calculated for the X variable. The differences between species 1 and 2 on traits Y and X isolate the evolutionary changes that have occurred on the branches a and b. All of the evolutionary changes along the branch leading to the ancestor of those two species (branch e) is shared by the two species and thus subtracted out when the difference is found. The same set of contrasts can be calculated for species 3 and 4 (and so on for larger phylogenies). These contrasts are independent of those calculated for species 1 and 2 because the evolutionary changes that have taken place in branches c and d are independent of those in the first pair. Finally, in Fig. 2, it is possible to calculate a difference between the two nodes. This difference isolates the evolutionary changes that have occurred along branches e and f, and so is independent of the first two contrasts.

The test of convergence is to ask whether these pairwise differences in the two traits are associated. I have shown elsewhere (Pagel, 1993) that the correlation and regression coefficients describing these two sets of changes can be interpreted just the same as one might interpret the correlation across species, except that, unlike the simple correlation across species, the 'contrasts' explicitly measure derived states. Thus a positive correlation would mean that evolutionary changes to the trait

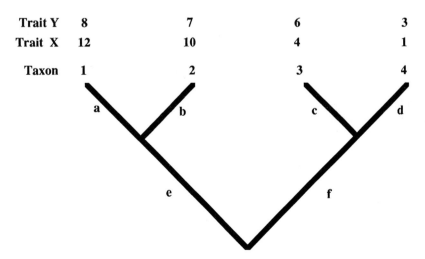

Figure 2 Idealized phylogeny of four species showing the values of two quantitative traits, Y and X.

X are directly associated with evolutionary changes to the trait Y. The logic of contrasts can also be applied to studies that include a categorical variable. For example, to study the association between lekking and sexual dimorphism one would find differences in the degree of sexual dimorphism between pairs of sister-species, one of which leks and one of which does not (apropos the above, one would actually be interested in comparing differences between males with differences between females in lekking versus non-lekking pairs). A significant positive correlation between changes to male size and changes in lekking would indicate that males are bigger in the presence of leks. The contrasts do not in themselves identify the ancestral condition, so one cannot be sure whether the presence of leks leads to larger males, the absence of leks leads to smaller males, or both. To determine this, one would have to reconstruct for every pair of sister-species what the probable ancestral condition was: lekking or non-lekking. In each case the difference indicates that one or both features is a new or derived state. Similarly, one cannot tell in this example whether changes to the degree of sexual dimorphism lead to the differences in lekking or vice versa, but such questions often also plague studies of single derived traits. The important points are that the measures are independent and derived.

The interpretation of the pairwise differences given above illustrates why the finding of an association between two such independent and derived measures is *prima facie* evidence for adaptation. To believe that the association merely indicates current utility is to believe that somewhere along the branch leading directly from the common ancestor of each sister-pair, the derived trait evolved for some other reason. Then, in each of these independent pairs the trait came to be used for its current purpose. This is of course possible, but requires quite unusual coincidences. Such a view may apply rather more strongly to a single apomorphy, where only one such coincidence is required.

But what about rhinoceros' horns? I return to this example because it illustrates both an apparent weakness in my argument, and the strength of the convergence approach. Imagine that there were many species of rhinoceros and several independent instances among them of the evolution of horns. Imagine further that in every case, rhinoceros' horns evolved as a means of defence against predators, but that they have also come to be used in mate choice. Thus, here is an example in which in each of several independent instances one of the functions of the derived trait is not the function for which the trait evolved. The convergence approach would measure pairwise differences in horn length (or some other feature of the horns) and pairwise differences in the strength of preferences for larger horns, and compute a correlation. If a significant correlation were found would not, then, this investigation conclude erroneously that the horns had evolved for reasons of mate choice and sexual selection? We must, here, separate the mechanical process of a method from the imagination of the individual investigator. On purely mechanical grounds, one conclusion from this correlation might be that the horns had evolved for mate choice. The same mechanical process applied to any one of these evolutionary events in isolation (that is, studied as a unique event) would also lead to the conclusion that the horn had evolved for mate choice. That is, applying the logic of Fig. 1, horns would be associated with mate choice in the derived condition, and the ancestral condition would by definition lack horns, and therefore would also lack the derived function of sexual attraction.

But the convergence investigation could also test whether evolutionary changes in horn size were correlated with evolutionary changes in the strength of predation pressure. It could additionally test via partial correlation whether variation in horn size was associated with predation pressure independently of variation in the strength of mate choice. If it were, as most models of sexual selection might expect (see section on 'Testing competing explanations' for a real example), the convergence approach could find the variation in horn size owing to natural versus to sexual selection. That is, the convergence method could identify and define separate derived components of the horn. The convergence approach would fail to separate these two explanations only under two circumstances. One is if the strength and direction of sexual selection were the same as that imposed by selection for repelling predators. Then the variation in horn size owing to one of the forces would be indistinguishable from that owing to the other. The other circumstance would arise if the horns evolved for predator defence but then, for some reason, did not evolve at all in response to sexual selection. There is no good reason to believe that this would happen unless even the smallest sexually selected change to the horn resulted in large loss of its capacity to repel predators. In any event, these latter difficulties also apply to the homology approach.

This example illustrates that if traits are properly defined, then it is reasonable under most views of evolution to expect that the original and current function of the trait are the same. It would be incorrect to say that the rhinoceros' horn was an adaptation to sexual selection. However, some of the size (or shape) of the horn may have originated in response to the extra force of sexual selection and it is correct to call that part an adaptation. Convergence methods based on pairwise contrasts are strongly placed to identify the separate components of a trait. To do so with a single apomorphy, one would need, in the case of rhino horns, a highly detailed and accurate optimality model relating the separate forces of natural and sexual selection to horn size. The model would make predictions about how much of the size (or shape) of the horn could be attributed to natural selection and how much to sexual selection. Comparison of the single horn to the predictions of the model could then be used to say whether the trait had evolved for one, the other, or both of the functions.

A further point about convergence is that repeated instances of convergence are not only persuasive as a group, but they also separately increase our confidence about the inference of adaptation in each of the homologies that together make up the convergences. Logically, of course, this assertion is false. Independent events are just that, and what happens in one has no bearing on what happens in another. Just because each of several convergent traits is an adaptation to 'X', does not prove that all such instances of convergence are adaptations to 'X': every blackbird I have ever seen was black, but the next one I see may be white. But convergence is different. Convergence indicates that similar selective forces have led to similar responses. In one sense, then, the cases of convergence are not independent because one set of responses to the selective force is more likely than others. The 'independent' cases actually share some sort of ability to respond to the selective force, by virtue perhaps of other similarities among them (and here I harken back to the arguments about inclusion criteria). Thus, convergences demonstrate how the same or similar explanation applies to each of several independent cases. Taken separately, each individual derived trait may admit a plausible and competing alternative explanation to the inference of adaptation. But it is unlikely that the same alternative will apply to each of the cases of convergence. Convergence thus can strengthen our confidence in one kind of explanation over the others.

TESTING COMPETING EXPLANATIONS

I shall continue my discussion in this section of alternative or competing explanations of a convergent trend. The null hypothesis in a study of convergence is that the character and the putative selective force (e.g. size and lekking), or the two sets of characters (age at maturity and body size) have evolved independently, the alternative hypothesis is that they are not independent. A significant correlation is evidence that independent evolutionary changes (i.e. 'contrasts' as defined in the previous section) in the selective force are associated with independent evolutionary changes in the trait, or that independent evolutionary changes in two traits are correlated.

Carpenter (1992: 192) asserts that in studies of convergence "significant correlation is always simply attributed to adaptation" and that the methods "do not really even attempt to measure corroboration" of the hypothesis of adaptation. Carpenter's assertions are false, but the topic of testing alternatives is worth pursuing.

Two issues are relevant in this context. One is whether the convergences making up the correlation are derived features and whether their current utility can be separated from adaptation. This was discussed in the previous section. The second issue concerns whether there are stronger null hypotheses than the absence of correlation against which to test the hypothesis of adaptation. Two sorts of explanation come to mind. One might be broad non-adaptive processes that can provide competing explanations for an observed correlation between two traits or between a trait and environmental feature. Another might be competing adaptive explanations. I consider these in turn below. In the next section I take up additional issues of corroboration and the inference of causality.

Forces such as genetic drift, allometry, pleiotropy, evolutionary lag and epigenesis, may all produce non-adaptive differences among populations. I shall, for purposes of this section, refer to these forces collectively as the 'theory of non-adaptation' because they are mechanisms by which organisms might differ but not for reasons related to natural selection. We want to investigate the theory of non-adaptation as a research programme for generating hypotheses to explain patterns of organismic diversity. Suppose the theory of non-adaptation were used to guide an investigation into variation in antler sizes in deer. The theory might assert that differences among species in antler size were not necessarily related to fitness or survival, but instead were caused by drift, allometry or pleiotropy. It is not my intention to caricature this view. Lewontin (1979) argued that because of the allometry of antler size and body size in deer it is "unnecessary to give a specific adaptive explanation for the extremely large antlers of large deer". Lewontin's phenomenological view does not of course rule out adaptation, but can 'explain' (in the statistical sense) the variation without reference to adaptation, even if it relies on an occult force, namely 'allometry'.

But antler size in deer also correlates strongly with the size of a species' social group. Male deer use their antlers to fight each other over access to females, so one inference is that antler size is an adaptation to male–male competition in these deer. Clutton-Brock, Albon & Harvey (1980) found that antler sizes of deer species not only vary with body size, but are larger in species with larger social groups, even after controlling for body size. The non-adaptive competing 'theory of allometry' is mute about this variation that is independent of body size. How about the other forces in the theory of non-adaptation, can they explain the variation that is independent of body size? Possibly the variation that is independent of body size represents genetic drift, or alternative adaptive peaks, or epistatic interactions, or time lags, or . . .? Each of these additional non-adaptive explanations could be invoked *ad hoc* species by species to explain some part of the remaining variance, but none could claim generality. However, the explanation of Clutton-Brock, Albon & Harvey was not *ad hoc*, and does claim generality. In fact, one could argue that the selective force Clutton-Brock *et al.* identified – male–male competition – has a stronger claim to being the principal determinant of antler size in deer than does body size.

Leaving aside for the moment the possibility in the Clutton-Brock *et al.* study that there is some other variable that correlates with social group size, it is simply not very parsimonious to explain a large number of different and independent instances of convergence between relative antler size and group size as instances of the trait arising for some other purpose then each coming to be used for male-male competition in just the right group size for the trait. To hold such a view one has to believe that a species evolves an antler size which then becomes more or less fixed and that the species subsequently finds itself coincidentally with the right group size to fit the trend, or that antler size is the principal determinant of group size. One might be inclined to speculate about such possibilities in the case of a single homologous trait, but not, as in the case of the antlers, for each successive enlargement or reduction in size of the trait among species: it is these repeated and independent changes in both the selective force and the trait that make the inference of adaptation so persuasive.

What I hope is clear from this example is that the theory of non-adaptation is not a general theory of complex organismic diversity. Indeed, it has no general mechanisms to be one. Drift, allometry and the like may be plausible explanations for a trait in isolated instances, but their force as explanations for convergence declines sharply with each new additional instance. Thus, one might be tempted to advert to a non-adaptive explanation such as genetic drift for any single difference between two populations or between two species, but doing so for broad variation in a trait among many species lacks plausibility and frequently requires one to invoke such fuzzy and typically untestable ideas as 'allometry' or 'pleiotropy' as explanations.

Merely showing that one trait changes when another is selected is not sufficient to explain an allometric trend via pleiotropy. One has to believe that the allometric trend is inviolable; that even if there was a phenotype of higher fitness it would not have arisen. This seems very unlikely in the case of many allometric trends where there is independent genetic variance within species in both characters. There is another peculiarity of the 'pleiotropy' view when given as an alternative non-adaptive explanation for allometric trends. It implies that, using antlers as an example, even if antlers were not adaptive the deer would still have them simply because of pleiotropy. Antlers presumably are costly to produce and to carry around, and may get tangled in trees. So, it is difficult to imagine that they would remain unchanged if they served no purpose. Is it not simply much more parsimonious to believe that they are adaptive? After all, the females do not have them. What seems a much more plausible explanation for pleiotropy or allometry is that they arise because selection favours sets of traits that bear certain relationships to each other. The 'allometry' or pleiotropy' then is a consequence of selection, not a cause of the observed phenomenon. For these reasons the theory of non-adaptation does not constitute a progressive programme of research, rather it is doomed to nibble away at the edges.

Carpenter (1992: 192) attempts a statistical version of the 'weak corroboration' criticism. Carpenter seizes on what he takes to be the assumption of some statistical methods that "the expected association between traits is modeled for a given phylogeny as a process of Brownian motion [!!!]. Of course, nothing is learned from the finding or not of a significant association with this method; presumably it is already known that characters did not evolve by this process" (exclamation points and brackets in the original). But Carpenter is wrong about his assumption, and this nullifies his conclusion. I find this such a common misunderstanding that I will spend some time explaining why it is wrong.

Statistical methods do not model 'the expected association between two traits as a process of Brownian motion'. Statistical methods such as correlation only presuppose that the data are normally distributed, or in the case of regression only that the residuals from the regression line are normally distributed. The methods do not make any assumptions about how normal distributions come about. Brownian motion is a process (one of many) that can generate normal distributions and so is often invoked for convenience. What this all means is that, *contra* Carpenter, the failure to find a significant association has nothing to do with whether the characters evolved by Brownian motion. The failure to find a significant association suggests only that evolutionary changes in one variable have been independent of evolutionary changes in another. The finding of a significant association indicates that the changes in one variable have been associated with changes in the other. Either is quite an important conclusion, one which presumably we did not know before we calculated the correlation, and which say nothing about Brownian motion.

An analogy may make this argument clearer. Consider this caricature of Carpenter's view that exposes its flaw. 'The process of reconstructing on a tree the ancestral character states of a trait, is modelled using parsimony rules [!!!]. Of course, nothing is learned from the finding that we can or cannot reconstruct unambiguously with these rules the ancestral character states of a trait; presumably it is already known that characters do not evolve according to parsimony.' Parsimony (or any other set of rules) is followed as an ideal that can be used to reconstruct ancestral character states. I do not think anyone seriously believes that all characters always follow parsimony. But that misses the point: one does not reconstruct ancestral characters to test parsimony anymore than one calculates a correlation as a way of testing whether a character evolved by Brownian motion.

An informed version of Carpenter's 'weak corroboration' criticism is that our belief in the hypothesis of evolution by natural selection would be strengthened if we could reject competing hypotheses that made stronger claims than the null hypothesis. I have some sympathy with this philosophical position. In fact, it is one of the strengths of the convergence approach. Rigorous 'null' hypotheses or competing adaptive theories are frequently available for the same data. Balmford, Thomas &

Jones (1993) for example, make careful predictions about the size and shape of birds' tails on the basis of an aerodynamic model. The model predicts what natural selection might be expected to produce in the absence of sexual selection. Against this 'null' model the authors derive predictions from the theory of sexual selection about which species should deviate and how they should deviate from the null model. The data support these predictions.

Another example comes from a study of two competing functional explanations for variation among organisms in the size of the nuclear genome (Pagel & Johnstone, 1992). The size of the nuclear genome varies about 80 000 fold across eukaryotes. One explanation (Cavalier-Smith, 1985) provides a functional link between the size of the nuclear genome and the volume of cytoplasm. Another explanation suggests that much of the nuclear DNA is pure junk that functions only to improve its own fitness, and thereby will accumulate in the genome until the costs to the organism become too great. Both explanations predict a relationship between genome size and generation time: species with longer generation times are expected to pay lower costs of having excess DNA, and longer generation times are correlated with cytoplasmic volume. We were able to remove the variation in genome size that is correlated with cytoplasmic volume, and show that the remaining variation still correlated with generation time. However, the reverse was not true, providing support for the junk-DNA explanation.

Convergence, then, does allow one to test competing explanations for a hypothesized adaptive trend, or to exclude on logical grounds many of the non-adaptive forces that might apply in single instances of a homology.

CORROBORATION AND THE INFERENCE OF ADAPTATION AND CAUSATION

I discuss in this section differences between the homology and convergence approaches in the ways that they form causal inferences. Many of the points made here draw upon points I have made in previous sections, but the emphasis here is on what might be termed the epistemology of the two approaches.

The data set with the homology approach often consists of a single evolutionary event (the evolution of the derived trait) that may be shared by several species (Fig. 1). The key point for purposes of understanding inference is that only one evolutionary event separates the ingroup and outgroup. In principle, then, any variable that differs between the two groups and which is shared by the members of the ingroup that have the derived trait is potentially an explanation for the evolution of the derived trait. This is not a strong position of inference because closely related species tend to be similar on many aspects of their morphology, behaviour and environments, and thus there will often be more than one potential explanation for the derived feature. The inferential weaknesses of the homology approach are not confined to competing selectionist explanations. I argued that the theory of non-adaptation does not provide very many plausible alternatives to natural selection when the phenomenon to be explained is the repeated convergence of a trait with some feature of the environment. However, it is to single evolutionary events that the theory of non-adaptation will apply most plausibly, if it has any plausible application.

These weaknesses mean that the homology approach must rely on other forms of corroboration than the statistically improbable repetition of the convergence approach. Corroboration may be sought in finding other parts of the picture that make sense or are even expected given the idea one wants to test. This sort of corroboration takes the form of a narrative or listing of other facts deemed relevant to the story. I find many such narratives persuasive. But narratives are difficult to judge logically. Who decides what facts to include and what facts not to include? How should the myriad facts be weighted in one's mind, especially given that many of them will not be independent (closely related species tend to come with suites of common characters). There is no objective way to determine when a narrative is sufficient or complete, short of direct evidence for adaptation. Instead, one stops when one is satisfied that no other explanation is as plausible as the adaptive explanation. This is reasonable.

A different form of corroboration is to test a transformation series, in which a set of predictions is made about the order of changes to several or more traits. The degree of match between the predicted and observed orderings when the traits are reconstructed on to the cladogram is used as a measure of the strength of corroboration of the idea. Like the convergence approach this form of corroboration rests on improbability. In this case the improbability is that the observed ordering should match the predicted ordering. One can even attach probabilities to it. For example three derived features can appear in six equally likely orders under the null hypothesis that they bear no relation to each other. The probability that a predicted and observed ordering match, then, is 1/6 by chance.

Carpenter (1989) tested a four-step transformation series to investigate West-Eberhard's (1978) theory of the evolution of eusociality in wasps. West-Eberhard suggested that primitively solitary bees first evolve a casteless form of polygyny. This is followed by a caste-containing group with only partial reproductive division of labour. A eusocial stage with greater division of labour is hypothesized to follow. The test of this transformation series then proceeds by reconstructing changes in these traits onto the cladogram of a suitable group. Carpenter found some but not all of the predicted transformations in the Vespidae, concluding (1989: 142) "this test corroborates several of the transitions proposed by West-Eberhard (1978), but indicates that others probably did not occur". Is this evidence for a match or a mismatch of the model with the observed data? How do we judge mismatches? Do they falsify the entire model, or only part of it? Do we suspect the data?

These may seem harsh questions to raise about a careful empirical study. However, in purely logical terms much of the inferential strength of the transformation series approach resides in what the theory under test rules out: given a four-step transformation series, there are twenty-four possible outcomes. West-Eberhard's theory boldly rules out all but one. A match between the predicted and the observed series is arresting in such cases because it is unlikely to have occurred by chance. Mismatches are much less arresting unless one particular mismatch has been predicted from some other theory. However, given the uncertainties about the reconstruction of ancestral characters (even if they fit unambiguously, they are an inference), and the possibility of unknown or extinct species that might alter the reconstructions, it may be unclear whether to reject the original theory or that deriving from the reconstructed data. In such instances one must seek out additional information that supports one or the other views, and Carpenter does just this in his article to support the validity of his reconstructions.

The data set for a convergence approach consists of many independent replications of the phenomenon of interest. This frequently makes possible a number of things that can increase our faith in one sort of explanation over another. Competing explanations and confounding variables can be tested, or excluded as implausible. I have argued repeated convergence strengthens our belief in the inference of the origin and function of a trait in individual cases of homology. Thus, were it possible to test a transformation series in each of several independent lineages, at least some of the inferential difficulties that may arise from a test of a single transformation series would disappear.

This comparison may be unfair because some things have happened only once. The convergence approach can only study things that have happened more than once. One cannot study the evolution of birds' wings with the convergence approach, although as I mentioned earlier one can study quite precisely the reasons for variation in birds' wings (e.g. Balmford *et al.*, 1993). This should not be seen as a criticism of convergence approaches, any more than to say that one cannot apply the homology approach to analogous traits. One does not, normally, view the heavens through a microscope. But this does not render microscopes useless or the heavens uninteresting. The two methods of inference have different applications. Similarly it has been suggested (Coddington, 1990; Carpenter, 1992) that the homology approach may overturn some things that the convergence approach calls adaptations, but that the convergence approach will not overturn any conclusions of the homology approach. I know of no evidence to support this assertion. But more to the point it is quite clearly a sociological and not a logical statement: its truth or falsity relies upon the practitioners and not the methods themselves.

Coddington (1990: 384) questions convergence, saying that "the strength of the convergence approach is proportional to the number of independent gains, transformations, or losses on the cladogram. This result is at least ironic: the worse the fit of the character is to the cladogram, the better the chance it has to be accepted as an adaptation". This criticism may be slightly overstated. To a cladist, a labile character poses grave difficulties because the more times a character changes the less likely one is able to reconstruct its ancestral conditions with certainty. However, to a functionalist interested in convergence a complex character that has evolved many times independently provides much more information for studying its environmental and other correlates than a character that has evolved only once. I have slightly fiddled this comparison by assuming that it is known that the trait has evolved more than once. But this will often be apparent even when the phylogenies are not known with certainty, either because of morphological differences among the analogous traits, or because the convergent trait is not found in species closely related to those with it.

CONTROLLING FOR PHYLOGENETIC EFFECTS

There is much confusion about how comparative methods for convergence make use of phylogenetic information. There is a widespread belief that they 'control' for phylogeny by somehow removing from the data variation associated with phylogeny.

There also seems to be a belief that variation that is associated with phylogeny somehow represents phylogenetic 'inertia', 'constraint' or variation that is otherwise not maintained by selection. I present here a very brief history of methods for convergence in hopes to dispel these beliefs. Fuller accounts can be found in Pagel & Harvey (1988), Harvey & Pagel (1991) and Pagel (1992). Newer comparative methods make use of all of the variation in the data set, all of it is treated as relevant to the hypothesis of adaptation, and yet these methods still take account of similarity among species that is correlated with common ancestry in such a way as not to overestimate the amount of evolutionary innovation.

In early comparative studies of convergence, correlations across species between two traits or between a trait and an environmental feature were frequently used. Quite apart from any of the concerns about identifying ancestral versus derived states, and separating current utility from adaptation, these correlations implicitly treat species as independent data points. This is equivalent to the assumption that each species has independently evolved its character. For anyone believing in descent with modification, this assumption is wildly implausible. Closely related species tend to be similar. It is not because they each have independently evolved similar traits, but rather because they share much of their evolutionary history. This is not an argument that phylogeny constrains evolution, merely that similar species living in similar environments will tend to make use of similar traits that may have evolved in a common ancestor. The upshot is that with respect to that part of their evolutionary history that is shared, species cannot be treated as independent with respect to the evolutionary hypothesis.

Some of the first attempts to take into account similarity owing to shared phylogenetic history made use of nested analysis of variance (Crook, 1965; Clutton-Brock & Harvey, 1977), and what might be termed phylogenetic subtraction methods (Stearns, 1983). The nested analysis of variance was used to identify the taxonomic level in a Linnean classification at which one could assume independence. Thus, for example if species were not independent perhaps genera or families were. If families were chosen as the level at which independence could be assumed, then all variability below the level of families was ignored for purposes of testing hypotheses, and the degrees of freedom for statistical tests were based on the number of families represented in the data. The logic is as follows. Treating each species as independent would count the evolutionary changes leading to families as an independent event in each species descended from a particular family. In the context of Fig. 2, treating species as independent would count twice the evolutionary changes that occurred in branch e, and similarly those in branch f. Using the family as the level of analysis, this variation counts only once. Phylogenetic subtraction methods (Stearns, 1983), although having the same goal, act on the data in precisely the opposite manner. These methods make the assumption that it is variation at higher taxonomic levels that should not be counted when comparing differences among species. Phylogenetic subtraction methods remove from the species data point variation owing to differences among orders and families, leaving only variation among genera and species within genera – exactly that which was ignored by the nested analysis of variance.

The nested analysis of variance and the phylogenetic subtraction methods both suffered from ignoring useful variation in the data, and for not having a way to produce independent data points. Treating, for example, families as the unit of analysis merely shifts to a higher taxonomic level concerns about non-independ-ence. Families will tend not to be independent for all the same reasons as species are not. Conversely the phylogenetic subtraction method must account for why it treats differences among orders and families as irrelevant to questions about adaptation. Wanntorp *et al.* (1990), for example conclude that most variation in life history can be attributed to descent with modification and body size. But what does such a statement imply about evolution and adaptation? Do these authors believe that differences in life history tactics between ungulates and rodents are not adaptive simply because they are correlated with body size and phylogeny?

The solution to these problems was first suggested by Felsenstein (1985) who showed how, given a phylogenetic tree of species, the set of pairwise comparisons between each pair of sister-species and pairs of higher nodes on the tree defined a group of mutually independent comparisons. These comparisons are the same as the contrasts I described earlier in connection with Fig. 2. The difference between two sister-species captures the evolutionary divergence between them since they split from their common ancestor. This will be independent of the evolutionary divergence between other species-pairs, between pairs of higher nodes, or between a species and a node. Differences between higher nodes on the tree capture earlier evolutionary divergences, that is, ones that preceded the evolution of the extant species. Precisely how to form these comparisons and to estimate appropriate values to place at higher nodes is the subject of much research and discussion (Felsenstein, 1985; Grafen, 1989, 1992; Pagel & Harvey, 1989, 1992; Harvey & Pagel, 1991; Pagel, 1992). However, the key point is that these new comparative methods based on contrasts make use of all of the variation in the data set, and have a way in principle of dividing up the data into a number of independent pieces of evolutionary change, each of which bears on the hypothesis of correlation between two traits. Nothing is removed, subtracted, or treated as irrelevant to the hypothesis of adaptation.

SUMMARY

I have argued in this chapter that most of the common criticisms of convergence methods for studying adaptations are either overstated, are sociological rather than methodological statements, or apply with equal force to homology approaches. Convergence can be defined, and rational criteria for the inclusion and exclusion of species from a study of convergence can be developed. I showed how in any particular study these criteria follow naturally from the idea one wants to test. Convergence studies are frequently criticized for failing to distinguish current utility from adaptation. However, given a proper definition of the trait, the current utility of a trait is very likely to be the function for which it arose. This is because it is unlikely that the same trait would repeatedly arise for some other purpose, then come to be used for its present purpose. Homology studies are plagued by the problem of current utility versus adaptation as badly if not worse than studies of convergence. Convergence allows one to test competing non-adaptive and adaptive explanations, and to rule out many possible confounding variables. I argued that these alternatives are much more likely to apply to a single case than to multiple cases of convergence. Finally, I put forward the argument that given certain assumptions about diversity, repeated instances of convergence actually increase our confidence in the inference of the origin and function of individual homologies. This form of inference is unique to the convergence approach.

ACKNOWLEDGEMENTS

I thank Paul Eggleton and Dick Vane-Wright for inviting me to the symposium that resulted in this book, and Jonathan Coddington for discussion of many of the points I raise in this chapter. Work on this chapter was supported by a Science and Engineering Research Council Grant (UK), No. GR/F 98727.

REFERENCES

BALMFORD, A., THOMAS, A.L.R. & JONES, I.L., 1993. Aerodynamics and the evolution of long tails in birds. *Nature, 361*: 628–631.

BROOKS, D.R. & McLENNAN, D.A., 1991. *Phylogeny, ecology, and behavior. A research program in comparative biology.* Chicago: University of Chicago Press.

CARPENTER, J.M., 1989. Testing scenarios: wasp social behaviour. *Cladistics, 5*: 131–144.

CARPENTER, J.M., 1992. Comparing methods [Review of P.H. Harvey & M.D. Pagel, 1991. *The Comparative Method in Evolutionary Biology*, Oxford University Press]. *Cladistics, 8*: 191–196.

CAVALIER-SMITH, T., 1985. Eukaryotic gene numbers, non-coding DNA, and genome size. In T. Cavalier-Smith (Ed.), *The Evolution of Genome Size*, 69–103. New York: John Wiley.

CLUTTON-BROCK, T.H. & HARVEY, P.H., 1977. Primate ecology and social organization. *Journal of Zoology, 183*: 1–33.

CLUTTON-BROCK, T.H., ALBON, S.D. & HARVEY, P.H., 1980. Antlers, body size, and breeding group size in the Cervidae. *Nature, 285*: 565–567.

CODDINGTON, J.A., 1988. Cladistic tests of adaptational hypotheses. *Cladistics, 4*: 3–22.

CODDINGTON, J.A., 1990. Bridges between evolutionary pattern and process. *Cladistics, 6*: 379–386.

CODDINGTON, J.A., 1992. Avoiding phylogenetic bias [Review of P.H. Harvey & M.D. Pagel, 1991. *The Comparative Method in Evolutionary Biology*, Oxford University Press]. *Trends in Ecology and Evolution, 7*: 68–69.

CODDINGTON, J.A. 1994. The roles of homology and convergence in studies of adaptation. In P. Eggleton & R.I. Vane-Wright (Eds.), *Phylogenetics and Ecology*, 53–78. London: Academic Press.

CROOK, J.H., 1965. The adaptive significance of avian social organization. *Symposia of the Zoological Society of London, 14*: 181–218.

FELSENSTEIN, J., 1985. Phylogenies and the comparative method. *American Naturalist, 125*: 1–15.

GRAFEN, A., 1989. The phylogenetic regression. *Philosophical Transactions of the Royal Society of London, Series B, 326*: 119–156.

GRAFEN, A., 1992. The uniqueness of the phylogenetic regression. *Journal of Theoretical Biology, 156*: 405–423.

HARVEY, P.H. & PAGEL, M.D., 1991. *The Comparative Method in Evolutionary Biology.* Oxford: Oxford University Press.

HOGLUND, J., 1989. Size and plummage dimorphism in lek-breeding birds: a comparative analysis. *American Naturalist, 134*: 72–87.

LEWONTIN, R.C., 1979. Sociobiology as an adaptationist program. *Behavioral Science, 24*: 5–14.

MAYNARD SMITH, J. & HOLLIDAY, R., 1979. Preface. In J. Maynard Smith & R. Holliday (Eds), *The Evolution of Adaptation by Natural Selection*, v-vii. London: The Royal Society.

PAGEL, M., 1992. A method for the analysis of comparative data. *Journal of Theoretical Biology, 156*: 431–442.

PAGEL, M., 1993. Seeking the evolutionary regression coefficient: an analysis of what comparative methods measure. *Journal of Theoretical Biology, 164*: 191–205.

PAGEL, M. & HARVEY, P.H., 1988. Recent developments in the analysis of comparative data. *Quarterly Review of Biology, 63*: 413–440.

PAGEL, M. & HARVEY, P.H., 1989. Comparative methods for examining adaptations depend on evolutionary models. *Folia Primatologica, 53*: 203–220.

PAGEL, M. & HARVEY, P.H., 1992. On solving the correct problem: wishing does not make it so. *Journal of Theoretical Biology, 156*: 425–430.

PAGEL, M. & JOHNSTONE, R., 1992. Variation across species in the size of the nuclear genome supports the junk-DNA hypothesis for the C-value paradox. *Proceedings of the Royal Society, Series B, 249*: 119–124.

READ, A. & NEE, S., 1991. Is Haldane's rule significant? *Evolution, 45*: 1707–1709.

STEARNS, S.C., 1983. The influence of size and phylogeny on patterns of covariation in life history traits in mammals. *Oikos, 41*: 173–187.

WANNTORP, H.E., BROOKS, D.R., NILSSON, T., NYLIN, S., RONQUIST, F., STEARNS, S.C. & WEDELL, N., 1990. Phylogenetic approaches in ecology. *Oikos, 57*: 119–132.

WEST-EBERHARD, M.J., 1978. Polygyny and the evolution of social behavior in wasps. *Journal of the Kansas Entomological Society, 51*: 832–856.

The roles of homology and convergence in studies of adaptation

JONATHAN A. CODDINGTON

CONTENTS

Keywords: Adaptation – comparative methods – cladistics – homology – convergence – philosophy.

Abstract

The study of adaptation traditionally has proceeded under either of two modes, here termed the homology approach and the convergence approach. In recent years, both approaches have benefited greatly by using cladistics to define homology and homoplasy (convergence is one kind of homoplasy) as alternative explanations of pattern in a comparative data set. The homology approach treats adaptation as one potential causal explanation of synapomorphy among many. It tests the assertion that natural selection predominantly determines biological pattern (Darwin's theory) by evaluating data on the performance, utility, or function of a homologous (hence historically unique) trait under the twin strictures of an adaptive hypothesis and optimization on the cladogram. It uses data on current utility in the test, is rooted in the natural history of the case and makes falsifiable claims about particular instances. The results pertain only to the case or clade studied. Except when summed, such results are unlikely to test evolutionary 'law' or to establish overarching evolutionary pattern. The method is best used to investigate historical events perceived to be of exceptional interest or importance.

The convergence approach forgoes detailed study of particular cases to reach for statistically significant correlations between classes of non-homologous events and nearly always attributes 'significant' results to one common cause – natural selection. Neither the homology of the 'trait' under study nor monophyly of the clade circumscribe its application. It is best used to establish evolutionary laws unbounded by the particulars of history. Some subjectivity in application and the separation from the particulars of the individual cases composing the correlation poses problems for the method. Criteria governing the hypothesis (whether that under test or the null), the definition of the phenomenon and taxon sampling need clarification.

INTRODUCTION

This paper seeks to differentiate and reconcile what may appear to be two different approaches to phylogenetically based research on adaptation. They ask similar questions, have different goals, use different methods and focus on different kinds of data. One is the explanation of unique events within lineages. The other is the explanation of correlations among similar events across lineages. The former emphasizes the analysis of evolutionary novelties (apomorphies), the latter empha- sizes the analysis of coincidences (homoplasies, convergence). Both approaches are basic to the 'comparative method'. While both approaches have their own methodo- logical concerns, my intent here is to contrast and analyse their respective strengths and weaknesses, paying particular attention to the logical interrelationships.

The analysis of unique events seeks to explain single, perhaps major features of evolution (Wanntorp, 1983; Padian, 1985; Greene, 1986; Coddington, 1988, 1990; Carpenter, 1989, 1991; Donoghue, 1989; Lauder, 1990; Wanntorp et al., 1990; Baum & Larson, 1991). It studies evolutionary homologues in their phylogenetic and ecological/functional contexts, and places great emphasis on empirical assessments of the function or utility of the trait. In general, such evolutionary events are easy to delimit and richly detailed. The price is that they are unique, or nearly so. Their investigation is peculiarly open-ended, and often frustratingly inconclusive.

Correlations among non-homologues, on the other hand, are convergences, parallelisms, replicates, not unique (Ridley, 1983; Huey, 1987; Pagel & Harvey, 1988; Sillén-Tullberg, 1988; Bell, 1989; Burt, 1989; Grafen, 1989, 1992; Gittleman & Kot, 1990; Maddison, 1990; Martins & Garland, 1991; Garland Harvey & Ives, 1992; Gittleman & Luh, 1992; Pagel, 1992; Sillén-Tullberg & Moller, 1993; reviewed in Harvey & Pagel, 1991). Correlation establishes a statistical pattern, but necessarily at great remove from the biology of any given instance. In general, this approach places greater emphasis on statistical patterns than the quality of data relevant to performance, function or the utility of traits. Correlations therefore can have a shallow quality – at worst just two series of numbers that barely achieve significance through rejection of a rather prosaic null hypothesis. Such is the price, but the payoff can be general trends verging on evolutionary law. For reasons that will become clear later, I call the study of historical uniques the 'homology' approach and the study of replicated events the 'convergence' approach.

An example may clarify the distinction. The homology approach might focus on an amazing thing, for example the origin of the orb as an architectural pattern in spider webs (Coddington, 1986a, 1986b, 1991). The convergence approach instead might focus on amazing coincidences, such as mimicry of ants by many different groups of spiders (Oliveira & Sazima, 1984; Oliveira, 1988). However different the two interests may seem, both depend on an appreciation of the amazing. The former concerns a startlingly unique event for which several alternative hypotheses are plausible, the latter concerns a improbably large set of (unique) events that possibly result from a common cause, perhaps selection to reduce attacks by predators.

Both approaches have always depended on taxonomy to structure and define the questions asked, but taxonomy has not always been equal to the task. Not until the mid 1970s did taxonomists decide to focus on one central scientific problem – reconstructing phylogeny. Since then, non-evolutionary and intuitive taxonomy has largely been replaced by cladistics, here construed as the reconstruction of the evolutionary history of lineages and traits by quantitative means. The resurgence of interest in history's role in biological pattern has more to do with the methodological innovation, that is cladistics, than it has with biologists 'remembering' or 'rediscovering' the importance of history. Prior to cladistics, taxonomic products were so often a matter of opinion, flawed and skewed by misrepresentation of lineage (paraphyly or polyphyly versus monophyly), misleading evidence (synapomorphy versus 'grades' or symplesiomorphy) and real versus imaginary categories (conflation of real taxa with artificial rank in the Linnaean hierarchy), that their reliability for structuring evolutionary research was, at best, hard to assess. Cladistics clarified these confusions, and two of the three problems mentioned above are now widely appreciated by non-systematists. The use of classically defined taxonomic ranks as objective and comparable categories to guide evolutionary comparisons remains an area of confusion, but it is as destined for the scrap pile of bad ideas as paraphyly or symplesiomorphic groupings. Part of this chapter will attempt to explain why the latter point is so. Even though all of the lessons of cladistics have not yet been learned, cladograms are now widely acknowledged as central to investigations or explanations of evolutionary history (Huey, 1987; Wanntorp *et al.*, 1990; Brooks & McLennan, 1991; Harvey & Pagel, 1991)

Both kinds of research mentioned above, the explanation of amazing things and the explanation of amazing coincidences, agree that cladograms are a prerequisite – the more detailed, complete and accurate, the better. The investigation of coincidences is necessarily statistical, and that point of view has questioned or even despaired of the 'scientific' study of uniques because they provided no replication (Lauder, 1982; Mitter, Farrell & Weigmann, 1988; Harvey & Pagel, 1991; Gittleman & Luh, 1992). In the convergence approach, phylogeny is something to be 'removed', 'extirpated' or 'controlled' because resemblance due to common ancestry is held to be non-selected.

However, statistics and 'science' are not synonymous. While commonplace statistics may be inapplicable to unique events, science is not merely the study of numerical distributions. Many valid scientific results are not testable statistically. Although many points of view on what constitutes 'science' are possible, good science uses null versus alternative hypotheses, some criterion of improbability (what it means to be amazing), unbiased observation and falsifiability or at least testability. Rejection of a null hypothesis by unbiased observation that remains testable and potentially falsifiable through additional observation (replication) seems enough like science to me. In what follows, I will compare and contrast each approach to the study of adaptation, seeking to identify its strengths and weaknesses, and how and whether these are appreciated by practioners of either the homology or the convergence approach.

DEFINING ADAPTATION

I take adaptation in the most rigorous sense to mean an apomorphic feature that evolved in response to natural (or any other kind of) selection for an apomorphic function (Fig. 1). As such adaptations are always apomorphies, though not all apomorphies are adaptations. Rigorous adaptational hypotheses include specific descriptions of the nature of the selection presumed to have operated, because the core criterion of adaptation is function (Greene, 1986; Coddington, 1988; Lauder, 1990; Baum & Larson, 1991). Williams (1966) felt strongly that the scientific study of adaptation reduced in large measure to careful analysis of function (in his terms as opposed to mere 'effect'). He thought the study of adaptation needed much improvement, constituted a distinct speciality and required a special name – teleonomy.

To this crucial link to function, cladistics has added the additional criterion of novelty or apomorphy. Apomorphy is a relative concept; all characters are apomorphic at some restricted level, and all except the latest are also plesiomorphic at a more inclusive level. Like apomorphy and homology, adaptation is a relative concept. Bird and bat wings are not homologous unless homologous as vertebrate forelimbs, thus specifying the level of apomorphy. It is as nonsensical to say that wings are an adaptation as it is to say that they are homologous. Compared to what? Because adaptation is a bridge that links evolutionary pattern and process, complete hypotheses of adaptation require comparative statements on both process (derived function) and pattern (apomorphy). Although tiresome to specify all the necessary components of an adaptive hypothesis, such details ought to be implicit in the hypothesis. With respect to the phylogeny depicted in Fig. 1, a complete adaptive hypothesis is of the form: the derived trait M_1 arose at time (t) in the stem lineage of taxa C, D and E via selection for the derived function F_1, with respect to the primitive trait M_0 with primitive function F_0, which still persist in taxa A and B. The main intent of this definition is to distinguish adaptation clearly from current utility, as historically viewed. Like homology, adaptation is a directed, polarized concept. Current utility, on the other hand, includes both the function for which a feature was built (what Williams (1966) called 'purpose'), as well as its current use (what Williams (1966) called 'effects').

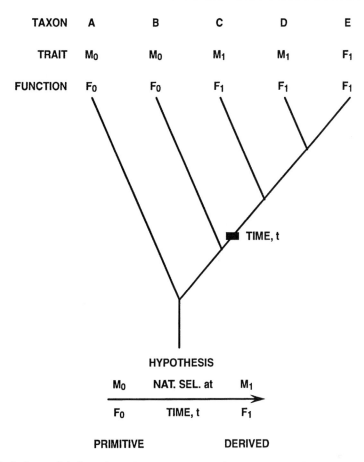

Figure 1 Cladistic model for adaptation. An apomorphic trait M_1 is built and maintained by natural selection for its function F_1 as compared to the plesimorphic trait and function (after Coddington, 1988).

An adaptational hypothesis therefore must link an observed pattern to a specific cause. The cause is particular in the case of unique events (though it may represent an instance of a common evolutionary trend) and it is common cause in the case of coincidences. If the sequence of the supposedly correlated traits is an essential part of the hypothesis (*if* A, *then* B), joint presence of A and B is insufficient to establish the relation. A must have evolved before B for the hypothesis to be supported (Greene, 1986; Donoghue, 1989; Baum & Larson, 1991). However, if A is extremely widespread in a lineage, by chance alone it will tend to pre-date the evolution of the rarer B (Maddison, 1990). An apomorphy that arose for other than the reason specified in the adaptive hypothesis may still be apomorphic and it may still be an adaptation, but the adaptive explanation was wrong. Evolutionary convergence due to several clearly distinct and logically different causes (different kinds of selection or non-selective causes such as linkage or various kinds of constraint) are not jointly evidence for adaptation, even if they all result in the 'same' startling phenotype. In most cases, distinct causes map clearly to distinct adaptational hypotheses. Pooling convergences that result from different kinds of selection is just a logical mistake. It is a mistake about evolutionary process, logically similar to the most common mistake about evolutionary pattern – distinguishing true from false homology. Greater precision in identifying and distinguishing evolutionary processes as causes can only help to connect the explanation of pattern to process, just as great precision in identifying pattern has been recognized as essential (Farris, 1988; Nixon, 1991; Maddison & Maddison, 1992; Swofford, 1993).

For example, a classic evolutionary problem concerns the evolution of 'dwarf' males in many lineages independently. 'Dwarf' already implies that selection acted to reduce male size. If 'dwarf' is operationally interpreted as the male–female size ratio, it is ambiguous whether the size change occurred in males, females or both sexes. Derived giant females (fecundity-driven selection?) could as easily be the answer as dwarf males (mating success/mortality-driven selection?). In the spider genus *Nephila* (Tetragnathidae: Nephilinae), in which female mean body length is about 40 mm and males about 11 mm, sexual size dimorphism conventionally is interpreted as dwarf males (e.g. Vollrath & Parker, 1992). Explanations consequently tend to emphasize evolutionary change in males (Vollrath, 1980; Elgar, Ghaffar & Read, 1990). In tetragnathids most likely to be outgroups to the Nephilinae, females and males average perhaps 6 mm and 5 mm respectively. (Other nephiline genera such as *Nephilengys* and *Herennia* are also dimorphic in size, which means the generality of the hypothesis, as usually treated in the literature, is also mistaken; dimorphism probably arose in the common ancestor of these nephilines, not de *novo* in *Nephila*.) It appears that nephiline males are not 'dwarves' but larger relative to their likely ancestors, while females are much larger. Hypotheses in this case may need to explain female giantism, not male dwarfism, or perhaps selection for increased size in both sexes as the cause of size dimorphism (Cheverud, Dow & Leutenegger, 1985). In most other dimorphic spider lineages (*Gasteracantha, Mastophora, Latrodectus, Tidarren, Misumena*) the males are indeed small relative to likely outgroups, and female size has not changed dramatically. In still other lineages, such as mammals, large males are derived. Conflating distinct changes as 'dimorphism' hides rather than elucidates selective mechanisms. Other kinds of 'ratio' characters may suffer from the same confusion of mechanism and effect.

The above definition of adaptation may seem unrealistically rigorous. Admittedly it sets a high standard, but clearer and higher standards have often been called for in work on adaptation (e.g. Williams, 1966; Gould & Lewontin, 1979). Perhaps it is best seen as an upper bound, a realizable ideal that may only be attained under ideal circumstances. Debating the lower bound seems fruitless, since it is never clear whether one is merely remarking on evidence for adaptation in a noncommittal fashion, or coyly making the claim itself. The line can be drawn in many places.

This definition of adaptation can be relaxed in two main ways. An adaptational hypothesis that is consistent with the above but declared to be untestable by comparative data on current utility might still be an adaptation. This situation may frequently occur when a synapomorphy of a large clade is proposed to have originated as an adaptation (= built by natural selection for a specific function), but now is thought to be maintained by some other agent than selection, e.g. ontogeny, epistasis, linkage or lack of genetic variation. By definition, data on current utility of the derived and plesiomorphic versions of the character will not falsify or strongly test the adaptive origin. However, this lack of testability does not mean the hypothesis is false.

The criteria for a 'common' cause at work in all instances of the pattern can also be relaxed. Relating correlation to cause is an old problem, but in my view correlation severed from its causal underpinnings breaks a necessary link. If the causes responsible for the set of distinct observations are very different, the analysis is fundamentally flawed. If mere correlation is accepted as being as close to demonstrating 'adaptation' as we can ever get, then correlation and adaptation become nearly synonymous. Defining the hypothesized cause clearly and distinctly, i.e. the nature of the selection, is one way to distinguish the claim of correlation from the more onerous claim of adaptation. Making this distinction is often trickier than it seems.

HOMOLOGY AND HOMOPLASY ARE COMPLEMENTARY

It is important to understand the provenance of the data required for either the homology or the convergence approach. For either approach, the data are conceptually a matrix or table of taxa by traits, characters, conditions or values. In what follows, I assume these traits are discrete (qualitative), but they may be continuous (quantitative) as well. Variation in this matrix is interpreted either as homology or homoplasy. Homology is similarity that is due to common ancestry. Homoplasy is similarity that is not. Homoplasy has three explanations: independent gain of the trait; independent or secondary loss; or observer error. In most real examples some variation is equivocal, meaning that although the total amount of homoplasy is known, it can be allocated to characters in various equivalent ways.

Both the homology and the convergence approach allocate variation in the matrix to either explanation based on the fit of the data to a cladogram. It is perhaps not widely enough appreciated that for any trait in the matrix, variation can be wholly attributed to homology or in large part to homoplasy, depending on the cladogram chosen. More succinctly, every character in a matrix is completely consistent with some genealogy (Farris, 1983). What then guides the choice of the cladogram that will determine the evidence for or against adaptation? In cladistics maximum parsimony thus far has been the most common criterion; homology is assumed until proven otherwise. All variation that can be explained as homology is so explained; maximizing homology (or minimizing homoplasy) specifies the cladogram of choice. *Homology is the null hypothesis of cladistics.* Variation that cannot be explained as homology is then termed homoplasy, and as many *ad hoc* hypotheses of convergence or loss as necessary are made to explain the distribution of inconsilient traits on the cladogram. Least squares regression is similar. As much variation as possible is attributed to the independent variable by the method and the remainder is termed 'error' or residual variation. Homoplasy is the 'error' term in cladistic analysis; the cladogram is the best-fit hypothesis given the data and a criterion of fit. *Parsimony is only a criterion of fit* and remains silent on whether the amount of homoplasy found is large or small, just as a least squares fit is agnostic on how much variation is explained or not.

A more important and subtler point about the use of the homology hypothesis in cladistics is that the method is also silent on whether congruence in data (no homoplasy) is due to homology, parallelism, chance, coincidence, earth history, drift or a lucky mistake. The method makes only the positive statement that given a specific cladogram, some similarities must be explained as homoplasy (Farris, 1983).

In an exactly similar fashion for least squares fits, complete prediction of variable A by variable B for any given ordered pair is silent on whether the expected causal relationship in fact underlies the results. Such an agreement between prediction and observation in any given instance might well be due to chance.

Given a cladogram that specifies how variation in a matrix is allocated between homology and homoplasy, it is significant that the data favoured by the homology and convergence approaches are complementary (Fig. 2). Each approach discards or ignores that component in the data prized by the other approach.

The homology approach looks at each evolutionary change as a unique event that requires its own explanation. Even if the trait evolves more than once on the cladogram, each instance is unique and therefore has its particular historical explanation. The method is largely silent on the significance of 'traits' that have evolved frequently (show high homoplasy), or at least it does not demand that traits be so. Even if the apomorphy is part of an impressive set of coincidences, arguing from this general trend to a specific instance of the trait may be a weaker inference than direct assessment of functional or performance data on that trait. If a coin comes up heads 49 times in a row and is tossed again, the trend predicts heads again. But if one is offered the choice of peeking at the result, is it not more secure to do so than to bet on the trend?

In contrast, the convergence approach discards homology and views homoplasy as the interesting and evidentially powerful source of information. Each convergent instance increases the coincidence, degrees of freedom and therefore statistical power – hence 'more' is always 'better'. This concern with sample size is well-founded (Ridley, 1989). The original and long-recognized problem in counting individual taxa as independent data points was precisely that it artificially inflated degrees of freedom and produced unacceptable Type I error rates (rejecting a true null). For the convergence approach homology not only is not evidence of adaptation, it is misleading and the largest source of error. Homology is 'phylogenetic inertia' or 'constraint', which should be removed or statistically 'extirpated' from a

	HOMOLOGY APPROACH	CONVERGENCE APPROACH
HOMOLOGY	**ADAPTATION?** (Alternatives exist and are likely)	Inertia, constraint, remove, extirpate!
HOMOPLASY	Error, secondary loss or gain, minimize "ignore"	**ADAPTATION** (Alternatives not commonly considered)

Figure 2 The homology and convergence approaches to studying adaptation, caricatured as extremes. They use complementary aspects of data, and in general regard the discarded portion as error or something to be minimized.

data set. How to do this properly is currently the focus of much research (Bell, 1989; Grafen 1989, 1992; Gittleman & Kot, 1990; Burt, 1989; Maddison, 1990; Pagel, 1992). Like any statistical test, the convergence approach assumes multiple independent events drawn indiscriminately or randomly from the set of all relevant comparisons. Traits that evolved once or only a few times are beyond the method.

A COMMON MODEL TO TEST HYPOTHESES OF ADAPTATION

In attempting to establish natural selection as the predominant cause of evolutionary pattern both approaches implicitly assume that selection explains change. In the literature, the hypothesis of adaptation usually seems to attribute evolutionary change to directional selection. It is peculiar that both the homology and the convergence approach focus so exclusively on change. In the limit, no change at any genetic locus is improbable (the second law of thermodynamics, if for no other reason) but very rapid change at any given locus is also unlikely. Either extreme requires explanation (Fig. 3). If 'directional' selection explains change, then 'stabilizing' selection explains stasis (Fig. 4). Extraordinary stasis is fully as interesting as extraordinary change (Levinton, 1983), and perhaps more common. For a null model like Brownian motion or Markov processes, testing for improbable stasis should make the test two- rather than one-tailed. Although the statistics to detect unusual change (reviewed in Harvey & Pagel, 1991) have attracted much attention lately, stasis has been of less interest. Adaptation has not always been so exclusively linked to change (Stebbins & Ayala, 1981). Williams (1966: 54) thought it "... unfortunate that the theory of natural selection was first developed as an explanation for evolutionary change. It is much more important as an explanation for the maintenance of adaptation". In contrast, the homology approach is not so dependent on finding aberrant frequencies of change, whether high or low. It relies on evidence specific to individual cases to test an adaptive hypothesis.

Figure 5 extends Fig. 4 to illustrate further the differences between the homology and convergence approach. A series of species (or taxa of any rank) occupy the tips

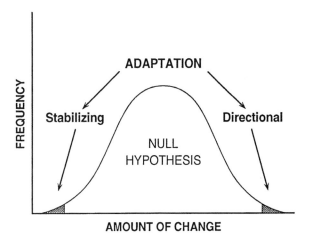

Figure 3 Both extremes in the frequency distribution of observed number of changes per traits may be rare compared to some null hypothesis. Both require explanation.

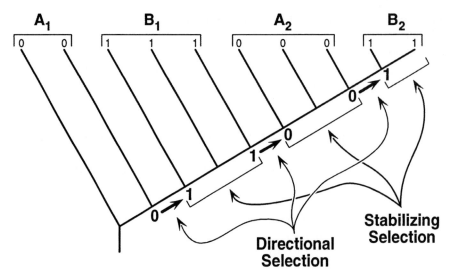

Figure 4 A cladogram for 10 taxa and one character with two states, coded as 0 and 1, each of which has evolved twice. Selection can explain both the origin and the persistence of novelties.

of the branches and display one of two conditions, A or B for 'Character 1' (initially coded as 0 and 1, respectively). A second trait, 'Character 2' also has two conditions, C and D, coded as for the first character. The traits could be continuous variates (Maddison and Maddison, 1992), but it does no violence to the logic to treat the simpler discrete case. For Character 1 on this tree three evolutionary events took place. If the tree is rooted properly, trait 'B' evolved twice (B_1 and B_2), indicated by the bold arrows, and trait 'A' has as well, once on the cladogram visible to us (A_2), and once somewhere below the displayed portion of the tree (A_1). Both conditions show homoplasy. The A at the bottom of the tree is not the same historically as the

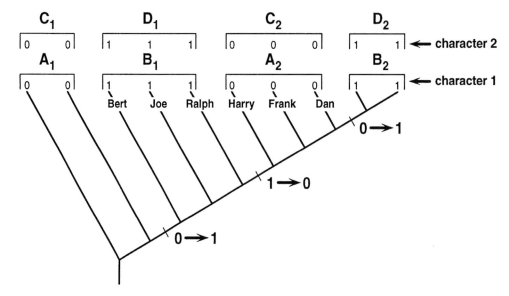

Figure 5 As in Fig. 4, but with taxa named and two perfectly correlated traits mapped on the cladogram.

A towards the top, and nor are the Bs. Even if they are phenetically and genetically identical (such as aligned, identical nucleotides), the homoplasies represent distinct historical events. To represent the convergence approach realistically Fig. 5 should be larger with A and B evolving 'n' times, but twice suffices to make the point. The varying ways to interpret the patterns of 0s and 1s for either character underlie all the differences between the homology and the convergence approaches.

With regard to the homology method as applied to character 1, Fig. 5 shows three events requiring explanation as indicated in the figure by bold arrows. The As and Bs, although similar, evolved independently and are not identical. The fact that B_2 is like B_1 or A_2 like A_1, though intriguing and worth investigating, is not *necessary* to test the hypothesis of adaptation. Perhaps B_1 and B_2 are only superficially similar (observer error). The tendency of systematists is to minimize homoplasy, and unproblematic cases of coding 'blunders' due to superficial similarity are usually re-coded as a matter of course.

With regard to the convergence method as applied to character 1, Fig. 5 shows two events, the parallel evolution of both A and B. The naive worker sees five instances of B, one for each taxon with B, whether B_1 or B_2, and improperly assumes inflated degrees of freedom. The sophisticate sees two evolutionary events with at best one degree of freedom and puts little emphasis on the pattern.

In summary, cladograms parse trait variation into either homology or homoplasy. If traits are continuous, it is still a question of dependent or independent change in two lineages. Both homology and homoplasy are interesting evolutionary patterns, both require a comparative method for analysis. The homology approach focuses on explaining individual changes. The convergence approach seeks a large ensemble of similar events because high levels of homoplasy improve the ability to detect correlated change. From a strictly phylogenetic point of view this goal is odd, as most systematic data sets are constructed to avoid homoplasy rather than to seek it out. For the convergence approach, the worse the trait fits the cladogram, the more powerful the test. If homology and homoplasy are complementary, what are the strengths and weaknesses of each?

THE HOMOLOGY APPROACH

In evolution, historical uniques are of at least two kinds, apomorphies and lineages. Together they form what O'Hara (1992) has called the evolutionary chronicle – what happened during evolution, rather than why. The chronicle is simply evolutionary 'history', as commonly construed.

Homologies are logical individuals

What is the scientific status of historical uniques? Here it becomes useful to borrow a metaphysical distinction from the philosophy of science. Philosophers define historical uniques as logical individuals as opposed to logical classes (e.g. Popper, 1965); these distinctions can also be applied within evolutionary biology (Ghiselin, 1974). Biological examples of logical individuals are the orb-weaving spider family Uloboridae (a taxon), or spider spinnerets (a set of synapomorphies). Logical individuals are considered to be 'spatio-temporally' limited, meaning that they originated in a single place and at a single time, will come to an end and will never recur. Individuals are most easily defined ostensively, by pointing to them, by the thing itself. Classes on the other hand are atemporal logical sets sharing some defining characteristic. Examples of classes are 'red' fruit, polygyny, aposematism, mimicry, dwarfism, gynodioecy, metabolic rate, GC–AT ratio, home range, body size, etc. All past and future taxa can be scored for such categories, although sometimes a little reflection is necessary ('fruits' are progeny; polygyny and gynodioecy refer to patterns among heterogametic organisms; mimicry and aposematism may require that the communication channel be specified, etc.). The homology approach studies individuals, but the convergence approach studies classes.

Is the study of logical individuals (whether apomorphies or lineages) less falsifiable or scientific than the study of 'replicated' events? The classic models of the scientific method applied to physics, where one hydrogen atom is the same as another, and new hydrogen, identical to all other hydrogen that has ever existed, could be created at will. In contrast, many well-corroborated scientific theories concern historical uniques, e.g. continental drift, the Pleistocene ice ages or even the Big Bang. The Universe may collapse and produce another Bang in accordance with timeless physical laws, but it will be a different Bang.

Evaluating hypotheses about historical uniques is not therefore necessarily statistical in the sense that repetitive events are required. Instead we recognize that such hypotheses could be wrong (are falsifiable) but that they have been corroborated by testing multiple independent deductions against facts. Each hypothesis is richly detailed and thus offers many points where correspondence to fact can be tested. This potential multiplicity of deductive statements provides replication analogous to 'frequency' in statistics. To the extent that hypotheses survive many tests, we place more confidence in them. Other examples of logical individuals are the Yucatan bolide impact, Gondwanaland, the taxa Araneae or Vertebrata (or any other taxon), or the vicariance of ancient African and South American biotas, all relatively well-corroborated scientific theories. Historical uniques are complex (another typical difference between individuals and classes), and multiple deductions testing explanatory theories are usually feasible. In sum, if studying historical uniques is untestable science because uniques happened only once, then study of lineages (taxonomy) is as unverifiable as that of apomorphic characters, and the entire comparative method will founder for lack of a phylogenetic framework.

Given that hypotheses concerning historical uniques are 'scientific,' how does it apply to the study of unique adaptations? With respect to Fig. 5, let us say the adaptive hypothesis is that trait A_2 of Character 1 arose in the most recent common ancestor to Harry, Frank and Dan because natural selection acted on trait B_1 in the outgroup (the most recent common ancestor of Harry $et\ al.$ and Ralph) for a derived function that had some average effect on fitness, given the environment. The data stipulate that Bert, Joe and Ralph have B_1, and that Harry, Frank and Dan have A_2. That B_1 arose in an ancestor basal to Bert or that descendants sharing a common ancestor with Dan evolved B_2 from A_2 is largely irrelevant to the time span in which A_2 evolved and persisted. These latter patterns concern different historical events.

Testing unique adaptations

How can this scenario be tested? There are a number of ways, most of which are straightforward, intuitive and merely pull together earlier threads of thought into a more rigorous cladistic context (Greene, 1986; Coddington, 1988, 1990; Donoghue, 1989; Lauder, 1990; Losos, 1990; Baum & Larson, 1991). First the effect of B_1 on Ralph's fitness can be assessed, thus characterizing the plesiomorphic selective context in which the feature evolved. Likewise the performance of the novel trait in the derived context can be assessed, and the consistency of these two measures of current utility or performance with the adaptive hypothesis can be assessed. The adaptive hypothesis predicts a significant performance advantage for the derived trait. The hypothesis is testable because it could fail – if the performance advantage is equivocal or opposite to that expected, the adaptive hypothesis has failed this particular test.

If an adaptive hypothesis withstands one test under the homology approach, it can be exposed to further tests that focus on more specific aspects of the natural history or ancillary deductions from the adaptive hypothesis. As the intersection between a narrative scenario and its biological and phylogenetic context grows, the initial adaptive hypothesis is necessarily elaborated and refined, which of course implies further tests. However examples of such work are sparse, especially at low taxonomic levels where the plesiomorphic selective context is most likely to persist (Brooks, O'Grady & Glen, 1985; Basolo, 1990; Losos, 1990; McLennan, 1990; Bjorklund, 1991). The corroboration of the adaptive hypothesis is roughly a function of the number of tests it has withstood weighted by their severity. As in the case of continental drift or the monophyly of spiders, there is little profit in trying to assign a quantitative probability to the hypothesis. It seems more candid to assess corroboration directly.

Assumptions of the homology approach

The homology approach makes several major assumptions. Perhaps the largest is that the current natural history of Ralph and Harry adequately reflect the context in which Character 1 actually changed. This *ceteris paribus* assumption can itself be tested in other taxa with B_1 and A_2. These tests use the notion of falsifying taxa, the cladistic neighbours of Ralph and Harry, to offer additional, taxon-based tests that bear on the theory. These predictions state that Bert and Joe, which retain trait B_1 by descent, can also be compared to Harry as well as to Frank and Dan for the same functional contrasts applied to Ralph and Harry. In the best possible case, multiple independent tested predictions offer the same sort of scientific corroboration that replication provides in a statistical test of correlation.

If no performance-based test of the adaptive hypothesis can be developed, but the cladogram, character polarities and predicted scopes and sequences of events are sound, then at least the plausibility of the adaptive hypothesis has been established (Greene, 1986; Donoghue, 1989; Baum & Larson, 1991). Such assessments could be considered either minimum requirements (Coddington, 1988) or a set of minimal tests for an adaptive hypothesis.

Strengths and limitations of the homology approach

As explained more fully below, historical uniques are usually less ambiguous and more independent of the observer's point of view than are convergences. Problems of trait definition and taxon sampling are not as troublesome, because the cladogram, as given in the analysis, specifies both homology and monophyly. Historical uniques are closely tied to an observable, bounded natural history context, where many different comparisons can be brought to bear, resulting in a system of high empirical content. Alternative adaptive and non-adaptive explanations are equally as rich and testable as adaptation. While adaptations are apomorphies, not all apomorphies are adaptations. Adaptive and non-adaptive hypotheses competing to explain multiple independent comparisons offers a powerful framework in which to evaluate different explanations. In contrast, if no alternative hypotheses are available or are considered, the ability to reject the adaptive hypothesis is weakened. It persists by monopoly rather than competitive merit.

Finally, only the study of historical uniques addresses the important and interesting question of evolutionary innovation, and whether or how innovations have shaped evolutionary history (Coddington, 1988; Wanntorp *et al.*, 1990). Every clade displays unique synapomorphies that are interesting evolutionary patterns and events (e.g. Coddington & Levi, 1991) and that demand explanation. If the corpus of explainable phenomena is limited to those that have happened a statistically tractable number of times, our understanding of the major features of our world will be incomplete indeed.

In the extreme in which features are truly unique, this singularity is the greatest limitation of the approach. Even assuming the adaptive hypothesis has survived test repeatedly it is necessarily a particular case – not general and not elucidating general evolutionary trends. Second, as in any empirical test, support for predictions deduced from the adaptive hypothesis can be equivocal. Third, the degree of corroboration or test is not easily quantified (although the hard 'significance' levels offered by statistics can be quite misleading if the assumptions that they require are not critically assessed (Wenzel and Carpenter, Chapter 4)). Fourth, all other things are rarely equal, which questions the basic premise that the derived and plesio-morphic functions have persisted sufficiently 'unchanged' to the present to validate investigation of past events with current data. Fifth, the notion that cladistic pattern ought to be linked to performance data or demonstrated current utility or function (Gould & Vrba, 1982; Lauder, 1990), that adaptation has something to do with selection, and that selection is demonstrable in the wild (Arnold, 1983; Endler, 1986), makes the implicit assumption that performance data available in the ecological 'here and now' can be optimized on a cladogram just as morphology can be (Baum & Larson, 1991). One must assume that selection is responsible for the origin and the maintenance of the adaptation (Coddington, 1988), and furthermore, that the nature of the selection is the same throughout. The variety of evolutionary mechanisms that explain stasis often make such a uniformitarian stance problematic. However, all branches of historical biology use the logic of character optimization (Swofford and Maddison, 1992) to elucidate past history, e.g. biogeography, morphology, physiology, behaviour, gene evolution, etc. There is no logical obstacle to using the same logic with performance data. Optimizing performance data proceeds on the same epistemological basis as optimizing anything else. A demonstrated performance advantage in all relevant modern taxa applies equally well to their most recent common ancestor.

THE CONVERGENCE APPROACH

The convergence approach uses the same data and cladistic model as the homology approach (Figs. 2, 4–5), but emphasizes correlations between multiple, independent evolutionary events. Assumptions peculiar to the convergence approach have been covered by Pagel (Chapter 2); and Wenzel and Carpenter Chapter 4). Applying commonplace statistical notions to hierarchical systems can be difficult (e.g. Maddison, 1990; Gittleman & Luh, 1992; Grafen, 1992; Pagel, 1992), but characters in Fig. 5 A and B appear to correlate with C and D. Let us assume that after a thicket of methodological and statistical problems have been solved (reasonable and appropriate null hypotheses, appropriate degrees of freedom, branch lengths, non-independence of observations, polytomies, specifying ancestral values, etc. (Pagel, Chapter 2; Wenzel and Carpenter, Chapter 4), a significant correlation has resulted. The correlation is intriguing, but can we conclude adaptation?

Homoplasies are logical classes

As mentioned above, the most fundamental difference between the homology and the convergence approaches is that homologies are logical individuals whereas homoplasies are logical classes. Philosophers of science hold that scientific laws apply to classes, but not to individuals (Popper, 1965; Hull, 1974). By 'law' I mean something of the form, 'if the following situation applies, then the following specific consequences are predicted to occur or to have occurred in a large number of cases'. Most of the classical subjects treated by the convergence approach are evolutionary laws *in statu nascendi*. Recent examples are the relation between investment in male sexual display and polygyny, sex ratio and sociality, aposematism and gregariousness, mating systems and resource monopolization, sexual size dimorphism and mating strategies/age to first maturity, home range and body size, or velocity and optimal physiological temperature. Note that many of these traits are nearly universal themselves (classes) – all organisms have had and will have metabolic rates, temperatures, sizes, home ranges, life histories, ages, resources, and most have had and will have breeding systems, population structure and sex ratios. Attempts to explain pattern in these nearly universal attributes of life are a hallmark and strength of the convergence approach. The homology approach rarely considers traits of these sorts because they are widespread, immune to ostensive definition and vary continuously. To take just the first example mentioned above, demonstrating truly widespread correlation between the evolution of males with marked secondary sex characters and polygynous mating systems would be a significant achievement (Sillén-Tullberg & Moller, 1993). To the extent that studies of these two traits across clades supports the generalization, the law (or tendency or trend) holds. Elaborating and testing such generalizations is the proper business of the convergence approach.

If the above distinction is correct, then hypotheses about homoplasies are *necessarily* about classes. In their most falsifiable form, adaptive hypotheses under the convergence approach treat the phenomena under study as true universal classes. Several interesting topics suggest themselves. If the convergence approach treats classes, then the generality of the adaptive hypothesis may not be fully unveiled in a test that is arbitrarily limited to particular historical uniques. In other words, the true scope of the hypothesis may go unrecognized. Second, logic requires that any character fulfilling the definition of the phenomenon be included in the class. Limitations imposed on the convergence approach by characters as historical uniques concerns the definition of the phenomenon or class (class membership). Third, arbitrary inclusion and exclusion of taxa or lineages as historical uniques poses questions about sampling bias among the instances used to test the hypothesis. Fourth, if significant correlation is always, or nearly always, interpreted as adaptation, may it be falsified?

In contrast, the scope of the hypothesis, the phenomenon under study and the taxa to which it applies are straightforward in the homology approach as it is limited to historical uniques, although the assessment of the evidence is sometimes complex.

Scope of the hypothesis

As an example, consider the relation of antler size in cervids with breeding system (Clutton-Brock, Albon & Harvey, 1980). At one level this concerns only the range of one putative unique (antlers) within another (cervids). More profoundly, it concerns investment in male sexual display and its relation to male–male competition as influenced by breeding system, a hypothesis particular neither to antlers nor cervids (nor to bovids, butterflies, angel fish, jumping spiders, primates, bower birds, etc.), but perhaps to a set as large as all primitively biparental organisms. Obviously one could include a range of metazoan taxa other than cervids. Theories of sex and sexual selection are general (Maynard-Smith, 1978). If these imply trends within cervids that hold, so much the better, but the cervid case is just a particular set of instances. Game theory and evolutionarily stable strategies are other general theories in biology that explain patterns in competition reasonably well (Maynard-Smith, 1982). If given certain initial starting conditions, *Agelenopsis* spiders in desert communities corroborate the predictions of game theory (Riechert, 1986), again so much the better, but *Agelenopsis* is again just an instance. In contrast, evolutionary patterns in life histories which were once thought to be generally the result of selection are now often considered to be effects largely of descent with modification and body size (Wanntorp *et al.* 1990).

The scope of the phenomenon

Clarity about the scope of the phenomenon is important because it defines the universe under study. The intended universe dictates the sampling programme and analysis. In Fig. 5, A_1 and A_2 are linked by the defining attributes of the class, not by homology. The same holds for the characters B, C and D, of which B_1 and B_2 through D_1 and D_2 are instances of phenomena that fulfill the class definitions. Definitions of classes of homoplasies can be made precise and objective, but those definitions often are arbitrary in an evolutionary sense in a way that homology is not. Objectivity of definition is readily achieved, but non–arbitrariness is more difficult. To return to antlers as an example of investment in male sexual display in cervids, equivalent examples of such investment should not be permanently excluded from the test without justification that derives from the hypothesis itself, rather than from *ad hoc* reasons. If the sample is limited to antlers despite other traits in the same organisms that have been affected by the same sort of selection and could equally well be studied, the sample bias should be justified or at least acknowledged.

Taxon sampling

In the same sense that a study arbitrarily limited to a single character system (historical unique) can bias the definition of the traits under study, arbitrarily limiting the taxa studied to a particular clade (another kind of historical unique) may bias the study as well (Coddington, 1992). Given the rigorously parametric statistical world-view that the convergence approach adopts (Felsenstein, 1985; Harvey and Pagel, 1991; Pagel, Chapter 2; Wenzel and Carpenter, Chapter 4), the set of taxa should represent unbiased samples of all members of the class, which, as noted above, could be something like all taxa that lek (for one trait) compared to all taxa that show sexual dimorphism (for the other). If the sample is limited to a particular lineage, the limitation should stem from a valid *a priori* reason related to the hypothesis. One good reason to limit the sample to a particular lineage (e.g. Aves, Mammalia, Araneae, Insecta, Solananceae, or some other named phylogenetic node bequeathed to us by our culture) is because the apomorphies of the lineage justify the limitation. By apomorphies here I mean not just that small subset of 'characters' used by systematists to identify the taxon, but rather the total synapomorphic biological and ecological nexus that may characterize the lineage and which may include anything from habitat preferences to metabolic rate.

For an example that such considerations are not picayune, consider the analysis of the effect of phytophagy on insect diversification rates used as a test case of adaptive zones by Mitter, Farrell & Wiegmann (1988). I choose this example because it is one of the best comparative studies published to date and because it dealt with many of the thorny issues endemic to clade-based work on convergence. Strong, Lawton & Southwood (1984: 15) had proposed that "life on higher plants presents a formidable evolutionary hurdle, that most groups of insects have conspicuously failed to overcome. Once the hurdle is cleared, however, radiation may be dramatic". Mitter *et al.* (1988) found that in 11 out of 13 cases of phytophagous:non-phytophagous pairs of sister-groups, the phytophagous lineages contained at least twice as many species as the non-phytophagous lineage (a statistically significant association). Phytophagy seems to promote diversification within insects.

How do scopes of hypothesis, phenomenon and taxa relate to such a question? The adaptive zone argument concerned shifts between fundamentally different feeding zones. Thus one might wonder why just 'to' phytophagy instead of 'from' phytophagy as well? Why not detritivory to or from predation or parasitism, or phytophagy to predation, or at a lower level sucking plant juices to sucking apocynaceous plant juices? These questions concern the scope of the hypothesis.

Second, Mitter *et al.* (1988) followed Strong *et al.* (1984) in defining higher plant feeding as "feeding on the living tissues of higher plants, which excludes algal and other non-tracheophyte feeders, wood borers, nectar feeders, and species that use dead plants and leaf litter as food. Insects feeding on developing seeds are included, but those that take only shed seeds are excluded. Pollen feeding is also excluded, even though it is often hypothesized to be transitional to phytophagy in the strict sense . . .". In addition they excluded taxa in which the adult, but not the larvae, feed on higher plants because such species depend less on the adult diet than the larval diet. Mitter *et al.* (1988) were uncomfortable with the somewhat arbitrary nature of the definition of the trait 'higher plant feeding', but accepted it as the classical formulation favoured by students of plant–insect interactions. It contains some notable inclusions and exceptions that serve to limit the scope of the phenomenon in curious ways.

Third, this general hypothesis on adaptive zones is bounded by taxa: 'higher plants' and insects. Why just these two nodes in the cladogram of life? Did the choice make a difference, was it 'conventional' in the sense that Western culture has long chosen to name these nodes as taxa, or was it to test the sense of past and present workers that if the question was limited in roughly this way that an impressive trend would be disclosed? The number of species in the lineages ranged from 1 (Joppeicidae, sister to the phytophagous Tingidae, with *c.* 1800 spp.) to 130 000 (the phytophagous Phytophaga, sister to the 'non-phytophagous' Cucujoidea with *c.* 10 000). The sister-group pairs mainly were chosen from taxa carrying traditional Linnaean names and ranks, such as subfamilies, families and super-families. However, Insecta is a large clade, and conservatively at least several hundred thousand sister-group comparisons are potentially possible, not to mention Acari, Nematoda, Mollusca, various fungal and algal taxa, Aves, Mammalia and Teleostei as examples of other lineages that have been arbitrarily chosen to receive Latin names. These questions concern the scope and possible bias in sampling.

Finally, although Mitter *et al.* (1988) did their best to consider these biases and to make conservative choices so that minor changes in definition or scope would not affect the conclusions, if just two more comparisons had been against the trend (9 out of 13 instead of 11), the sign test would have been insignificant ($P > 0.133$).

If the core prediction of the adaptive zone hypothesis concerns invasion of formerly unoccupied zones, the sample would not be limited to a set of shifts that coincidentally includes just those taxa that tradition has conventionally recognized as 'megadiverse'. All pairwise feeding zone shifts could have been the universe, and one might well have concluded that only herbivory (and probably parasitism), not shifts to new 'feeding zones' *per se*, tended to influence diversification rate. The result actually obtained (Mitter *et al.*, 1988 claimed no more) may not test the fundamental hypothesis. Perhaps it is not a new zone, however empty, however formidable the barrier, but rather just particular zones that in fact drive diversification. Perhaps there is no general trend at all between invasion of new adaptive zones and increased diversification rate. In general, if we limit tests of general adaptive hypotheses to a set that includes mainly confirming instances, we are once again committing a sort of Panglossian, observer-biased mistake (O'Hara, 1992). Monophyletic groups often share coherent biologies, and if very diverse ones are compared to miscellanies defined as 'any biology but that one' adaptation will probably emerge triumphant. A slight change in the definition of higher-plant feeding would have altered the test in yet other ways. Why 'no' to pollen, wood and shed seeds? Why 'yes' to fruits, flowers and roots?

If mammals had been included, perhaps insectivory and granivory might have yielded significant trends; in mammals, phytophagy may not be the 'thing'. Maybe diversity in mammals is different from diversity in insects, but on what basis do we partition the general and testable hypothesis of adaptive zones into two lucky subsets, separately significant but jointly not? If it is because some of Life's disparities are obvious, suitably delimited, the sample may be biased.

As in the case of the definition of adaptation itself, these considerations seem, even to me, unrealistically high. I fully realize that data on phytophagous insects are preliminary, that other comparisons even within the definitions of the study can be made (the authors thought that phytophagy had arisen at least 50 times among extant 'orders'), and we are all just trying to add one brick more solid than the last to the wall of science. The point is not to set goals that no one can achieve, but rather to debate issues of rigor and power, and to investigate methodological issues that lurk beneath the surface. In day-to-day work on sets of replicated evolutionary events (homoplasy), it sounds odd to suggest that first the adaptive hypothesis should be clearly developed and described in detail, that the focal traits be defined without reference to homology, and that instances be sampled at random among traits and taxa to which the hypothesis applies. However, such procedures are fundamental if statistical 'significance' is to mean anything. Sampling effort could be allocated in a way that avoided the biases identified above. If nothing else, such considerations may force the real generality of the adaptive hypothesis into the open. As Williams (1966) said, "evolutionary adaptation is a special and onerous concept that should not be used unnecessarily." It is probably no accident that hypotheses such as sex-ratio theory, among the more synthetic process theories evolutionary biology has ever developed, are fully cognizant of their generality and acknowledge a much greater range of instances as potential falsifiers than typically do studies of convergent phenomena.

Falsifiability

Ignoring the above issues will have a predictable effect on adaptation as a general explanation for evolutionary pattern. Imagine a scenario in which significant correlation is found, but that on admitting either more traits (e.g. male plumage as well as antlers) or more taxa (Aves as well as Mammalia) to the analysis, the significance disappears. Two choices seem clear. We can protect the significance in the smaller data set by *ad hoc* exclusion of the contrary data, based on appeals to historical uniques (either (1) traits or (2) lineages). Examples of this dilemma already exist. Höglund (1989) studied the association between lekking and sexual size and plumage dimorphism in those 11 families of birds that lek and found no significant association between the two traits. Harvey & Pagel (1991) reported Höglund's conclusions, but singled out the data just for grouse and pheasants (Tetraonidae) where there was a 'perfect' association. Can the impressive association in Tetraonidae be protected and if so on what grounds? A hidden third variable or factor coincident with tetraonids is required. Such *post hoc* procedures are problematic in hypothesis testing. Do the apomorphies of Tetraonidae in some way differentiate their leks from others? Höglund pointed out that if males lek on the ground (as do grouse, pheasants and some other birds) as opposed to aerial or arboreal leks, size dimorphism is significantly associated. However, this observation casts great doubt on the effect on lekking *per se* on size dimorphism, and suggests consideration of all ground-lekking taxa before drawing the conclusion, not just tetraonids. Put negatively, delimiting convergence studies by vaguely justified claims about the uniqueness of lineages or characters seems like the worse form of *ad hoc* appeal. All lineages and characters are unique.

At the other extreme, if new data are added without prejudice to the growing evidence on an adaptive hypothesis, the 'significance' of an association may wink on and off as new data are compiled. Examples of this pattern in correlations in medical and epidemiological research are commonplace (e.g. cholesterol). Larger sample sizes (more taxa) test correlations better, all other things being equal. Faced with protecting odd bits of significance here and there by *ad hoc* appeals, or adding new data to the pile, I generally favour the second. It is through bolder and more powerful tests of adaptational hypotheses that real progress towards evolutionary generalities will be made, not through a gerrymandered series of small studies. In sum, the convergence approach should strive to avoid situations in which significant correlation is inevitably interpreted as adaptation.

Taxonomic ranks

A final, unsolved problem for the convergence approach concerns the use of taxonomic ranks. De Queiroz and Gautier (1992) and O'Hara (1992) have explained in some detail the mismatch between the current Linnaean system and real phylogeny. O'Hara in particular emphasizes how rank (fundamentally an expression of the observer's biased point of view) distorts understanding of history. By judicious lumping and splitting, severely asymmetric pectinate phylogenies can be transformed into the more balanced dichotomies predicted by equable rates of stochastic change (the molecular 'clock') and visa versa. The Linnaean hierarchy is already a viciously symmetrical model because it has only seven major ranks into which thousands of cladogenetic events must be compressed. Each newly corroborated cladogram of any size identifies dozens to hundreds more new taxa that all merit a rank in the system. But orders are already bursting with families, themselves stuffed with egregious numbers of equally ranked genera. Increasing the number of ranks available will not solve the problem. Though well-intended, 'Gigapicaorders' are a bad joke. The Linnaean model is so hopelessly limited that every category must be used within every lineage, which just goads workers to count or compare families within orders, or genera within tribes, or even 'five' kingdoms within life itself. In stark contrast, cladistic analysis often results in rather asymmetric topologies, though whether more or less than that expected under some null model is debatable (Slowinski & Guyer, 1989).

The Seven Ranks of Life skew not only names but thought and analysis. The empirical excess of asymmetric cladograms may be bad news for the molecular clock, Brownian motion and Linnaean hierarchies if that excess turns out to be real rather than artifactual (Shao & Sokal, 1990; Slowinski, 1990). Simulations of evolution only with topologies balanced to mimic taxonomies (Gittleman & Luh, 1992) almost certainly mislead for the most asymmetric resolutions implied by the polytomies. Figure 6a illustrates that taxonomies, interpreted phylogenetically, usually consist of reasonably balanced polytomous nodes. Even assuming that each polytomy subtends a monophyletic group, the taxa composing the polytomy are differentially related. Given no branch length information and polytomies, several statistical approaches recommend considering the ancestor as the mean of its descendants (Burt, 1989; Gittleman & Luh, 1992; Pagel, 1992). If a continuous variate

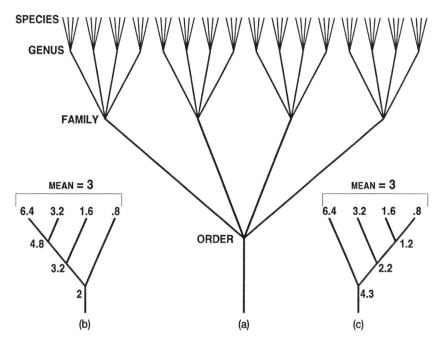

Figure 6 Linnaean taxonomies can bias comparative tests. (a) The cladogram implied by a Linnaean taxonomic classification in which all ancestral taxa contain four descendant taxa. (b) A four taxon tree with trait values known only for tips. The root value is 3 if taken as the mean of the tips, but 2 if topology is given and ancestors are the mean of their descendants. (c) As in (b), but a very different topology subtending the terminals, and therefore the root value is 4.3 rather than 2. Topology can severely affect the estimated ancestral value.

with values for four taxon topologies as in Fig. 6b and c is considered in the absence of topological information, the mean of 3 for the tips on both topologies will be the same. When the topologies are available, approximations using the mean are seen to overestimate the true values at the ancestral node by 50% in Figs. 6b or underestimate it by 30% in Fig. 6c (calculating ancestral nodes means of descendant nodes in all cases). Whether 8-fold ranges in continuous variates for unresolved polytomies are realistic, or what proportion of all possible topologies yield seriously different estimates of the ancestral mean, is currently unknown. Studies of the sensitivity of convergence approaches to topology, null models of branch length, and differential lumping and splitting are urgently needed.

SUMMARY

Neither the homology nor the convergence approach offer a royal and exclusive road to demonstration of adaptation, but perhaps proof in these cases is not obtainable. The former method requires a strong connection to trait function, because it is only the careful comparison of observed contrasts in performance or biological fact to prediction that provides the opportunity for the hypothesis to fail. Hidden third variables, ambiguous results, and events subsequent to the origin of the adaptation all conspire to cast doubt on the conclusion. Rigor in demonstrating adaptation resides in the hope that particular scenarios must, through every test, become more detailed, and soon, like the proverbial liar, acquire enough rope to hang themselves. Textbook examples of adaptation will be those that continue to accumulate the rope but skillfully avoid the noose.

The convergence approach will usually precede the homology approach insofar as each instance contributing to a correlation awaits the study of its unique history. It seems uncontroversial to suggest that closer attention to the instances of a correlation can weaken or strengthen the relationship as a whole. In the limit, use of the homology approach on each instance could falsify enough cases to exclude adaptation as the explanation for correlation, though presumably the statistical correlation would still remain. The converse seems possible, but less likely. Detailed natural history results that strongly favour adaptation are unlikely to be controverted simply because the example conflicts with a more superficial correlation. Clearly this point is arguable (Pagel, Chapter 2). While perfect correlations that cannot adduce plausible evidence on function will always be an incomplete argument for adaptation (unless correlation and adaptation are the same thing), cases in which no evidence on function support the correlation seem unlikely, especially as it is usually anecdotal functional evidence that suggests the correlation in the first place.

Like the old children's game in which scissors cut paper, paper covers stones and stones break scissors, an obvious circularity and complementarity directs comparison of these two ways of investigating adaptation. If in a given study, the methods conflict but have available the best data according to their own lights, homology may win pitched battles with convergence by falsifying particular instances. However, the assumptions involved in such a thought experiment seems far-fetched. The homology approach, on the other hand, always loses the war to achieve general explanations of evolutionary pattern.

Both the homology and the convergence approach are alike in attempting to devise more rigorous and more powerful methodologies to the study the historical effects of natural selection. In this review, I have deliberately caricatured the two views in order to clarify their fundamental differences and the implications of those differences for the study of adaptation. As a practical matter, however, it cannot be emphasized too strongly that these methods are complementary. A large number of individuals cases, each tested by the homology approach and together making an impressively general evolutionary trend as tested by convergence approach, is obviously the best result. A blend of these approaches will mean meticulous characterization of the kind of selection thought to have resulted in the putative adaptation, because clarity of the hypothesis (or lack of it) underlies most of the sampling or bias problems the convergence approach faces. To evade those problems, limits on the scope of problems addressed by the convergence approach (traits, taxa) ought to devolve from a carefully restricted hypothesis, rather than loose or sloppy testing of a very general hypothesis. The key feature, then, is the precision of the adaptive hypothesis. With either approach one should always guard against substituting a claim of adaptation for a simpler, less mystical conclusion (current utility on the one hand, correlation on the other). The study of adaptation enters the truly evolutionary dimension when a study transcends the demonstration of these simple goals.

ACKNOWLEDGEMENTS

I thank Paul Eggleton and Dick Vane-Wright for inviting me to the symposium, and to Jim Carpenter, Bill DiMichele, Michael Donoghue, John Gittleman, Mark Pagel, John Wenzel and anonymous reviewers for comments on a draft of the manuscript. Work on this project was supported by the Biological Diversity Program, the Lowland Tropical Ecosystems Program and Scholarly Studies Grants 1233S008 and 1233S148 of the Smithsonian Institution.

REFERENCES

ARNOLD, S., 1983. Morphology, performance, and fitness. *American Zoologist, 23*: 347–361.

BASOLO, A.L., 1990. Female preference predates the evolution of the sword in swordtail fish. *Science, 250*: 808–810.

BAUM, D.A. & LARSON, A., 1991. Adaptation reviewed: a phylogenetic methodology for studying character macroevolution. *Systematic Zoology, 40*: 1–18.

BELL, G., 1989. A comparative method. *American Naturalist, 133*: 553–571.

BJORKLUND, M., 1991. Evolution, phylogeny, sexual dimorphism, and mating system in the grackles (*Quiscalus* spp., Icterinae). *Evolution, 45*: 608–621.

BROOKS, D.R. & McLENNAN, D.A., 1991. *Phylogeny, Ecology and Behavior.* Chicago: Chicago University Press.

BROOKS, D.R., O'GRADY, R.T. & GLEN, D.R., 1985. Phylogenetic analysis of the *Digenea* (Platyhelminthes: Cercomeria) with comments on their adaptive radiation. *Canadian Journal of Zoology, 63*: 411–443.

BURT, A., 1989. Comparative methods using phylogenetically independent contrasts. In P.H. Harvey & L. Partridge (Eds), *Oxford Surveys in Evolutionary Biology*, (6), 33–53. Oxford: Oxford University Press.

CARPENTER, J.M., 1989. Testing scenarios: Wasp social behavior. *Cladistics, 5*: 131–144.

CARPENTER, J.M., 1991. Phylogenetic relationships and the origin of social behavior in the Vespidae. In K.G. Ross & R.W. Matthews (Eds), *The Social Biology of Wasps*, 7–32. Ithaca, NY: Cornell University Press.

CHEVERUD, J.M., DOW, M.M. & LEUTENEGGER, W., 1985. The quantitative assessment of phylogenetic constraints in comparative analysis: sexual dimorphism in body weight among primates. *Evolution, 39*: 1335–1351.

CLUTTON-BROCK, T.H., ALBON, S.D. & HARVEY, P.H., 1980. Antlers, body size, and breeing group size in the Cervidae. *Nature, 285*: 547–565.

CODDINGTON, J.A., 1986a. The monophyletic origin of the orb web. In W.A. Shear (Ed.), *Spiders: Webs, Behavior, and Evolution*, 319–363. Stanford, CA: Stanford University Press.

CODDINGTON, J.A., 1986b. Orb webs in non-orb-weaving ogre-faced spiders (Araneae: Deinopidae): a question of genealogy. *Cladistics, 2*: 53–67.

CODDINGTON, J.A., 1988. Cladistic tests of adaptational hypotheses. *Cladistics, 4*: 1–22.

CODDINGTON, J.A., 1990. Bridges between evolutionary pattern and process. *Cladistics, 6*: 379–386.

CODDINGTON, J.A., 1991. Cladistics and spider classification: Araneomorph phylogeny and the monophyly of orbweavers (Araneae: Araneomorphae; Orbiculariae). *Acta Zoologica Fennica, 190*: 75–87.

CODDINGTON, J.A., 1992. Avoiding phylogenetic bias. *Trends in Ecology and Evolution, 7*: 68–69.

CODDINGTON, J.A. & LEVI, H.W., 1991. Systematics and evolution of spiders (Araneae). *Annual Review of Ecology and Systematics, 22*: 565–592.

DE QUEIROZ, K. & GAUTIER, J., 1992. Phylogenetic taxonomy. *Annual Review of Ecology and Systematics, 23*: 449–480.

DONOGHUE, M.J., 1989. Phylogenies and the analysis of evolutionary sequences, with examples from seed plants. *Evolution, 43*: 1137–1156.

ELGAR, M.A., GHAFFAR, N. & READ, A.F., 1990. Sexual dimorphism in leg length among orb-weaving spiders: a possible role for sexual cannibalism. *Journal of Zoology (London), 222*: 455–470.

ENDLER, J.A., 1986. *Natural Selection in the Wild.* Princeton, NJ: Princeton University Press.

FARRIS, J.S., 1983. The logical basis of phylogenetic analysis. In N.I. Platnick & V.A. Funk (Eds), *Advances in Cladistics, 2,* 7–36. New York: Columbia University Press.

FARRIS, J.S., 1988. Hennig86, ver. 1.5. Microcomputer program. Available from author, 41 Admiral St, Port Jefferson Station, New York 11776, U.S.A.

FELSENSTEIN, J., 1985. Phylogenies and the comparative method. *American Naturalist, 125*: 1–15.

GARLAND Jr, T., HARVEY, P.H. & IVES, A.R., 1992. Procedures for the analysis of comparative data using phylogenetically independent contrasts. *Systematic Biology, 41*: 18–32.

GHISELIN, M.T., 1974. A radical solution to the species problem. *Systematic Zoology, 23*: 536–544.

GITTLEMAN, J.L. & KOT, M., 1990. Adaptation: statistics and a null model for estimating phylogenetic effects. *Systematic Zoology, 39*: 227–241.

GITTLEMAN, J.L. & LUH, H.K., 1992. On comparing comparative methods. *Annual Review of Ecology and Systematics, 23*: 383–404.

GOULD, S.J. & LEWONTIN, R.C., 1979. The spandrels of San Marco and the Panglossian Paradigm: a critique of the adaptationist programme. *Proceedings of the Royal Society of London, Series B, 205*: 581–598.

GOULD, S.J. & VRBA, E.S., 1982. Exaptation – a missing term in the science of form. *Paleobiology, 8*: 4–15.

GRAFEN, A., 1989. The phylogenetic regression. *Philosophical Transactions of the Royal Society of London, Series B, 326*: 119–157.

GRAFEN, A., 1992. The uniqueness of the phylogenetic regression. *Journal of Theoretical Biology, 156*: 405–423.

GREENE, H.W., 1986. Diet and arboreality in the emerald monitor, *Varanus prasinus,* with comments on the study of adaptation. *Fieldiana Zoologica, 31*: 1–12.

HARVEY, P.H. & PAGEL, M.D., 1991. *The Comparative Method in Evolutionary Biology.* Oxford: Oxford University Press.

HUEY, R.B., 1987. Phylogeny, history, and the comparative method. In M.E. Feder, A.F. Bennett, W. Burggren & R.B. Huey (Eds), *New Directions in Ecological Physiology,* 76–98. Cambridge: Cambridge University Press.

HULL, D.L., 1974. *Philosophy of Bological Science.* Englewood Cliffs, NJ: Prentice Hall.

HÖGLUND, J., 1989. Size and plumage dimorphism in lek-breeding birds: a comparative analysis. *American Naturalist, 134*: 72–87.

LAUDER, G.V., 1982. Historical biology and the problem of design. *Journal of Theoretical Biology, 97*: 57–67.

LAUDER, G.V., 1990. Functional morphology and systematics: studying functional patterns in an historical context. *Annual Review of Ecology and Systematics, 21*: 317–340.

LEVINTON, J.S., 1983. Stasis in progress: the empirical basis of macroevolution. *Annual Review of Ecology and Systematics, 14*: 103–137.

LOSOS, J.B., 1990. Ecomorphology, performance capability, and scaling of West Indian *Anolis* lizards: an evolutionary analysis. *Ecological Monographs, 60*: 369–388.

MADDISON, W.P., 1990. A method for testing the correlated evolution of two binary characters: are gains or losses concentrated on a certain branch of a phylogenetic tree? *Evolution, 44*: 539–557.

MADDISON, W.P. & MADDISON, D.R., 1992. *MacClade, ver. 3.0.* Sunderland, MA: Sinauer.

MARTINS, E.P. & GARLAND Jr, T., 1991. Phylogenetic analysis the correlated evolution of continous characters: a simulation study. *Evolution, 45*: 534–557.

MAYNARD-SMITH, J., 1978. *The Evolution of Sex.* Cambridge: Cambridge University Press.

MAYNARD-SMITH, J., 1982. *Evolution and the Theory of Games.* Cambridge: Cambridge University Press.

McLENNAN, D.A., 1990. Experimental investigations of the evolutionary significance of sexually dimorphic nuptial colouration in *Gasterosteus aculeatus* (L.): the relationships between male colour and female behaviour. *Canadian Journal of Zoology, 68*: 484–492.

MITTER, C., FARRELL, B. & WIEGMANN, B., 1988. The phylogenetic study of adaptive zones: has phytophagy promoted insect diversification? *American Naturalist, 132*: 107–128.

NIXON, K.C., 1991. Clados. Program and documentation. Distributed by the author, P.O. Box 270, Trumansburg, NY 14886, U.S.A.

O'HARA, R.J., 1992. Telling the tree: narrative representation and the study of evolutionary history. *Biology and Philosophy, 7*: 135–160.

OLIVEIRA, P.S., 1988. Ant-mimicry in some Brazilian salticid and clubionid spiders (Araneae: Salticidae, Clubionidae). *Biological Journal of the Linnaean Society, 33*: 1–15.

OLIVEIRA, P.S. & SAZIMA, I., 1984. The adaptive bases of ant-mimicry in a neotropical aphantochilid spider (Araneae: Aphantochilidae). *Biological Journal of the Linnaean Society, 22*: 145–155.

PADIAN, K., 1985. The origins and aerodynamics of flight in extinct vertebrates. *Paleontology, 28*: 413–333.

PAGEL, M.D., 1992. A method for the analysis of comparative data. *Journal of Theoretical Biology, 156*: 431–442.

PAGEL, M.D. & HARVEY, P.H., 1988. Recent developments in the analysis of comparative data. *Quarterly Review of Biology, 63*: 413–440.

PAGEL, M.D., 1994. The adaptionist wager. In P. Eggleton & R.I. Vane-Wright (Eds), *Phylogenetics and Ecology*, 29–51. London: Academic Press.

POPPER, K.R., 1965. *Conjectures and Refutations: the Growth of Scientific Knowledge*, 2nd edition. New York: Harper & Row.

RIDLEY, M., 1983. *The Explanation of Organic Diversity: the Comparative Method and Adaptations for Mating.* Oxford: Oxford University Press.

RIDLEY, M., 1989. Why not to use species in comparative tests. *Journal of Theoretical Biology, 136*: 361–364.

RIECHERT, S.E., 1986. Spider fights: a test of evolutionary game theory. *American Scientist, 47*: 604–610.

SHAO, K. & SOKAL, R.R., 1990. Tree balance. *Systematic Zoology, 39*: 266–276.

SILLÉN-TULLBERG, B., 1988. Evolution of gregariousness in aposematic butterfly larvae: a phylogenetic analysis. *Evolution, 42*: 293–305.

SILLÉN-TULLBERG, B. & MOLLER, A.P., 1993. The relationship between concealed ovulation and mating systems in anthropoid primates: a phylogenetic analysis. *American Naturalist, 141*: 1–25.

SLOWINSKI, J.B., 1990. Probabilities of n-trees under two models: a demonstration that asymmetrical interior nodes are not improbable. *Systematic Zoology, 39*: 89–94.

SLOWINSKI, J.B. & Guyer, C., 1989. Testing the stochastity of patterns of organismal diversity: an improved null model. *American Naturalist, 134*: 907–921.

STEBBINS, G.L. & AYALA, F.J., 1981. Is a new evolutionary synthesis necessary? *Science, 213*: 967–971.

STRONG, D.R., LAWTON, J.H. & SOUTHWOOD, T.R.E., 1984. *Insects on plants: community patterns and mechanisms.* Cambridge, MA: Harvard University Press.

SWOFFORD, D.L., 1993. PAUP (Phylogenetic Analysis Using Parsimony), ver. 3.1.1. Microcomputer program available from the author, Illinois State Natural History Survey, 172 Natural Resources Bldg., 607 E. Peabody, Champaign, IL 61820, U.S.A.

SWOFFORD, D.L. & MADDISON, W.P., 1992. Parsimony, character-state reconstruction and evolutionary inference. In R.L. Mayden (Ed.), *Systematics, Historical Ecology, and North American Freshwater Fishes*, 186–223. Stanford, CA: Stanford University Press.

VOLLRATH, F., 1980. Why are some spider males small? A discussion on *Nephila clavipes*. In J. Gruber (Ed.), *Verhandlungen. 8. Internationeler Arachnologen – Kongress abgehalten an der Universitat fur Bodenkultur Wien, 7–12 Juli 1980*, 165–169. Vienna: Egermann.

VOLLRATH, F. & PARKER, G.A., 1992. Sexual dimorphism and distorted sex ratios in spiders. *Nature, 360*: 156–159.

WANNTORP, H., 1983. Historical constraints in adaptation theory: traits and non-traits. *Oikos, 41*: 157–160.

WANNTORP, H.E., BROOKS, D.R., NILSSON, T., NYLIN, S., RONQUIST, F., STEARNS, S.C. & WEDELL, N., 1990. Phylogenetic approaches in ecology. *Oikos, 57*: 119–132.

WENZEL, J.W. & CARPENTER, J.M., 1994. Comparing methods: adaptive traits and tests of adaptation. In P. Eggleton & R.I. Vane-Wright (Eds), *Phylogenetics and Ecology*, 79–101. London: Academic Press.

WILLIAMS, G.C., 1966. *Adaptation and Natural Selection.* Princeton, NJ: Princeton University Press.

CHAPTER

4

Comparing methods: adaptive traits and tests of adaptation

JOHN W. WENZEL & JAMES M. CARPENTER

CONTENTS

Keywords: Adaptation – phylogeny – evolution – statistics – models – crypsis – wasps.

Abstract

Studies of adaptation rely increasingly on examination of traits that appear to be correlated in several taxa. Many statistical methods have been proposed to provide a probability value for the observed associations based on a presupposed model of the evolution. Flaws in these methods include inappropriate substitution of statistical uncertainty for logical uncertainty, the *a priori* characterization of the unknown evolutionary process under examination, axiomatic definitions of the traits themselves and inadequate representation of the unique historical relationships that are the crux of the study. Specific tests are discussed to demonstrate their unreasonable foundations and limited validity. An example is offered to show that the relationship between traits in given taxa is better understood by direct examination of specifics than by statistical description of generalities.

Phylogenetics and Ecology
ISBN 0-12-232990-2

INTRODUCTION

This chapter discusses the ways in which recent statistical approaches can account for and eliminate differences due to ancestry from studies of adaptation. Many of these methods do not require a phylogenetic hypothesis, and most rely upon specific assumptions about the process of evolution that produced the data under consideration. Only two chapters dissent that these methods are somehow lacking. Coddington (Chapter 3) takes a philosophical approach to the epistemological issues, while this chapter will discuss specifically what he calls the "thicket of methodological and statistical problems" that remain unsolved. Several chapters in this book correctly state that the 'homoplasy' approach (statistical examination of repeated events belonging to a predefined class) and the 'homology' approach (direct examination of unique events plotted on a phylogeny) do not attempt to do the same thing. This suggests that they should be able to coexist, each doing what it does best. However, we will argue that, coexistence or not, the homoplasy approaches available today do not fulfil their ambitious promises.

The objections made here will often resemble warnings that appear in introductory statistical texts, and some readers may be moved to cite this property as reason to dismiss our arguments. In fact, our objections are all the more severe because they derive from common knowledge. A real defence against our criticism must come from scientific argumentation rather than a glib dismissal based on 'everyone already knows that', or the suggestion that 'it's been said before'. Those who have worked to develop these methods will be disappointed to see that we do not propose remedies for the problems we see, but as such that does not invalidate our observations. Neither will an attack on the homology approach constitute a defence of the homoplasy approach. Authors may acknowledge "the arbitrariness of the starting points and the unrealistic assumptions of most comparative analyses" (Pagel & Harvey, 1992: 427) and argue that if we insist on strict criteria we will hinder the understanding of empirical phenomena (Pagel, Chapter 2), but an adequate defence for the statistical methods promoted in this book and elsewhere is not to be found in such arguments.

STATISTICS AND HISTORY

The homoplasy approach as discussed in this book portend to be more scientific by being more statistical. This chapter will show that in actuality, these may be statistical in the worst sense, substituting mathematical formalism for biological study. Most of the methods generally proceed in the following manner: first a null or prior distribution is imagined; second, the observed data are compared with an expectation under the prior distribution; finally, if the observations depart significantly ($P < 0.05$) from the expectation, adaptation is considered the probable cause for the disparity. This procedure contains both biological and statistical fallacies.

The biological problems arise from reliance upon a model of randomness, equal probability, or some other process understood through sampling theory, as a possible explanation for the association found among certain traits in given taxa (e.g. Ridley, 1983; Felsenstein, 1985; Grafen, 1989, 1992; Harvey & Pagel, 1991;

Gittelman & Luh, 1992; Pagel, 1992; Pagel & Harvey, 1992). Implicit in this perspective is that the world is composed of two complementary parts: adaptation versus a chosen model producing the null expectation (generally one supposed to include historical phylogenetic effects). This is sometimes made explicit, as when Gittelman & Kot (1990: 227) write that their method is "effective in partitioning trait variation into adaptive and phylogenetic components". But in addition to mere chance, factors known to allow departure from 'expectation' include epistasis (for within population comparisons), genetic hitchhiking or linkage disequilibrium (between populations), developmental constraints (between species), earth history (between entire faunas), and no doubt other influences that may stand independently from specified adaptive regimes. Only when biologically meaningful premises exclude other possibilities does a rejection of expectation suggest adaptation. Otherwise, such a conclusion is suspect. 'Significance' means only that the process supporting the null hypothesis is an unlikely explanation for the data, given that the assumptions are valid. If the null model is based on random sampling from a distribution of normal variates, then such a process is the only one we have rejected. This says nothing about adaptation *per se*, only that our data are unlikely to be derived from such normal variates sampled randomly. Testing the model means that 'expectation' is actually the centre of the study, not 'adaptation'.

Models may be designed so that they deal with uncertainty that is tractable statistically (see below), giving rise to probability statements. But why are we to believe that an assumption of randomness, equal probability, or another model including uncertainty, is appropriate in the first place? There are few naturalists who would regard the complexity of nature as a constellation of random or equally probable traits. Even creationists who reject the evolutionary model of natural hierarchies see strong pattern in the world. Some of the creationists claim that life is *too* orderly to be explained by evolution, repeating William Paley's 200-year-old argument that the purposeful design of nature demonstrates the existence of a Creator. So what does it mean to find that something is different from 'random expectation'? Certainly not the same as the question of whether dark blue automobiles are significantly more likely than random to be in collisions. Most every car is independent, as is most every collision, whereas our taxa and our traits are not, nor do we know how to account for their relationships in the process of assigning probability. Probabilistic data, like those from genetics, bring statistics to biology, but evolutionary history is not a set of data from a probabilistic model.

Statements of 'probability' come in two forms, those determined by frequency and those determined by logic (Popper, 1968). Statistical analysis is at home only with the former. Statistical statements concern repeatable events and rely upon sampling from a known population. When we say that a given bottle *probably* contains 75 redeyed flies out of 100, we make a statistical statement presupposing that we know all the factors that bear upon the problem except for deviations from expectation due to random chance. Random deviation itself is modelled by Student's t distribution or some other scheme, and is therefore far better understood than a chaotic process in which anything could happen, such as there being no flies at all. Repeated sampling tests the statement of probability. By observing directly and repeatedly the facts of automobile accidents, an insurance firm establishes this sort of probability to calculate automobile accident premiums.

In constrast, logical statements do not sample repeatedly the phenomenon in question, but rather use other data contained in the predicates of the problem. When we say that a free fly *probably* escaped from the bottle when the cap fell off, we mean that such an event is indicated but cannot be confirmed. The observation that another fly escapes under similar conditions is still only a logical scenario that shows 'how possibly' the first fly came to be free, not true repeated sampling of the event in question. Most evolutionary hypotheses are logical 'how possibly' arguments that rely on data which are connected indirectly to a historical event that is no longer observable (O'Hara, 1988).

Some argue that statistical approaches are simply ways of making decisions in the face of uncertainty. This is true, but these approaches are not applicable to all facets of uncertainty, particularly historical uncertainty. A satisfactory statistical statement about the 'probable' origin of the free fly would require a great deal of measurement and assumption, the uncertainty of each contributing to the frailty of the ultimate model. We would have to build correct frequency models to describe flies originating from closed bottles, from open bottles, from outside the building, from a bottle left open last week, or perhaps the fly has been there all along, etc. What is most immeasurable is the frequency associated with the actual fly and actual bottle in question; they are unique. The uncertainty in this sort of model is nothing like that found in Student's t, and is therefore best addressed logically (for instance, by deciding to forget about the possibility that the fly was there all along) regardless of our efforts to assign probability to the observation in question.

The inappropriateness of statistical models for historically unique events is shown clearly by a statistical question about an historical statement. *Statement*: William defeated Harold at Hastings in 1066. *Question*: What was the mean? A statistical examination is clearly absurd. Either the event happened as described ($P = 1.0$), or it did not ($P = 0$). There is no population from which we can draw repeated samples to evaluate a mean, variance, skew and hence an expectation that is testable according to formal statistics. Other battles are not repetitions of the one in question and they are alike only in some very limited way. Even if it is *logically expected* that Norman invaders would not defeat a larger force of Anglo-Saxon defenders, the historical fact is unique and has a probability $P = 1.0$. It happened.

CONTEXT OF THE STUDIES

Two practical problems arise from the general tendency for adaptive scenarios to: (1) define the traits axiomatically; and (2) circumscribe the problem so that what appears to be a generally informative statement is, in fact, narrow. In the first case, suppose we are interested in whether polygamous birds are more sexually dimorphic than monogamous birds. We begin by dividing the natural variation into complementary units: monogamy versus polygamy and monomorphic versus dimorphic. Clearly, the continuum of the real world does not fall easily into these discrete classes. The American robin, *Turdus migratorius*, would most likely be called monomorphic, yet any backyard birder can spot the male's blacker cap and redder breast on active birds at ten metres. Recently, high levels of parental uncertainty have challenged the utility of the dichotomy of 'monogamous' versus 'polygamous' (Gibbs *et al.*, 1990; Gowaty & Bridges, 1991a,b). If assignment of a species to either of the classes is somewhat arbitrary, then so are the data on which the test is based. Even if there is a consensus among researchers as to what goes where, division of continuous variables into two discrete halves is a poor starting point for arriving at generalities that truly have the probabilities ascribed to them. Of course, this problem obtains for the homology approach also, but cladistic analysis assigns expectations of either zero or one, which are more easily testable than the incremental probabilities of the homoplasy approaches.

One solution to the problem of breaking up continuous variables is simply to use them in a continuum. This is popular for traits that are logically continuous, such as home range size, body size, age at first reproduction, etc., and the values seem to lend themselves to some form of regression analysis. These analyses have their own set of assumptions, however. Evolution must reflect a metric quality such that the evolutionary difference between 20 and 17 is the same as between 17 and 14, and that 17 is evolutionarily intermediate between 20 and 14, or at least this must hold for some transformation of the data. It is statistically reasonable to take the logarithm of home range areas measuring l000, 100 and 10 to obtain 3, 2 and 1, but is it biologically reasonable to say that the intervals are evolutionarily the same? Does evolution require that an animal cannot go from a home range of 10 to one of 100 without passing through a range of 50? Such transformations and their implicit assumptions may or may not be reasonable, and their validity should be explicitly defended. If they are derived empirically, they may be more interesting than the correlation itself. This point is not lost on enthusiastic supporters of statistical methods, but it is sometimes understated in a quaint way. For instance, after recommending that contrasts between species be standardized according to trans-formed branch lengths, including the possibility of using different lengths or transformations for different variables, Garland, Harvey & Ives (1992: 24) caution that "use of different branch lengths for the dependent and independent variables would, of course, complicate interpretation of the slope for a set of standardized independent contrasts". Complicate indeed, for what do these transformations say of our presuppositions regarding the evolutionary process underlying the variously transformed data?

As for circumscribing problems, consider a query as to why the nightjar flies at night when the majority of birds are diurnal. The phylogenetic answer ('because it is a species of Caprimulgiformes') sounds somewhat tautological and not very satisfactory. The ecological answer ('because it eats flying insects') seems not to rely on phylogeny, infers a reasonable causal relationship and sounds much more informative. But each of these three latter impressions is mistaken. The question itself is framed by historical, phylogenetic reference because we have already defined the problem as one regarding birds only. We are not considering all flying things. Rather, we have limited ourselves to a phylogenetic unit where we expect to find coherent patterns in the data. As for causality, if we were to consider all birds that eat flying insects, any association with nocturnal habit will disappear. Even if the relationship did hold, we would have only demonstrated a pattern in the data, and nothing about the process (adaptation) by which things became as they are. For example, the features discussed might be the result of something wè have not considered (perhaps good binocular vision) rather than being linked directly. Lastly, 'eats flying insects' is far less informative as a class of things than 'Caprimulgiformes'. None of the objections to the ecological explanation mean that the inferred causality is necessarily false, but even in situations of carefully circumscribed traits and taxa there is no way to prove that it is true. Assigning a probability is still no more convincing than the logical arguments that an indicated causal mechanism is at work.

The correlations assessed with myriad procedures fail to explain adaptation. The functional evidence necessary to demonstrate adaptation is rarely supplied by either systematists or ethologists (Lauder, 1990), but several authors (Cracraft, 1981; Coddington, 1988, l990a; Baum & Larson, 1991) have emphasized that such evidence is required because the correlations may have multiple possible causes. Some methods that attempt to dispense with phylogenetic hypotheses (below) do not treat the evolutionary direction of trait associations (i.e. which trait came first), although recent work now recognizes that this is critical for inference of selection and causality (Donoghue, 1989; Hoglund, 1989; Harvey & Pagel, 1991). All of the methods face the ironic situation where the worse the fit of the traits to the cladogram (i.e. the more homoplasious the association), the greater the significance of the correlation (Coddington, 1990a, 1992). Unique adaptations (uniquely derived characters) are not amenable to investigation with this approach because they represent a sample size of one (Carpenter, 1992a; Coddington, Chapter 3), and therefore have no degrees of freedom. Such associations as forked hairs among all bees (one event in the proto-bee) and pollen carriage will always be outside the province of the methods.

ASSUMPTIONS OF THE MODELS

The main problem with any general method is that the assumptions that make sense in one context may seem absurd in another. Consider the extreme of the treatment of every terminal taxon as an independent data point (Fig. la). We assume that each terminus is equally likely to display a given trait, as the tick marks show. But of

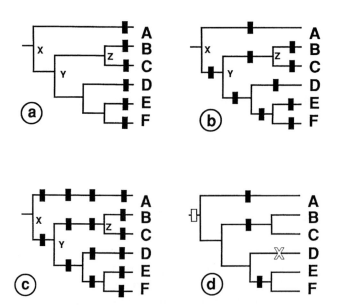

Figure 1 Alternative expectations for trait distributions. See text for explanation.

course, B and C might both have a trait if it was found in their mutual ancestor Z, so the occurrences of the trait are not really independent. This is the situation that has inspired most of the methods designed to factor out phylogenetic information, but it represents only one of many possible pitfalls.

The methods that have been proposed to deal with non-independence range from those that attempt to remove phylogenetic effects without requiring a phylogenetic hypothesis to those that use a cladogram to specify 'phylogenetically independent' contrasts. An example of the former is the use of nested ANOVA, with taxonomic rank in the Linnaean hierarchy treated as a categorical variable. Data for taxa at the rank which accounts for the largest percentage of the overall variance are assumed to be independent; covariation among taxa at lower ranks is attributed to common ancestry. More or less the opposite approach is taken through autoregression or autocorrelation (Cheverud, Dow & Leutenegger, 1985; Gittleman & Kot, 1990) in which expected correlation (or 'phylogenetic connectivity') between traits is related to rank: in the example, species within the same genus are assumed to have a connectivity of 1.0, species in the same family get 0.5, same superfamily yields 0.33, etc. However, both schemes suffer in that taxa given the same rank are related differentially, so that non-independence still obtains. Finding the largest variance at the level of families does not demonstrate their mutual independence because some families are very much more closely related to each other than they are to other families. Differential relatedness is just as serious for the autocorrelation method, but now the statistics rely upon quantitative values that originate in the investigator's imagination as opposed to being measured data, regardless of what values one chooses for connectivity.

Any procedure dependent upon existing Linnaean taxonomy to measure phylogenetic 'distance' will fail in unpredictable ways (Carpenter, 1992a). Authors of these methods would more likely prefer to include phylogenetic structure if it were available and only use taxonomic rank as a necessary surrogate. However, the gross misinformation of the Linnaean system motivates some systematists to recommend abandoning it even in taxonomy (De Queiroz & Gauthier, 1992). It is hard to know how badly taxonomy fails as a substitute for phylogeny, but it will range from good performance under the (very rare) situation with a perfectly symmetrical tree when taxa bearing equal rank also have equal age, to very bad when assuming that the aardvark (unique genus, family, order) is somehow comparable to speciose families or orders of mammals. The larger the tree, the worse these methods will fare since taxa are not necessarily comparable in any way other than rank, and many presently recognized taxa are certainly unnatural. Because these methods are intended to cover several levels of the hierarchy, they appear to be in danger of being most used where they are least secure.

Other, more subtle problems abound even when consulting a cladogram directly. For instance, it might seem logical to assume that the traits in question should have an equal probability to map to any internode, in which case we have the expectation illustrated in Fig. 1b. This is the foundation of Maddison's (1990) test, by which one character is mapped on the tree as a given, and the occurrence of the second character is presumed to be equally likely on every internode, providing an estimate of the likelihood that the two characters should co-occur at random. This is one of the best tests available because of its relatively modest assumptions and its explicit use of empirical data regarding one character and the cladogram.

Unfortunately, the computational complexity is burdensome and required Maddison to simulate the results of his own test rather than calculate them. The method is not powerful because it applies only to two binary variables and a fully resolved cladogram, and the crucial predicate of the method remains unjustified: the use of a random model. In any case, taxon A has been evolving as long or longer than any other, so should not that internode weigh at least as much as any other path from root to tip? Then we would not have four times as much likelihood of plotting something between X and F as we have between X and A just because of the internodes.

Making all root to tip paths equal, we get the expectation of Fig. 1c, akin to the assumptions of the molecular clock. Note that we have another problem with arbitrariness around internal nodes in that the paths from Y to B or C should have three ticks just as do the paths from Y to D, E and F. We may not know whether to expect two ticks between Y and Z (as illustrated), or two from Z to each B and C. Also note that in some sense, we may sample phenomena of internode X–Y five times as much as X–A simply because B, C, D, E, F are all downstream from the former and only A is downstream from the latter. Despite incorporation of branch lengths, we still fail to offer equal probability. Harvey & Pagel (1991) and Pagel (1992) use branch lengths (such as those derived from study of molecular data) to standardize changes in two characters under the assumption that most evolution occurs on the longest branches. However, this yields spurious correlations precisely because the branches vary, producing high significance when neither character changes (Maddison & Maddison, 1992: 315). In this case the P value is no improvement over the simple observation that two invariant characters are found together. Harvey & Pagel (1991: 100) propose to avoid the problem by restricting the analysis to only those branches where changes occur. This recommendation to crop data "remains incompletely justified" (Maddison & Maddison, 1992: 315). In any case, Peterson & Burt (1992) found that such a model was useless because the basic assumptions were clearly violated by the data. They plotted habitat use and breeding system on a tree (produced by 29 allozyme loci for 11 populations of three jays) and found a significant negative correlation between allozyme branch length and ecological character changes. Evolution does not always go according our expectations, and there is no *a priori* way to tell when our assumptions are wrong. In any event, branch length information is usually unavailable and the molecular clock is untenable (review in Carpenter, 1990) – unless we admit that there is a different clock for every trait in every species.

Grafen's (1989) complex phylogenetic regression is essentially a modification of the situation shown in Fig. 1c, in that all paths from a node to its various terminals are equal. Nodes derive a 'height' (value) as one less than the number of species (terminals) distal to that node. A branch derives its length from the difference in height between the basal and apical nodes defining it. These path lengths may then be stretched or compressed differentially, but ultimately they represent the expected covariance or variance for comparisons, depending upon whether the branches represent common ancestry or not. The method requires that these measures accurately reflect all phylogenetic effects; but relative, sampled branching order clearly does not.

Pagel (1992) offers a method for continuous characters using branch lengths supplied by another data set (e.g. molecular) to generate expected variance for traits in question. The method is intended to be flexible with respect to the assumptions of the model and is offered specifically as a generalization of Felsenstein's (1985) model such that polytomies are allowed. The method performs well (that is, produces accurate probability statements) with simulated Brownian motion data on known dichotomous trees (Martins & Garland, 1991). Presumably, it is already known that characters are not interrelated by this process, so a rejection of the model with real data is not particularly informative (Carpenter, 1992a,b). The performance of the method with respect to its advertised use on polytomies is not yet known. In fact, the method relies in part upon arbitrary resolution of the polytomy. Pagel (1992: 434, 435) warned readers that "these decisions are not likely to be reliable, and thus the comparisons derived from them are likely to introduce much incorrect information". If poor resolution of the cladogram is problematic, instability of the cladogram is a disaster for probability models. Phylogenies can be wrong or incomplete for many reasons. We all agree that extinction does occur and our living taxa are not a complete representation of the whole evolutionary history of any group. What if we discover an extinct taxon on that long internode X–A in Fig. 1? Do we now have two internodes, or still just one? What if A, B, C, D, E, F are all genera, and then we decide that we really want to look at species? What if A is a genus with many species, like oaks, and the others are genera with relatively few species, like beeches and chestnuts? Now the cladogenesis is mostly on the X–A lineage, which was previously monobasic. Adding branches to a cladogram is easy enough, but figuring out what that means for probability models is not. If the real root turns out to be somewhere in the D, E, F clade then everything we have done with respect to statistical expectations is uninterpretable, even if the branching diagram changes in no other way. Although the methods discussed here are not concerned with reconstructing the phylogeny *per se*, it is important to note that many will be highly sensitive to modest error in the cladogram, such as re–rooting (which, cladistically, affects only the polarity of traits between the old and new roots). If these 'what ifs' are not reason enough to reject the methods, they demonstrate the highly restricted and provisional nature of conclusions based on homoplasy.

In addition to these problems of topology and expectations, the exact optimization of traits upon a given topology affects how we view the problem in the first place. This indicates one of the major problems with Maddison's (1990) test: the observed frequency of character state changes can not be reproduced without arbitrarily neglecting more parsimonious optimizations. It is easy to construct situations in which 'randomly' assigned traits fall so that some of the random patterns will not be recognized as different from more parsimonious interpretations. If the three black tickmarks evolved 'randomly' (independently), as shown in Fig. 1d, such that five of six species display it, would we ever see it as convergence? Not necessarily, because we would likely interpret it under an accelerated transformation, shown by the white marks, to indicate a single synapomorphy for the whole group and a reversal in the tip of the tree. The question might never arise in the first place because it could be taken as an issue about a unique trait for D, instead of a triple derivation. These are very different statements about the pattern of evolution, and there has to be a reason for us to reject the simplest explanation in favour of one more complicated. The problems we choose to address are not only completely dependent upon the exact topology of the cladogram, they are dependent upon the particular optimization of the characters.

Readers who think we have constructed this example as an esoteric cavil should study the explanations given by Harvey & Pagel (1991: 90, ff.) for Maddison's (1990) method. We offer alternative optimizations that show their example to be lacking (Fig. 2). Nine cladograms of four taxa with two evolutions of a trait are shown to illustrate all possible ways to map two occurrences (black tick marks). Harvey & Pagel (1991) conclude that there is a 1/9 chance of seeing the model, given dual origin of the trait. Of these nine, cladograms 1, 4 and 9 can be interpreted to show a single synapomorphy for a clade (white tick marks) rather than evidence of dual origin (white tick marks). Cladograms 7 and 8 can be interpreted to show a loss of trait 1 (white X) rather than a derivation. Five of their nine cladograms do not give unambiguous evidence of dual origins based on the presence/absence of a trait in the terminal taxa. Thus, *given evidence of two origins*, may not the probability of observing the dual origin of the model be 1/4 rather than 1/9? Parsimonious interpretations must predominate, otherwise what would make dual origins preferable to an infinity of less parsimonious ad hoc alternatives? Why not examine how we could have had three origins and one loss to produce the model? Oddly, Maddison & Maddison (1992) propose exactly that under the name 'compensation'. "If you need to generate 6 changes on the clade selected, then by having a compensation of 1 or 2, one would throw down 7 or 8 actual changes, and this would yield a better chance of parsimony reconstructing 6 than if you had thrown down only 6 actual changes" (Maddison & Maddison, 1992: 312). They caution that this "could

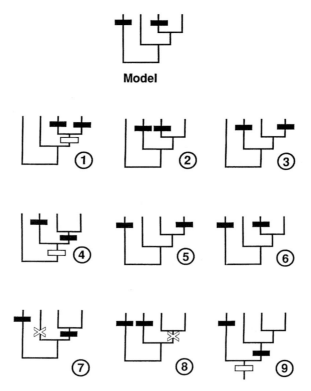

Figure 2 Modified from Harvey & Pagel (1991: box 4.1). Black tick marks show their explanation of possibilities, white tick marks are alternative, more parsimonious explanations not discussed by these authors.

affect probabilities", and indeed in the example offered below, the effect of 'compensation' is to shrink a probability value from 0.016 to 0.001. Some may feel that the distinction is unimportant at this level of high significance, but there is no argument around the fact that compensation means that the student is no longer modelling the data he thinks he is examining. This represents one example of why general tests as such cannot be valid.

If the preceding arguments demonstrate problems with statistical analysis of a single trait, covariation for more than one trait can only be more problematic. The occurrences of the relevant variables may be too few, some are not independent and the methods have not been extended to multistate variables. The methods are lacking in power and their applications are limited (Carpenter, 1992a). Significance values may be entirely and obviously the product of biased sampling (Coddington, 1992).

The complex relationships between traits and taxa are understood only through formal cladistic analysis. A surprising denial of this obvious truth rings throughout the assumptions of various statistical 'comparative methods'. Purvis (1991) claims that ancestral node values for quantitative traits may be estimated by this equation (cited in Gittleman & Luh, 1992):

$$X_k = \frac{\sum\limits_{i}^{m}(1/v_i)X_i}{\sum\limits_{i}^{m}(1/v_i)} \tag{1}$$

where X_i are traits of descendants, v_i are branch lengths and m is the number of descendants. Evidently, Pagel (1992) is willing to rely on this also. This method generates annectent values intermediate between the extremes found in terminal taxa. This property corresponds nicely with other assumptions about the evolution of quantitative traits (above), but has the undesirable property of generating states in order to satisfy an *a priori* decision about the path of evolution. Even more extreme than Purvis' assumptions (above), Harvey & Pagel (1991) argued that when both branch length and ancestral values are unknown, one can assume that the ancestor is the mean of its descendants. Imagine a situation in which we have sister species, one large and one small, and the two successive outgroups are both large. Acccording to the recommendations above, we deduce that the ancestor of the two sisters was intermediate between them, the ancestor of that species and the outgroup was intermediate between them, etc. generating states that do not exist today and for which there is no empirical evidence. The possibility that ancestral species were big and one of our sister-terminals got small is ruled out by the model, although it is logically better supported by the data and is commonly observed in reality. It is too much to believe that the great homogenizing effect of Purvis' equation embodies the known record of strange and unpredictable patterns of elaboration and loss, of radiation and extinction. This is not to say that the model may never be adequate, but there is no way to tell when it is failing.

Authors often make explicit statements regarding the bold assumptions that replace uncertainty about the evolutionary process, but these do not necessarily improve their believability. Grafen (1992: 417) stated: ". . . it is neither desirable nor possible to use the data in the analysis to estimate all the branch lengths. Most of the information must be simply assumed, when possible on the basis of other types of evidence." It sounds as if he recommends that we should make up the data we need whenever they are lacking. Citing this proceedure with approval, Gittelman & Luh (1992) stated: ". . . while the branch lengths generated from this process are arbitrary, nevertheless they retain statistically tractable results." Strangely, one of the things some workers will not assume is the very problem that provides the foundation of their work: historical effects. Grafen (1992: 417) said ". . . it will rarely be sensible to assume in advance that we know the strength of phylogeny." Harvey & Pagel (1991: 138) praise a maximum likelihood method because "it avoids altogether the problem of reconstructing ancestral character states." This is particularly shocking because all of these methods aspire to account for similarity due to ancestry. While advertising methods to study the effects of phylogeny, the authors appear eager to ignore it. Garland *et al.* (1992: 19) appreciate substituting taxonomy for phylogeny because "it does allow analysis of the data now, rather than waiting for actual phylogenetic information to become available." It would seem more prudent to recommend that, after assuming away the historical facts of evolution, one should not attach any statistical probability to the findings.

VALIDITY OF THE MODELS

As discussed in another context by Carpenter (1992b), significance relative to a null model of random (or other) correlation is at best weak corroboration for historical explanation: 'signficance' merely calls into question the particular model used. False premises do not lead to correct conclusions. Even if we fail to reject a given model, that does not mean that the model does apply (statistical Type II error).

Many statistical homoplasy approaches are akin to Felsenstein's (1978) 'maximum likelihood' method of phylogenetic inference (see in particular Harvey & Pagel, 1991; Pagel, 1992). Farris (1983) contrasted the fundamental logic of phylogenetic inference against Felsenstein's approach. He demonstrated that explanatory power is maximized by phylogenetic parsimony, therefore a more complex model-dependent approach is unnecessary. For Felsenstein's results to hold (in this case, failure of parsimony to estimate phylogeny consistently), the models must be defended as realistic – the very opposite of what Felsenstein maintains. Yet, conditions that critics of parsimony stress (high rates of change or highly disparate rates in different taxa) occupy a very minor fraction of biologically meaningful values (Albert, Mishler & Chase, 1992). As Farris (1983: 17) put it: "The statistical approach to phylogenetic inference was wrong from the start, for it rests on the idea that to study phylogeny at all, one must first know in great detail how evolution has proceeded. That cannot very well be the way in which scientific knowledge is obtained. What we know of evolution must have been learned by other means." Ten years later, this statement applies with equal force to the correlations of the 'comparative method' as embodied in the homoplasy approaches. Ironically, Felsenstein himself allowed (1979: 60) that maximum likelihood methods could not be used in practice, because the parameters of the model remain unknown.

To the (superficially reasonable) argument that we should use our present knowledge of evolutionary process, the response is that such 'knowledge' comprises empirical claims. Building a process model amounts to presupposing empirical propositions. Herein lies the flaw: the results cannot question presuppositions of a model. Cladistic parsimony combined with the most modest assumptions imaginable for an evolutionary process (descent with modification and cladogenesis) offers the possibility of testing process theories that are more detailed than parsimony requires. When parsimonious results explain the data as well as or better than the expection under some model, they supply evidence against that model. This asymmetry between the ability of models versus cladograms to test a process theory is what underlies the success of cladistics – including applications such as the associations investigated by the 'comparative methods' of the homoplasy approach.

In some circles, ad hoc claims to knowledge are incorporated into models as fiats that will directly impede understanding nature. Common objections to cladistic study of behaviour rely on the impression that "behavior (movement) is an especially plastic aspect of phenotypes" (West-Eberhard, 1989: 252). But is it? How are we to know except by testing that hypothesis? If we assume that behaviour is plastic, we cannot conclude that our subsequent work demonstrated plasticity without being guilty of discovering our own assumptions. The dogma that behaviour is evolution-arily labile persists, despite many studies showing that a suite of behaviours does reflect hierarchical order of phylogeny (Jander, 1966; Chvàla, 1976; Mundinger, 1979; Lauder, 1986; Carpenter, 1987; Grimaldi, 1987; McLennan, Brooks & McPhail, 1988; Miller, 1988; Coddington, 1990b; Prum, 1990; Maurakis, Wollcot & Sabaj, 1991; Wenzel, 1993; see also reviews in Wenzel, 1992; and De Queiroz & Wimberger, 1993). Sometimes the behavioural traits are more conservative than morphology (Coddington, 1990b; de Queiroz & Wimberger, 1993; Wenzel, 1993). Yet, the *status quo* lives on, because despite recognized demonstrations of phylogenetic stability in presumably rapidly evolving traits (Cheverud *et al.*, 1985; Höglund, 1989; Björklund, 1990), investigators return to their assumptions to escape the rejection of their attractive hypotheses, as when Webster (1992: 1636) declares "It is likely that these researchers [Cheverud and Björklund] would have reached different conclusions had they not automatically assumed associations with phylogeny to represent a non-adaptive phylogenetic constraint." Webster invokes a postulate of lability in order to reject separate demonstrations of stability. What growth in our understanding do we expect from this avenue? We should be glad that Cheverud, Höglund and Björklund did as they did, for it is only by challenging a hypothesis with the possibility of rejection that we will ever discover that, for instance, behaviour is not really that different from morphology after all.

AN EXAMPLE

The example offered below serves to demonstrate that the relationship between traits and taxa is better understood by direct examination of specifics than by statistical description of generalities.

Social wasps are often easily recognized by striking coloration, and they are the putative models for large guilds of harmless mimics (Richards, 1978). Yet, the mass of resources represented by the defenceless brood tempts monkeys (Boinski & Timm, 1985), birds (Winsdor, 1976; JWW personal observations) and other verte-brates to remove wasp nests from frail supports (such as leaves or terminal twigs) and consume the brood away from the original location. Perhaps because of this, the nests built by many social wasps are cryptic and difficult to see against a natural background. In the subfamily Polistinae (Vespidae), the brood comb of *Ropalidia carinata* is composed of lobes, resembling pinnate leaves of the forest understory. *Mischocyttarus punctatus* dissolves the hexagonal pattern of the comb entirely to produce cells arranged linearly, resembling a slender dead twig. Similar architectures are found in some species of *Polistes, Belonogaster* and *Ropalidia*. Some species of *Brachygastra* and *Charterginus* build a covered nest of dark paper but paint the underside white so that the nest is dark (like the forest floor) if viewed from above, but bright (like the sky) when viewed from below. Nests of some species of *Protopolybia* are green in colour, like the leaves themselves, while others span gaps in curled leaves with a sheet of pure salivary secretion, free of wood pulp and dried into a transparent film that is truly invisible. This transparent envelope has evolved separately in *Ropalidia opifex*, which also nests on leaves.

The most variable and cryptic nests are built by colonies of *Leipomeles dorsata*. Swarms always build initial combs on the midrib of a leaf and conceal these with an envelope that may be a simple funnel-shape or may be flattened against the underside of the leaf. The surface of the envelope may be smooth or with acute ridges, with or without a longitudinal furrow, coloured white, yellow, grey or green, with or without blots of constrasting wood pulp to disguise the outline or make it resemble the substrate. In the most striking forms, the nest is flattened against the

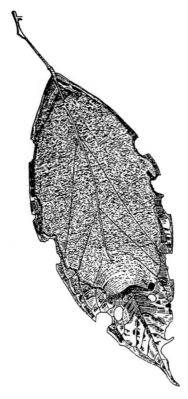

Figure 3 *Leipomeles dorsata* nest, founded on the underside of a leaf, modified from Williams (1928: plate 29, fig. 221). Brood combs on the midrib are concealed by a paper envelope that may employ three elements to remain cryptic: blots of coloured pulp interrupt the uniform appearance; V-shaped ridges suggest the leaf veins; a longitudinal, central furrow matches the placement of the midrib.

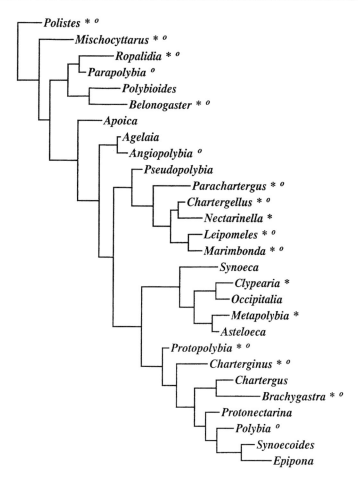

Figure 4 Cladogram for the genera of Polistinae. The character matrix of Carpenter (1991) has been modified by synonymy of *Pseudochartergus* with *Protopolybia* (Carpenter & Wenzel, 1989) and incorporation of more detailed data on nest architcture (Wenzel, 1993) and larvae (Kojima, personal communication). Approximate analysis of 97 characters resulted in 41 cladograms (length 314, consistency index 0.48, retention index 0.70); successive weighting resulted in the cladogram shown. The tree differs from that of Carpenter (1991: fig. 1.5) in resolving, and partly rearranging, the relationships of the genera in the branch subtending *Protopolybia* and *Epipona*. Genera with cryptic nests mentioned in the text are marked with an asterisk; those that include species that prefer to nest on leaves and twigs are marked with a superscript '*o*'.

leaf, a series of ridges suggest veins of the leaf, a central furrow suggests the midrib of the leaf, and bits of coloured pulp reinforce the image of the veins or suggest lichens. At a glance, these nests look like nothing more than a slight irregularity or a dead region of an otherwise ordinary leaf (Fig. 3).

Social wasp nests illustrate Darwinian adaptation through both specific function (Jeanne, 1975) and wide-spread convergence (Wenzel, 1991) of many architectural details. Has brood predation selected for crypsis among nests constructed on leaves and terminal twigs? The proposal has intuitive appeal, and the appearance of the nests seems to argue in favour of it. The examples cited above are scattered widely across the phylogeny of Polistinae (Fig. 4), which might suggest generality. Most of the genera contain both species that have cryptic nests and those that do not, suggesting that crypsis has evolved independently many times. Only a few genera that build cryptic nests (*Nectarinella*, *Metapolybia* and *Clypearia*) prefer trunks and major limbs. If we compare genera that build on twigs and leaves against those that build only on other substrates, we will likely inflate our estimates of 'independent' evolutions, as shown above. Because we have a cladogram and two characters that can be considered binary (nests on twigs and leaves versus not, cryptic architecture versus not), we could employ Maddison's (1990) test. We would start by plotting the character 'nests on twigs and leaves' as a given, optimizing it across ancestral nodes. In this case, equivocal branches would not be relevant because there are no taxa that are cryptic, not on twigs or leaves, and immediately descended from an equivocal ancestor. Then we observe how many derivations of cryptic nests there are, and how many occur in lineages that nest on twigs and leaves (counting *Parachartergus* to *Nectarinella* as a single clade we get ten for the former, but eight for the latter). Implementing 1000 simulations of the concentrated changes test in MacClade, we find that the probability of getting eight of ten changes associated with lineages nesting on twigs or leaves ranges from $P = 0.016$ (plotting ten changes, no optimization) to $P = 0.001$ (optimizing the reconstructions to maximize derivations and using a compensation value of three), a broad range but certainly significant. Therefore, we conclude that nest crypsis is associated with construction on twigs and leaves, and logically infer that selection has repeatedly produced crypsis in taxa that nest in such sites. We may go further to say that *Leipomeles* nests look like leaves because they are built on leaves, and if they did not look like leaves more would get eaten.

The problem with this conclusion is that it only tells us (if somewhat more forcefully) what we observed to begin with; cryptic nests are associated more with leaves and twigs than with other sites. We do not know any more about the derivation of this putative adaptation than we knew before. Furthermore, functional definitions of behavioural traits are generally inappropriate for retracing (historical) patterns in evolution and rarely contribute to understanding the origin of the traits themselves (Cracraft, 1981, 1990; Wenzel, 1992). 'Cryptic nest' may be an ill advised starting point because we lump together such different things as a green nest and a lobate nest, claiming that they are the same for our purposes, and obscuring the fact that they are not the *same thing*. Only the close relatives of beautifully cryptic *Leipomeles* can contribute meaningful information regarding the question of how *Leipomeles* got where it is today and whether its appearance is related to the chosen nest site. *Protopolybia*, *Ropalidia* and the other genera mentioned above are too distantly related to *Leipomeles*, and nest crypsis in these genera is clearly not produced by the same mechanism.

If we want to understand architectural adaptation to nest site, we must examine it in detail rather than in general overview. Crypsis in *Leipomeles* is more than just a single trait. There are several components, some of which are shared with near relatives, and by studying them in their own right we will best understand the problem at hand. Therefore, we need to study how Leipomeles and her relatives have changed their ancestral ground plan to become what they are today.

Wenzel (1991) published an overview of the variation and evolution of nest architecture in the three subfamilies of social Vespidae, as well as a detailed study of the phylogenetic patterns of nest architecture in Polistinae (Wenzel, 1993). With the exception of a few new observations from recent field and museum work, most of the data for arguments we make below are available in those papers. Character-izations are simplified because variability of nest forms within a species far exceeds that of adult morphology, but a trait is considered to be part of the endowment of a species or genus if it appears as a regular part of the diversity of forms, even if it is not expressed in every individual. For example, if a species sometimes builds an envelope with ridges (when they nest on a flat surface) and sometimes does not (when they hang a nest from a twig), we consider these ridges to be a part of the natural endowment and ground-plan of the species even though not all nests display the trait. This is no different from considering A, B and O to be typical human blood types despite the fact that each of us has only one of them. It results in a generous characterization, but does not necessarily favour finding synapomorphy as opposed to homoplasy.

The cladogram we use for the near relatives of *Leipomeles* (Fig. 5) is derived from Carpenter's (1991) analysis of morphological data (and see Fig. 4). The best estimate for the primitive ground plan of the nest for this clade is a smooth, flattened dome, like certain modern species of *Parachartergus*, e.g. *P. fulgidipennis* or *P. colobo-perus*, or alternatively a smooth, simple, pendant flask, like modern species of *Angiopolybia*. These forms are built on terminal twigs, and surfaces such as large leaves, tree trunks and broad limbs.

A conspicuous difference between the envelopes of the basal taxa *Angiopolybia* and *Pseudopolybia* and those of the component subtending *Parachartergus* and *Leipomeles* is that the latter five genera commonly apply blots of coloured pulp to the envelope, disguising the otherwise monochromatic paper. The crypsis can be remarkable and many biologists mention it specifically (*Parachartergus*: Strass-mann, Hughes & Queller, 1990, *Marimbonda* and *Chartergellus*: Richards, 1978;

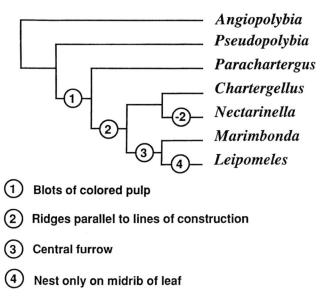

Figure 5 Cladogram for *Leipomeles* and near relatives, showing optimization of nest architectural traits illustrated in Fig. 3. The extant, relatively plesiomorphic genus *Angiopolybia* is depicted as a far outgroup rather than the hypothetical mutual ancestor of the 13 genera in the lower region of Fig. 4.

Nectarinella: Schremmer, 1977). This character is represented and optimized as (1) in Fig. 5.

When on broad horizontal surfaces, such as a large *Heliconia* leaves, *Leipomeles dorsata* builds a flask-shaped envelope similar to that of *Angiopolybia*, except that it is marked with simple, acute and widely separated circumferential ridges. When closely appressed to the underside of a small leaf, the envelopes of *Leipomeles* mimic leaf veins by having such ridges extend from the midline to the margin of the envelope (Fig. 3). These ridges are parallel to the lines of paper construction produced by successive addition of separate loads of wood pulp and may approximate the limit of construction on successive days. The ridges have an inverted V-shape, with the apex somewhat higher than the sides of the ridge when the nest is in its natural orientation. Similar ridges are also found in some *Chartergellus* and perhaps in *Marimbonda*. This character is illustrated as (2) with two steps in Fig. 5, showing a synapomorphy for the component including *Chartergellus* and *Leipomeles*, with a reversal in *Nectarinella*. It is also possible to optimize the trait with one derivation in *Chartergellus* and one in the ancestor of *Leipomeles* and *Marimbonda*, *Nectarinella* showing primitive absence. This latter possibility is not shown, but the distinction is irrelevant for the points made below.

The image of the midrib of the leaf is produced in *Leipomeles* envelopes by a central furrow running against the lines of paper construction from the uppermost margin of the envelope toward the nest entrance. Its location (even when weak) is marked by the apices of the V-shaped ridges (described above, Fig. 3). Loads of pulp applied in tight arches form the furrow and contrast with the nearly straight applications of pulp in the rest of the envelope. The same tight arches produce a weak furrow in the sister genus, *Marimbonda*. This is shown as (3) in Fig. 5.

Many of the 28 genera of Polistinae will build a nest under the shelter provided by a firm leaf, but nearly all of these also contain species that nest on twigs. *Leipomeles* is unusual in that it appears to nest only on leaves, the first combs usually placed on the midrib. This exclusive use of leaves is represented as (4) in Fig. 5. Notice that the stepwise evolution of the elements of crypsis all comes before the preference for nesting on the midrib of a leaf. We conclude that the design of the nest of *Leipomeles* evolved prior to the habit of nesting only on leaves.

Returning to the question 'Is cryptic nest form an adaptation for building on leaves?', the answer is 'no' for *Leipomeles*. We now see that the recent evolutionary innovation would have to be the nest site rather than the nest design. Field experiments exploring adaptations should focus on how nest location, not nest design, relates to survival. If nest design does influence survival, the relationship cannot be used to explain the origin of the present architecture as a result of the present nest site, for the causality (if any) must go the other way. The statistical methods that propose to study adaptation reveal only a correlation and cannot show that causality in *Leipomeles* is opposite to what we inferred.

The example above shows that *Leipomeles* (and perhaps the other members of the clade subtending *Parachartergus* and *Nectarinella*) would have to be ruled out if we intend to continue with the question of whether cryptic nests are adaptations to living on leaves and twigs. More properly, perhaps they should count against the hypothesized relationship because the crypsis evolved before the preference for nesting exclusively on leaves. The cryptic nests of *Polistes*, *Belonogaster*,

Mischocytarus, Ropalidia, Brachygastra, Charterginus and *Protopolybia* must each have passed their own trials like the one presented above if they are to be applicable to the general case. The significant results of Maddison's concentrated changes test (above) are misleading because adaptation is presupposed and the statistic evaluates whether crypsis is or is not associated with leaves and twigs. There is no way to reject the proposal of adaptation *per se*. Now that we have independently rejected it for *Leipomeles*, what does the test say about seven of ten derivations of crypsis associated with leaf- and twig-nesting? From the same simulation described above, the probability ranges from $P = 0.073$ to $P = 0.037$. If we take a conservative approach and say that there is no relationship, we might obscure a few cases in which adaptation is at work. Only by studying each case will we know the degree to which crypsis and nest site are linked. It is possible, perhaps even likely, that a relationship between two traits in some taxa is supported in an adaptive way while in others it is not, and that generalizations are meaningless (Coddington, 1988; Baum & Larson, 1991).

The study of *Leipomeles* demonstrates that if there is a causal relationship, it must be that recent selection pressure favoured changing the nest site rather than changing the nest form. Yet, this exercise does not by itself demonstrate that adaptation is involved. Adaptation, as a process, occurs within a species when polymorphisms have different fitness and one replaces another. Data to demonstrate the process of adaptation are available only from studies at the level of populations. Variation between individuals is the currency of adaptation, not variation between species. Therefore, phylogenetic study can refine questions of adaptation by rejecting certain false proposals, but even in the best case it cannot be used to demonstrate that a given proposal is true. To demonstrate the legitimacy of an adaptive scenario, biologists must collect data on individual organisms.

CONCLUSIONS

Biologists want to make evolutionary conclusions about 'laws' of biology when there are inadequate data to do so. Some mistakenly try to replace logical uncertainty with statistical uncertainty, others believe that statements made outside a method conferring statistical significance are unscientific. As we have argued above, frequency probabilities, such as those derived from experimentation, are out of place in problems of history, such as phylogenetic problems. Phylogenetic inference is not unscientific just because it is not accompanied by a statistical probability value.

If the phylogenetic relationships and character states were inadequately known to begin with, then they are still inadequate to support a conclusion after the statistical massage. Lead does not become gold. Some biologists will reply to our arguments by saying that it is better to proceed with admittedly weak techniques than to do nothing. We argue that it is better to do nothing than to misinform. It is foolhardy to begin with a foundation of assumptions that are unsupported and then assert that the resulting *P* value has proven (or disproven) something. There are now available enough 'comparative methods' for the homoplasy approach that the authors dispute the validity of competing models (Maddison, 1990, *versus* Sillén-Tullberg, 1988; Maddison & Maddison, 1992, *versus* Harvey & Pagel, 1991; Pagel & Harvey, 1992, *versus* Grafen, 1992; Peterson & Burt, 1992, *versus* Harvey & Pagel, 1991). To some extent the debates are sterile because a dispute about the legitimacy of one or another statistical test or expectation is not an improvement on our fundamental understanding of the biology in question. Nevertheless, all these authors would stand united in the claim that these methods allow us to look at broad patterns representing the selective regimes important to many taxa. However, this claim is hollow in that the tests are performed only after certain observations for certain taxa suggest adaptation. How would we interpret the results of such tests if we did not already believe that adaptation was a likely explanation? In short, nothing is gained by the statistical approach that is not evident from direct study of the taxa and characters using a cladistic perspective. Plotting traits on well corroborated trees and observing the pattern is the best method.

The arguments above all reduce to the critical necessity of a cladogram. The cladogram is the best statement about our understanding of the evolution of a given group, and it is not clear that there is a benefit from a 'test' which is by necessity less secure and less interpretable than the cladogram itself. Yet, the scepticism through which geneticists and behavioural ecologists view phylogenetic reconstructions is applied myopically to the work of the systematist and rarely to the statistical machinations of their own fields. When Crozier (1992: 1979) writes "although the use of cladistics is an advance over purely intuitive methods, it is here unaccompanied by statistical tests, so that one cannot tell which aspects of the various phylogenies presented are well founded as against being essentially conjectural," he explicitly declares that probability models are to be believed whereas parsimonious tree-building is suspect. In fact, Crozier got it backwards; the cladograms are based on observed presence/absence of traits in taxa, whereas the process theories leading to probability statements are conjectural. Pagel (1992: 441) observed that his method suffers from the crucial deficiency necessarily shown by all model-dependent approaches: if the predicates are incorrect, the conclusions can be wrong. As Pagel & Harvey (1992: 425) themselves put it: "Nature is not as simple as our models". Pagel (1992: 441) argued: "But this should not be taken as a criticism of the use of a model". Yes, it should.

ACKNOWLEDGEMENTS

Participation in the symposium and preparation of this paper was made possible by NSF grant BSR–9006102 to JMC, a Kalbfleisch Fellowship (AMNH) to JWW, and the American Museum of Natural History. We are grateful to J. Coddington, J. Cracraft., P. DeVries, D. Frost, M. Mickevich and M. Pagel for comments on the manuscript.

NOTE ADDED IN PROOF

Leroi *et al.* (1994, *American Naturalist 143*: 381–402) argue against comparative methods on grounds complementary to those above. They stress genetic mechanisms leading to correlation and affirm that data relevant to questions of adaptation are collected only within species.

REFERENCES

ALBERT, V.A., MISHLER, B.D. & CHASE, M.W., 1992. Character-state weighting for restriction site data in phylogenetic reconstruction, with an example from chloroplast DNA. In P.S. Soltis, D.E. Soltis & J.J. Doyle (Eds), *Molecular Systematics of Plants*, 369–403. New York: Chapman & Hall.

BAUM, D.A. & LARSON, A., 1991. Adaptation reviewed: a phylogenetic methodology for studying character macroevolution. *Systematic Zoology, 40*: 1–18

BOINSKI, S. & TIMM, R.M., 1985. Predation by squirrel monkeys and double-toothed kites on tent-making bats. *American Journal of Primatology, 9*: 121–127.

BJORKLUND, M., 1990. A phylogenetic interpretation of sexual dimorphism and in body size and ornament in relation to mating system in birds. *Journal of Evolutionary Biology, 3*: 171–183.

CARPENTER, J.M., 1987. Phylogenetic relationships and classification of the Vespinae (Hymenoptera: Vespidae). *Systematic Entomology, 12*: 413–431.

CARPENTER, J.M., 1990. Of genetic distances and social wasps. *Systematic Zoology, 39*: 391–397.

CARPENTER, J.M., 1991. Phylogenetic relationships and the origin of social behavior in the Vespidae. In K.G. Ross & R.W. Matthews (Eds), *The Social Biology of Wasps*, 7–32. Ithaca, NY: Cornell University Press.

CARPENTER, J.M., 1992a. Comparing methods. [Review of P.H. Harvey & M.D. Pagel, 1991, *The Comparative Method in Evolutionary Biology*, Oxford University Press.] *Cladistics, 8*: 191–195.

CARPENTER, J.M., 1992b. Random cladistics. *Cladistics, 8*: 147–153.

CARPENTER, J.M. & WENZEL, J.W., 1989. Synonymy of the genera *Protopolybia* and *Pseudochartergus* (Hymenoptera: Vespidae; Polistinae). *Psyche, 96*: 177–186.

CHEVERUD, J.M., DOW, M.M. & LEUTENEGGER, W., 1985. The quantitiative assessment of phylogenetic constraints in comparative analyses: sexual dimorphism in body weight among primates. *Evolution, 39*: 1335–1351.

CHVALA, M., 1976. Swarming, mating and feeding habits in Empididae (Diptera), and their significance in evolution of the family. *Acta Entomologica Bohemoslovaka, 73*: 353–366.

CODDINGTON, J.A., 1988. Cladistic tests of adaptational hypotheses. *Cladistics, 4*: 3–22.

CODDINGTON, J.A., 1990a. Bridges between evolutionary pattern and process. *Cladistics, 6*: 379–386.

CODDINGTON, J.A., 1990b. Cladistics and spider classification: araneomorph phylogeny and the monophyly of orbweavers (Araneae: Araneomorphae; Orbiculariae). *Acta Zoologica Fennica, 90*: 75–87.

CODDINGTON, J.A., 1992. Avoiding phylogenetic bias. [Review of P.H. Harvey & M.D. Pagel, 1991, *The Comparative Method in Evolutionary Biology*, Oxford University Press.] *Trends in Ecology and Evolution, 7*: 68–69.

CODDINGTON, J.A., 1994. The roles of homology and convergence in studies of adaptation. In P. Eggleton & R.I. Vane-Wright (Eds), *Phylogenetics and Ecology*, 53–78. London: Academic Press.

CRACRAFT, J., 1981. The use of functional and adaptive criteria in phylogenetic systematics. *American Zoologist, 21*: 21–36

CRACRAFT, J., 1990. The origin of evolutionary novelties: pattern and process at different hierarchical levels. In M.H. Nitecki (Ed.), *Evolutionary Innovations*, 21–44. Chicago: University of Chicago Press.

CROZIER, R.H., 1992. All about (eusocial) wasps. [Review of K.G. Ross & R.W. Matthews (Eds), 1991, *The Social Biology of Wasps*, Cornell University Press.] *Evolution, 46*: 1979–1981.

De QUEIROZ, A. & WIMBERGER, P., 1993. The usefulness of behavior for phylogeny estimation: levels of homolplasy in behavioral and morphological characters. *Evolution, 47*: 46–60.

De QUEIROZ, K. & GAUTHIER, J., 1992. Phylogenetic taxonomy. *Annual Review of Ecology and Systematics, 23*: 449–480.

DONOGHUE, M.J., 1989. Phylogenies and the analysis of evolutionary sequences, with examples from seed plants. *Evolution, 43*: 1137–1156.

FARRIS, J.S., 1983. The logical basis of phylogenetic analysis. In N.I. Platnick & V.A. Funk (Eds), *Advances in Cladistics, 2*, 7–36. New York: Columbia University Press.

FELSENSTEIN, J., 1978. Cases in which parsimony or compatibility methods will be positively misleading. *Systematic Zoology, 27*: 401–410.

FELSENSTEIN, J., 1979. Alternative methods of phylogenetic inference and their relationship. *Systematic Zoology, 28*: 49–62.

FELSENSTEIN, J., 1985. Phylogenies and the comparative method. *American Naturalist, 125*: 1–15.

GARLAND Jr, T., HARVEY, P.H. & IVES, A.R., 1992. Procedures for the analysis of comparative data using phylogenetically independent contrasts. *Systematic Biology, 41*: 18–32.

GIBBS, H.L., WEATHERHEAD, P.J., BOAG, P.T., WHITE, B.V., TABAK, L.M. & HOYSAK, O.J., 1990. Realized reproductive success of polygynous red-winged blackbirds. *Science, 250*: 1394–1397.

GITTLEMAN, J.L. & KOT, M., 1990. Adaptation: statistics and a null model for estimating phylogenetic effects. *Systematic Zoology, 39*: 227–241.

GITTLEMAN, J. L. & LUH, H.-K., 1992. On comparing comparative methods. *Annual Review of Ecology and Systematics, 23*: 383–404.

GOWATY, P. & BRIDGES, W.C., 1991a. Nestbox availability affects extra-pair fertilizations and conspecific nest parasitism in eastern bluebirds, *Sialis sialis. Animal Behaviour, 41*: 661– 675.

GOWATY, P. & BRIDGES, W.C., 1991b. Behavioral, demographic, and environmental correlates of extrapair fertilizations in eastern bluebirds, *Sialis sialis. Behavioral Ecology, 2*: 339–350.

GRAFEN, A., 1989. The phylogenetic regression. *Philosophical Transactions of the Royal Society of London, Series B, 326*: 119–157.

GRAFEN, A., 1992. The uniqueness of the phylogenetic regression. *Journal of Theoretical Biology, 156*: 405–423.

GRIMALDI, D.A. 1987. Phylogenetics and taxonomy of *Zygothrica* (Diptera: Drosophilidae). *Bulletin of the American Museum of Natural History, 186*: 103–268.

HARVEY, P.H. & PAGEL, M.D., 1991. *The Comparative Method in Evolutionary Biology*. Oxford: Oxford University Press.

HÖGLUND, J., 1989. Size and plumage dimorphism in lek-breeding birds: a comparative analysis. *American Naturalist, 134*: 72–87.

JANDER, U., 1966. Untersuchungen zur Stammesgeschichte von Putzbewegugen von Tracheaten. *Zeitschrift fur Tierpsychologie, 23*: 799–844.

JEANNE, R.L., 1975. The adaptiveness of social wasp nest architecture. *Quarterly Review of Biology, 50*: 267–287.

LAUDER, G.V., 1986. Homology, analogy, and the evolution of behavior. In M.H. Nitecki & J.A. Kitchell (Eds), *Evolution of Animal Behavior*, 9–40. New York: Oxford University Press.

LAUDER, G.V., 1990. Functional morphology and systematics: studying functional patterns in an historical context. *Annual Review of Ecology and Systematics, 21*: 317–340

MADDISON, W.P., 1990. A method for testing the correlated evolution of two binary characters: are gains and losses concentrated on certain branches of a phylogenetic tree? *Evolution, 44*: 539–557.

MADDISON, W.P. & MADDISON, D.R., 1992. *MacClade, version 3. Documentation.* Sunderland, MA: Sinauer.

MARTINS, E.P. & GARLAND Jr, T., 1991. Phylogenetic analyses of the correlated evolution of continuous characters: a simulation study. *Evolution, 45*: 534–557.

MAURAKIS, E.G., WOLLCOT, W.S. & SABAJ, M.H., 1991. Reproductive behavioral phylogenetics of *Nocomis* species groups. *American Midland Naturalist, 126*: 103–110.

McLENNAN, D.A., BROOKS, D.R. & McPHAIL, J.D., 1988. The benefits of communication between comparative ethology and phylogenetic systematics: a case study using gasterosteid fishes. *Canadian Journal of Zoology, 66*: 2177–2190.

MILLER, E.H., 1988. Breeding voacalizations of Baird's sandpiper *Calidris bairdii* and related species, with remarks on phylogeny and adaptation. *Ornis Scandanavica, 19*: 257–267.

MUNDINGER, P.C., 1979. Call learning in the Carduelinae: ethological and systematic consideration. *Systematic Zoology, 28*: 270–283.

O'HARA, R.J., 1988. Homage to Clio, or, toward an historical philosophy for evolutionary biology. *Systematic Zoology, 37*: 142–155.

PAGEL, M.D., 1992. A method for the analysis of comparative data. *Journal of Theoretical Biology, 156*: 431–442.

PAGEL, M.D., 1994. The adaptationist wager. In P. Eggleton & R.I. Vane-Wright (Eds), *Phylogenetics and Ecology*, 29–51. London: Academic Press.

PAGEL, M.D. & HARVEY, P.H., 1992. On solving the correct problem: wishing does not make it so. *Journal of Theoretical Biology, 156*: 425–430.

PETERSON, A.T. & BURT, D.B., 1992. Phylogenetic history of social evolution and habitat use in the *Aphelocoma* jays. *Animal Behaviour, 44*: 859–866.

POPPER, K.R., 1968. *The Logic of Scientific Discovery*. New York: Harper & Row.

PRUM, R.O., 1990. Phylogenetic analysis of the evolution of display behavior in the noetropical manakins (Aves: Pipridae). *Ethology, 84*: 202–231.

PURVIS, A., 1991. CAIC, version TEST. Program and documentation. Oxford: Department of Zoology, University of Oxford.

RICHARDS, O.W., 1978. *The social wasps of the Americas excluding the Vespinae*. London: British Museum (Natural History).

RIDLEY, M., 1983. *The Explanation of Organic Diversity: The Comparative Method and Adaptations for Mating*. Oxford: Oxford University Press.

SCHREMMER, F., 1977. Das Baumrinden-Nest der neotropischen Faltenwespe *Nectarinella championi*, umgeben von einem Leimring als Ameisen-Abwehr (Hymenoptera: Vespidae). *Entomologica Germanica, 3*: 344–355.

SILLEN-TULLBERG, B., 1988. Evolution of gregariousness in aposematic butterfly larvae: a phylogenetic analysis. *Evolution, 42*: 293–305.

STRASSMANN, J.E., HUGHES, C.R. & QUELLER, D.C., 1990. Colony defense in the social wasp, *Parachartergus colobopterus*. *Biotropica, 22*: 324–327.

WEBSTER, M.S., 1992. Sexual dimorphism, mating system and body size in New World blackbirds (Icterinae). *Evolution, 46*: 1621–1641.

WENZEL, J.W., 1991. Evolution of nest architecture. In K.G. Ross & R.W. Matthews (Eds), *The Social Biology of Wasps*, 480–519. Ithaca, NY: Cornell University Press.

WENZEL, J.W., 1992. Behavioral homology and phylogeny. *Annual Review of Ecology and Systematics, 23*: 361–381.

WENZEL, J.W., 1993. Application of the biogenetic law to behavioral ontogeny: a test using nest architecture in paper wasps. *Journal of Evolutionary Biology, 6*: 229–247.

WEST EBERHARD, M.J., 1989. Phenotypic plasticity and the origins of diversity. *Annual Review of Ecology and Systematics, 20*: 249–278.

WILLIAMS, F.X., 1928. Studies in tropical wasps – their hosts and associates (with descriptions of new species). *Bulletin of the Experiment Station of the Hawaiian Sugar Planters Association, Entomological Series, 19*: 1–179.

WINSDOR, D.M., 1976. Birds as predators on the brood of *Polybia* wasps (Hymenoptera: Vespidae: Polistinae) in a Costa Rican deciduous forest. *Biotropica, 8*: 111–116.

5

Phylogeny, evolutionary models and comparative methods: a simulation study

JOHN L. GITTLEMAN & HANG-KWANG LUH

CONTENTS

Keywords: Comparative method – evolutionary model – simulation – statistics.

Abstract

Many comparative methods are now (and will become) available for evaluating and removing phylogenetic correlation in cross-taxonomic traits. Methods differ in assumptions about the process of evolution, statistical procedures for extirpating phylogenetic correlation and tests of adaptive hypotheses. We simulate various conditions typical of comparative problems and then apply two extreme examples of comparative methods, independent contrasts and autoregression, to assess whether and how the methods themselves influence comparative results. Simulations include effects of sample size (10–90 species), level of phylogenetic correlation with traits, systematic information (taxonomic ranks *versus* phylogenetic trees) and assumed evolutionary model (gradual, punctuational or no model). Simulations involve traits randomly drawn from a normal distribution with a mean of 0.0 (variance of 1.0) and a bivariate (null) correlation with a mean of 0.0. Traits are then placed on a phylogeny, whereby we can view the evolutionary change of these traits along the branches. Each comparative method is applied to the evolving traits to expose how well a method removes phylogenetic correlation and hence reveals the null distribution. Autoregression yields better results with small sample sizes ($N < 10$) and, assuming a punctuational model of evolutionary change, with taxonomic rank data. The independent contrasts method performs better with larger samples ($N = 50$ or 90) and, assuming gradual evolution, with phylogenetic trees. The independent contrasts method is more sensitive to an assumed evolutionary model whereas autoregression is vulnerable to phylogenetic (distance) information.

INTRODUCTION

Haldane (1938: 79) wrote, "The analysis of interspecific differences is a prerequisite for any complete theory of evolution". The steady accumulation of comparative data and phylogenetic information along with greater understanding of the evolutionary process is testimony to this view. Nevertheless, with this knowledge comes an increasing awareness of what we do not know about comparative study. This is especially true in applications of the comparative method itself. Any comparative test of an ecological or evolutionary hypothesis requires an initial diagnosis of whether the data are statistically independent from phylogenetic relations (Felsenstein, 1985; Harvey & Pagel, 1991; Gittleman & Luh, 1992). Although not guaranteed, phylogenetic relations are often found and therefore comparative methods are needed to account statistically for, or partition out phylogenetic correlation prior to hypothesis testing. At least seven methods are currently available to carry out this task; detailed descriptions and applications of these methods are available elsewhere (for review see Harvey & Pagel, 1991; Gittleman & Luh, 1992). The purpose of this paper is to investigate, by quantitative simulation, the extent to which the comparative methods themselves might influence comparative findings.

We chose two methods, independent contrasts and autoregression, for simulation tests. These methods are used not because of their superiority over others but because of inherent analytical differences, thus perhaps exposing the range of effects, and their common use in the literature. The basic ideas behind each method are as follows.

The independent contrasts method derives from a model by Felsenstein (1985, 1988) that suggests calculating contrasts between pairs of species at each bifurcation in a phylogeny. Because these contrasts are at each node which link species that share an immediate common ancestry, essentially this is removing phylogenetic heritage and creating values that represent independent evolutionary events. If two traits are correlated (e.g. population density and body size), then a large calculated contrast with one trait at a given node should be associated with a large contrast in the other. As Felsenstein originally stated, a known phylogeny is central to using the independent contrasts method. Unfortunately, phylogenetic information is often lacking and therefore an estimation must be made. Two general approaches may be taken. First, when ancestral values are unknown but branch length information is available, Equation (1) may be used to estimate ancestral nodes: where X_i are trait descendants; v_i are branch lengths; and m is the number of descendants (Pagel, 1992). Second, when both branch lengths and ancestral values are unknown (i.e. the current state of affairs in many, if not most, comparative studies that must rely on species trait values and systematic information based on taxonomic rank), evolutionary change may be assumed to follow a punctuational model in which trait evolution only occurs at a node. Since the variance of change is unrelated to branch length, all branches are equal to each other (Harvey & Pagel, 1991). In other words, the estimated ancestral node is the mean of its descendants. The independent contrasts method has been usefully applied to a wide variety of problems (e.g. locomotor morphology in lizards: Losos, 1990; metabolic rate and life histories in birds: Trevelyan, Harvey & Pagel, 1990; sperm competition in birds: Møller, 1991; see also Harvey & Pagel, 1991) and a computer version is easily employed (see Purvis, 1991). Fundamental features of the technique are that it: (1) equally weights each (independent) evolutionary event; (2) explicitly uses phylogenetic (branch length) information for calculating expected contrast variances according to some model of evolution (e.g. gradual); and (3) uses all of the variation in traits throughout a phylogeny rather than partitioning certain phylogenetic and non-phylogenetic components, as in autoregression (see below).

$$X_k = \frac{\sum_{i=1}^{m}(\frac{1}{v_i})X_i}{\sum_{i=1}^{m}(\frac{1}{v_i})} \tag{1}$$

The autoregressive approach is based on a method developed by Cheverud, Dow & Leutenegger (1985) that considers trait variation in terms of a phylogenetic component and a specific component. The reasoning behind such trait division is based on similar models of quantitative genetics where phenotypic values are considered as additive genetic and environmental values. The statistical properties of the technique of Cheverud *et al.* (1985) is derived from the autocorrelation literature for analyzing problems of spatial pattern. For comparative study, the method takes the form:

$$\mathbf{y} = \rho\mathbf{W}\mathbf{y} + \epsilon \tag{2}$$

with $\mathbf{W}$ and $N \times N$ weighting matrix. The vector $\mathbf{y}$ of standardized trait values takes the linear combination $\rho\mathbf{W}\mathbf{y}$ as its phylogenetic component and the residual vector ϵ as its specific component. The autocorrelation coefficient, ρ, measures the correlation between the phenotypic trait vector $\mathbf{y}$ and the purely phylogenetic value $\mathbf{W}\mathbf{y}$. The residual ϵ depicts the independent (adaptive) evolution of each species. Empirical examples of the autoregressive approach include studies of body size sexual dimorphism in primates (Cheverud *et al.*, 1985; Ely & Kurland, 1989), brain size and life history patterns in carnivores (Gittleman, 1991, 1992, 1994), ornament dimorphism in birds (Bjørklund, 1990) and cooperative breeding in perching birds (Edwards & Naeem, 1993). A macro- or microcomputer version of the method is available from the authors.

Basic features of this method are that it: (1) calculates values for each taxonomic unit (typically species) under study; (2) retains flexibility to match systematic information (either taxonomic ranks or phylogenetic trees) with phenotypic similarity, particularly when applying a maximum likelihood procedure (Gittleman & Kot, 1990); and (3) assumes an autoregressive statistical process whereby comparative data are correlated with phylogeny and this correlation tends to fall-off with phylogenetic distance (Gittleman & Luh, 1992). As stipulated in the spatial processes literature (Cliff & Ord, 1981; Upton & Fingleton, 1985), autoregressive methods assume some correlation in the data and are designed to remove it. If phylogenetic correlation is not apparent, then the autoregressive method is likely to make errors. Therefore, a prerequisite for using the autoregressive method for comparative problems is diagnosing phylogenetic correlation in the data (Gittleman & Kot, 1990).

Both the independent contrasts and autoregressive methods involve particular kinds of data in comparative tests. These include quantitative (or continuous) traits, systematic information either in the form of taxonomic ranks or phylogenetic trees, levels of correlation between traits and phylogeny, and rate of trait evolution through phylogenetic time. The certainty of these comparative characteristics varies from empirically-observed factors, as with sample size or phylogenetic correlation of traits, to venturing into the unknown with evolutionary rate. Understanding the impact of these factors, on the effectiveness of our methods is critical. For example, evolutionary models are essential for testing comparative hypotheses (see Harvey & Pagel, 1991; Harvey & Purvis, 1992; Purvis, Gittleman and Luh, 1994), yet we rarely know anything about them for specific traits. The aim of the following simulations, therefore, is to investigate relative differences in results from each comparative method as contingent on how the methods use different kinds of comparative information.

METHODS

Our simulation studies involve a three-stage process: (1) creating different phylogenetic topologies; (2) simulating trait evolution along each phylogeny; and (3) using the simulated phylogenies and associated traits to test the robustness of each comparative method. The general sequence of these simulations is similar to Martins & Garland (1991), however, various methodological details are different so as to reflect more realistic comparative problems (see also Gittleman & Luh, 1992).

Phylogenetic trees

The simulated phylogenetic trees used in this study are generated by random joining (Maddison & Slatkin, 1991). Given a number of species, each simulation begins with the lowest taxa (i.e. species) and randomly groups taxa to form larger taxa in the next higher levels (i.e. genera) until only one taxon is formed. A random joining tree only provides information about the hierarchical taxonomic rank among species, genera, families, etc. It does not show the evolutionary distance between nodes. To generate random phylogenetic trees, we simply use the above taxonomic ranks and assign a random number to each branch as its length. One restriction is necessary for generating phylogenies: branch lengths between higher nodes must be longer than the branch lengths between lower nodes. This assumption is, however, suitable for either comparative method. In autoregression, it is assumed that nodes nearer to one another are more closely related. In the independent contrasts, based on a Brownian motion model, the expected variance along a branch is related to its length; thus a contrast between higher nodes will have a higher variance than a contrast between lower nodes.

On the simulations of phylogenetic trees we construct three sets of phylogenies which comprise 10, 50 and 90 species respectively. For each sample size, we generate 50 different trees.

In contrast to Martins & Garland (1991), we simulated multifurcating rather than bifurcating phylogenies. Multifurcating (or polytomic) trees were used because they more closely reflect the type of phylogenetic information typically available in the systematic literature, both from taxonomic rankings or phylogenetic trees. This issue is fundamental in comparing comparative methods (see also Purvis, Gittleman and Luh, 1994). The independent contrasts method calculates data values at each node in a tree rather than per species. As mentioned in the 'Introduction' (Pagel, 1992; see also Purvis, 1991, for implementation) presents a model for calculating contrasts from unresolved trees which essentially expands multiple branches from a node into two subgroups. These contrasts, as determined from the number of nodes in a tree, are then used for hypothesis testing and for setting appropriate degrees of freedom. Thus, with multifurcations, the independent contrasts method will retain fewer analysable values than the original (species) sample size. By comparison, auto-regression calculates data values for each species, thus providing $N - 2$ degrees of freedom where N represents the number of species.

Obviously, such disparate procedures for statistical tests will influence results between the two comparative methods. Rather than preset degrees of freedom in our simulations, we chose to calculate them according to typical systematic conditions (i.e. unresolved multifurcating trees) observed in comparative studies. Other simulation work has investigated the degrees of freedom issue by setting the number of nodes in bifurcating trees (Purvis, Gittleman and Luh, 1994).

Trait evolution and evolutionary models

Evolutionary changes of two continuous (quantitative) traits can be viewed as two random variables drawn from a bivariate normal distribution. The pattern of evolutionary change is determined by parameters of the probability distribution. The mean of each variable indicates the net change of evolution in each trait. The correlation of the bivariate distribution shows the general relationship of the two traits changing along a tree. All simulations of trait evolution assume that the two random trait changes are drawn from a bivariate normal distribution with means of zero (i.e. the evolutionary changes are neutral) and a correlation of zero (see Fig. 1).

Evolutionary models determine the way that traits evolve along the phylogeny (Martins & Garland, 1991; Gittleman & Luh, 1992). In this study we used two evolutionary models – gradual and punctuational. If simulated trait evolution follows a gradual manner, random trait change occurs at each time step (i.e. an arbitrary time scale) along the branch (see Fig. 2a). If simulated trait evolution follows a punctuational manner, random trait change only occurs at the splitting from a node (i.e. the speciation events) and branch lengths are not involved (see Fig. 2b). One simulation was structured as if there were no evolutionary model. In this case (see Fig. 2c), trait values are randomly placed on the tips with no evolutionary structure in relation to branches or nodes, as in the other evolutionary models. This case is

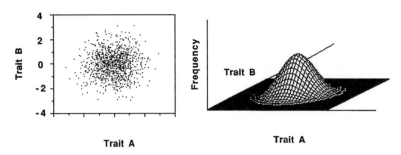

Figure 1 Illustration of simulated random traits A and B (left). Frequency of bivariate correlations are normally distributed with a mean of 0.0 (right).

also meant to represent the conditions of no phylogenetic correlation of traits (i.e. no relation between trait variation and phylogenetic structure), a condition verified by descriptive statistics (see below). It should be pointed out that in Martins & Garland's (1991) study, a simulated 'star' phylogeny was used to reflect no phylogenetic correlation. Branch lengths were extended to increase independence among contemporary tip species. Variance of the traits, however, remained proportional to the branch lengths. Therefore, some phylogenetic correlation was likely.

All trait simulations began at the initial base node (i.e. the original ancestor). We assigned two random numbers as two trait values of that ancestor, then random evolutionary changes were continuously added to the previous trait value until trait values of species (i.e. tips on the phylogeny) were obtained. For the gradual model of evolutionary change, the number of changes along any branch is proportional to that branch length (see Fig. 2a). Therefore, each offspring derived from one particular ancestor not only receives a copy of that ancestor's trait values but also the total evolutionary changes along the branch derived from the ancestor. For the punctuational model of evolutionary change (see Fig. 2b), the number of evolutionary changes is the number of speciation events from one ancestor. Each branch has only one trait change. The trait value of offspring is the summation of the ancestor's trait value and one trait change.

For each simulated phylogeny, 50 sets of two traits are generated.

Simulated comparisons

Given the above protocols, simulations involve three comparative problems which reflect various conditions and availability of information (see Table 1). (1) No evolutionary model is involved in the simulation of trait evolution. That is, no evolution occurs along the phylogeny: two sets of trait values are simply placed on the tips of phylogenies as representing the trait values of species. (2) Trait evolution

Table 1 Conditions for simulating comparative methods.

Condition	Model	Systematic information
No phylogenetic correlation	None (random)	Taxonomy/Phylogeny
Phylogenetic correlation	Gradual or Punctuational	Taxonomy
Phylogenetic correlation	Gradual or Punctuational	Phylogeny

(a) Gradual Evolutionary Model

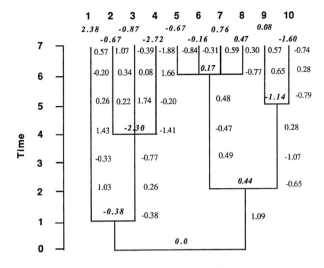

(b) Punctuational Evolutionary Model

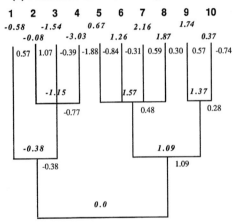

(c) No Evolutionary Model

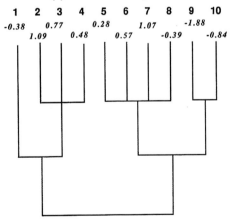

Figure 2 Three simulated evolutionary conditions for examining comparative methods. (a) Gradual evolutionary model where traits evolve in proportion to branch lengths. (b) Punctuational evolutionary model where traits evolve in relation to speciation events occurring at nodes. (c) No evolutionary model (also designates no phylogenetic correlation) where trait values are placed on species tips and not allowed to evolve along a phylogeny.

follows a punctuational model of evolutionary change. (3) Trait evolution follows a gradual model of evolutionary change. For each condition, comparative methods use either of two kinds of systematic information to analyse the simulated data, taxonomic ranks or phylogenies with branch length information.

Statistical analysis

Descriptive statistics in autoregressive method   To investigate whether and the degree to which descriptive statistics (i.e. Moran's I) are necessary for using the autoregressive method, we compare the effects of comparative tests from auto-regression with and without applying Moran's I statistics. For testing the autoregressive method without using a Moran's I, we simply apply the method regardless of whether or not simulated traits are phylogenetically correlated. In other words, all phylogenetic information is involved in the comparative test. In contrast, with applying Moran's I, a series of Is are calculated at different taxonomic levels and then plotted on a graph (a phylogenetic correlogram) to show where autocorrelation varies with taxonomic listing or phylogenetic distance. Four criteria are applied to the phylogenetic correlogram to detect the extent and location of phylogenetic correlation (see Gittleman & Kot, 1990, for details of this statistical procedure).

1. At least one of the z values is greater than 1.96 (i.e. observed phylogenetic autocorrelation at $P<0.05$).
2. The correlation coefficient in the correlogram must be less than -0.5, indicating a negative (autoregressive) slope.
3. Cut-off distance is set when z values drop to negative (i.e. data are not filtered through the method when uncorrelated).
4. No z values which are below the cut-off distance are above 1.96.

   In essence, these criteria permit a more effective application of the autoregressive method because it detects the level and location of phylogenetic correlation. Consequently, there is appropriate determination of what phylogenetic information to involve in a comparative test.

Statistical comparisons of methods   Two approaches are adopted for comparing comparative methods. First, to assess general patterns, frequency distributions are examined of bivariate correlation coefficients in each of the three evolutionary conditions. Second, to evaluate statistical power, observed differences in significance tests of correlation (rate of Type I error) are compared. Because of the different calculations of the degrees of freedom in both comparative methods, it is impossible to compare the robustness of two methods according to the frequency distributions. With independent contrasts, correlation is computed from contrasts at each node, so the degrees of freedom are set according to the number of nodes rather than total sample size. Conversely, with the autoregressive model, degrees of freedom are determined by the number of species. For each of different evolutionary models and different phylogenetic information, we count the number of correlation coefficients out of 2500 simulations which are significant at the 0.05 level (two-tailed t-test with different degrees of freedom).

RESULTS

We report results of simulations in two forms: (1) general frequency distributions for each evolutionary model, using taxonomic rank or phylogenetic tree information, after applying the independent contrasts or autoregressive methods; and (2) statistical (Type I) error rates for the same conditions. The general frequency distributions of each comparative method are shown against the null distribution, that is a normal distribution around a mean of 0.0. In general, when traits are evolving with an assumed evolutionary model, then the variance of the frequency distribution will be wider as expected by the traits evolving on a phylogeny, either along branches (gradual model) or nodes (punctuational model). In each figure we also present the distribution represented by analysing the species tips so as to designate the degree to which phylogenetic correlation is removed.

Prior to examining simulation results, it is important to assess whether the general condition of phylogenetic correlation is fulfilled. Moran's I statistics revealed that, out of 2500 simulations for each condition, the percentage of significant correlation is only found in 0.72–4.84 cases in comparison to the phylogenetically correlated assumption in which 22.2–85.68 cases were significant with phylogeny (Table 2). Thus, the assumption of phylogenetic correlation is satisfied.

Independent contrasts

In the first condition, with no evolutionary model (i.e. no phylogenetic correlation), the null distribution represented by species tips has a small variance, as expected given there is no phylogenetic correlation and an input (bivariate) correlation of 0.0 (Fig. 3). With small sample size ($N = 10$), the independent contrast method does a relatively poor job with both taxonomic rank and phylogenetic tree information. With larger samples ($N = 50$ or 90), contrasts appear to yield better results. Overall, though, it is important to recognize that in this situation with no phylogenetic correlation, the contrasts method does not seem to perform as well as species tips correlation. Assuming a punctuational model (Fig. 4) we first observe more variance in the overall distribution owing to phylogenetic correlation, as we expected (i.e. the traits are evolving in relation to the nodes on a given tree); also, note the greater

Table 2 Percentage of cases in which significant phylogenetic correlation was detected using the Moran's I statistic.

			Trait A	Trait B
	No model	$N = 10$	4.84	1.48
	(No phylogenetic	$N = 50$	3.20	0.72
	correlation)	$N = 90$	3.32	0.84
Evolutionary models	Punctuational	$N = 10$	31.52	22.20
		$N = 50$	80.56	81.36
		$N = 90$	84.28	83.84
	Gradual	$N = 10$	47.52	38.80
		$N = 50$	81.40	81.60
		$N = 90$	84.16	85.68

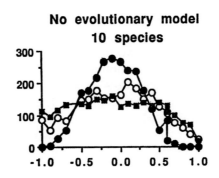

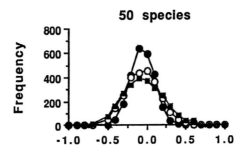

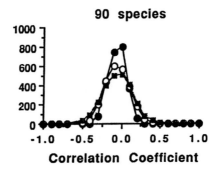

Figure 3 Assuming no evolutionary model of trait change, results of frequency distributions of bivariate correlations using the independent contrasts method on different sample sizes and systematic information. The null distribution of species tips is also shown. (—●—, Tips; —○—, Taxonomy; —■—, Phylogeny).

variance in small samples owing to the relatively unstable situation of correlations with few data points. Thus, we are comparing how well the contrasts method removes this correlation and reveals the real (null) distribution (i.e. a mean of 0.0). In each case, except possibly with the small sample size of 10 species, the contrasts method produces acceptable results. The independent contrasts method also appears to yield more acceptable error rates with taxonomic rank rather than phylogenetic tree information. Assuming a gradual model (Fig. 5), the frequency distributions again have a wide variance when phylogenetic correlation exists. The contrasts method also performs well, removing phylogenetic correlation and reveal-

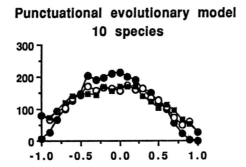

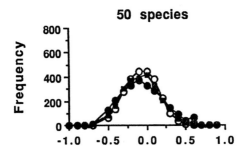

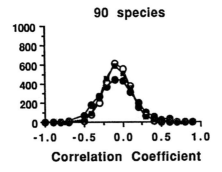

Correlation Coefficient

Figure 4 Assuming a punctuational evolutionary model, results of frequency distributions of bivariate correlations using the independent contrasts method of different sample sizes and systematic information. Comparable values are shown for analysing species tips data. (—●—, Tips; —○—, Taxonomy; —■—, Phylogeny).

ing the null distribution. However, in contrast to the punctuational condition, systematic information with phylogenetic trees appears to yield better results.

In Table 3 we present Type I error rates for each of the above frequency distributions. Results are given in relation to an alpha equal to 0.05, the error rate of the null distribution and standard of comparison. However, because our primary purpose in these simulations is to examine relative error rates in relation to different conditions and given that sample sizes are extremely high, we are also interested in rates approximating the 0.05 level. For comparison, we also present the error rates for analysing species tips.

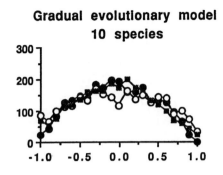

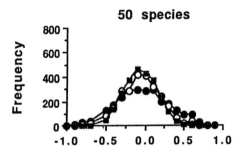

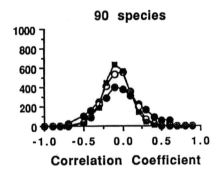

Figure 5   Assuming a gradual evolutionary model, results of frequency distributions of bivariate correlations using the independent contrasts method on different sample sizes and systematic information. Comparable values are shown for analysing species tips data. (——●——, Tips; ——○——, Taxonomy; ——■——, Phylogeny).

   A number of results are consistent with the patterns of frequency distributions. First, small sample sizes (<10 species) produce unacceptable error rates under all conditions. Second, when there is no phylogenetic correlation the independent contrasts only work (relatively) well with taxonomic ranks. Third, and perhaps most interestingly, the error rates between gradual and punctuational models emphasize the fact that independent contrasts are evolutionary-model-dependent and that the form of systematic information is tied to this model-dependence. In the punctuational model, when trait evolution relates to nodal change rather than branch lengths, taxonomic rank information elicits more acceptable error rates; conversely,

Table 3 Statistical results of comparing independent contrasts and autoregressive methods. Comparative results are reported as relative rates of Type I error ($\alpha = 0.05$). Comparable error rates are first given for species tips analysis and then second for the comparative method. For autoregression, values are given without and then with using Moran's I.

Evolutionary models	Comparative methods	Species number (df)	Phylogeny	Systematic information Taxonomy
No model	Independent contrasts	$N = 10$ (2–7)	0.064/0.131	0.064/0.075
		$N = 50$ (11–34)	0.048/0.112	0.048/0.057
		$N = 90$ (21–56)	0.035/0.111	0.042/0.061
	Autoregression	$N = 10$ (8)	0.064/0.090/0.037	0.064/0.090/0.037
		$N = 50$ (48)	0.048/0.051/0.048	0.048/0.050/0.048
		$N = 90$ (88)	0.035/0.036/0.036	0.042/0.044/0.044
Punctuational	Independent contrasts	$N = 10$ (2–7)	0.138/0.093	0.138/0.077
		$N = 50$ (11–34)	0.292/0.080	0.292/0.049
		$N = 90$ (21–56)	0.332/0.089	0.332/0.065
	Autoregression	$N = 10$ (8)	0.138/0.091/0.042	0.138/0.089/0.042
		$N = 50$ (48)	0.292/0.085/0.049	0.292/0.091/0.048
		$N = 90$ (88)	0.349/0.084/0.061	0.351/0.099/0.060
Gradual	Independent contrasts	$N = 10$ (2–7)	0.227/0.065	0.227/0.111
		$N = 50$ (11–34)	0.383/0.050	0.383/0.100
		$N = 90$ (21–56)	0.398/0.069	0.398/0.100
	Autoregression	$N = 10$ (8)	0.228/0.114/0.047	0.228/0.120/0.047
		$N = 50$ (48)	0.384/0.121/0.073	0.384/0.125/0.073
		$N = 90$ (88)	0.417/0.095/0.074	0.430/0.121/0.073

under a gradual model where trait evolution relates to branch lengths, phylogenetic tree information elicits more acceptable error rates.

Autoregression

As stated, descriptive statistics (i.e. Moran's I) are necessary prior to appropriately and effectively using the autoregressive method. Thus, in the following simulation results we first show effects of comparative tests from autoregression without examining the data, as in the previous independent contrasts simulations. We then present results, after applying criteria from diagnostic statistics, which show relative effects of evolutionary models and different kinds of comparative information.

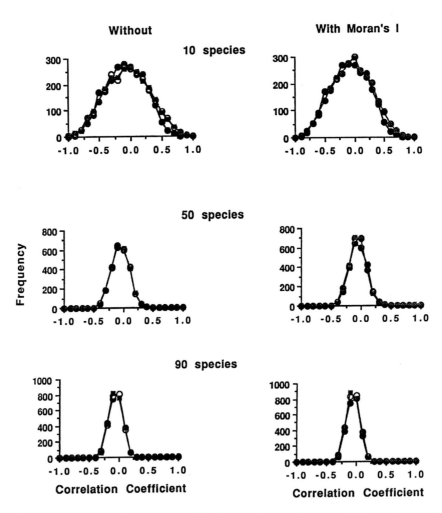

Figure 6 Assuming no evolutionary model of trait change (i.e., no phylogenetic correlation), results of frequency distributions of bivariate correlations using the autoregressive method on different sample sizes and systematic information. Results are given with (right) and without (left) applying the Moran's I statistics and cut-off criteria (see text). The null distribution of species tips is also shown. (——●——, Tips; ——○——, Taxonomy; ——■——, Phylogeny).

With general frequency distributions, the autoregressive method in the condition of no evolutionary model (or phylogenetic correlation) simply mimics the null distribution (Fig. 6), with little effect from either sample size or kind of systematic information (i.e. taxonomic *versus* phylogenetic). To an extent, these patterns are misleading because, as we will show in our statistical comparisons, the auto-regressive method may generate some phylogenetic correlation from the null distribution. Under the punctuational model, the autoregressive method performs equally well with systematic information based on either taxonomic ranks or phylogenetic trees (Fig. 7). More importantly, though, the autoregressive technique performs better after the Moran's I criteria are applied (and non-phylogenetically correlated data sets are excluded): there is a greater frequency distribution around the mean of 0.0 after executing the diagnostic statistics. With the gradual model (Fig. 8), as in the figures for the independent contrasts, the frequency distribution is wider

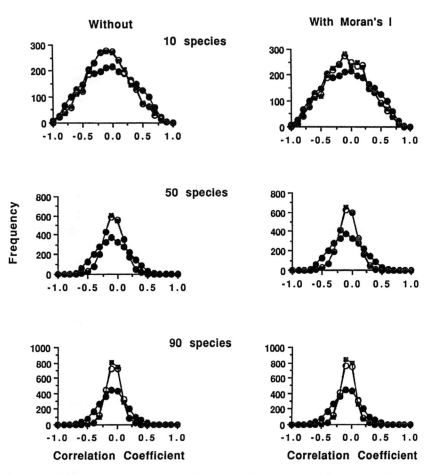

Punctuational evolutionary model

Figure 7 Assuming a punctuational evolutionary model, results of frequency distributions of bivariate correlations using the autoregressive method on different sample sizes and systematic information. Results are given with (right) and without (left) applying the Moran's I statistics and cut-off criteria (see text). Comparable values are shown for analysing species tips data. (—●—, Tips; —○—, Taxonomy; —■—, Phylogeny).

Gradual evolutionary model

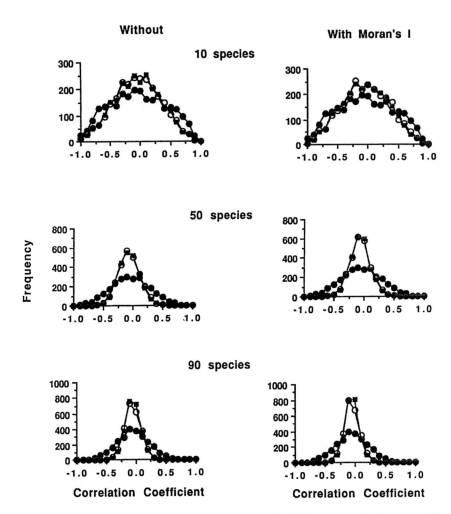

Figure 8 Assuming a gradual evolutionary model, results of frequency distributions of bivariate correlations using the autoregressive method on different sample sizes and systematic information. Results are given with (right) and without (left) applying the Moran's *I* statistics and cut-off criteria (see text). Comparable values are shown for analysing species tips data. (——●——, Tips; ——○——, Taxonomy; ——■——, Phylogeny).

due to increasing correlation between the traits as they evolve along the branches. The autoregressive method in the gradual condition performs about the same as with the punctuational model: different systematic information reveals similar patterns and application of the Moran's *I* is necessary for better results.

In Table 3 we present Type I error rates for each of the above frequency distributions. Results are given in relation to an alpha equal to 0.05, the error rate of the null distribution and standard of comparison; autoregressive performance is given with and without applying Moran's *I* criteria. For comparison, we also present the error rate of analysing species tips. When the Moran's *I* criteria are not applied to the simulated data, the autoregressive method only provides acceptable error rates under the condition of no evolutionary model; in this case, systematic data either in the form of phylogenetic trees or taxonomic ranks produce the same patterns. This is simply because there is little or no phylogenetic correlation at the tips (see above). But, even though error rates are acceptable, the autoregressive method generates higher error rates than in the original data unless the Moran's I is used.

Second, when the Moran's I criteria are applied and consequently non-phylo-genetically correlated data sets are excluded from analysis, the autoregressive method generally removes correlation under any condition. It seems to perform best with smaller sample sizes ($N = 10$ or 50) and under the condition of punctuational evolution. The larger samples ($N>50$), especially with the gradual model, produce some degree of error.

DISCUSSION

Comparing independent contrasts and autoregression: selecting a method

A comparative researcher is faced with making decisions about which comparative method to use based on the kind of information available. Comparing the relative results of these simulations with each method emphasizes this point. Thus we discuss the relative merits of each method in terms of information conditions.

Sample size When sample sizes are small ($N<10$), the autoregressive method will almost always yield better results, because of the method of calculating contrasts at each evolutionary event. If one must deal with small samples, especially when multifurcations are common, then autoregression is more appropriate. In general, though, comparative studies with such small samples are (probably) less likely anyway. With larger samples ($N = 50$ or 90), selection of an appropriate method then depends on access of systematic data and/or evolutionary model.

Taxonomic rank versus phylogenetic tree information Despite efforts to provide more accurate and detailed phylogenetic information, most comparative studies are still left with taxonomic rankings devoid of branch lengths. Given the current state of the art, both autoregression and independent contrasts yield acceptable rates within a punctuational framework. With the gradual model, autoregression may be more suitable with taxonomic data probably because of the maximum likelihood estimator.

With phylogenetic information containing branch lengths, the independent con-trasts method performs best under a gradual model and indeed outperforms the autoregressive method. The reason for this may be that with real branch length information the autoregressive method overcompensates the phylogenetic compon-ent and therefore makes errors in estimating the true bivariate relationship. With branch length information but with a punctuational model of evolutionary change, an unrealistic scenario, the autoregressive method is able to adjust to this situation.

Evolutionary model Deciding on an assumed evolutionary rate is guesswork in most comparative studies, thus it is difficult to draw conclusions of this factor based on access of real data. Nevertheless, assuming evolutionary rate is known, the independent contrasts is more appropriate with taxonomic rank information under a punctuational model and more appropriate with phylogenetic tree information under a gradual model. The autoregressive method is essentially model-independent in an evolutionary sense.

General results and conclusions of this study are summarized in a flow chart (Fig. 9), to aid in selecting an appropriate comparative method given different conditions of comparative tests. None of these guidelines is uncontroversial. Ironically, one of the most difficult decisions is what to do when there is no phylogenetic correlation. Our simulations suggest that it is probably (statistically) acceptable to simply analyse species tips; under similar conditions Martins & Garland (1991: Table 6) concur. Even so, there might be reasons for still applying a comparative method. For example, if there are *a priori* assumptions for phylogenetic correlation with a particular trait or if phylogenetic information is extremely questionable, then a comparative method should be employed alongside species analysis to assess statistical error. Furthermore, suggestions based on our simulations do not consider effects of topology on selecting appropriate comparative methods (see Purvis, Gittleman & Luh, 1994). Even more essential is the unknown of evolutionary model.

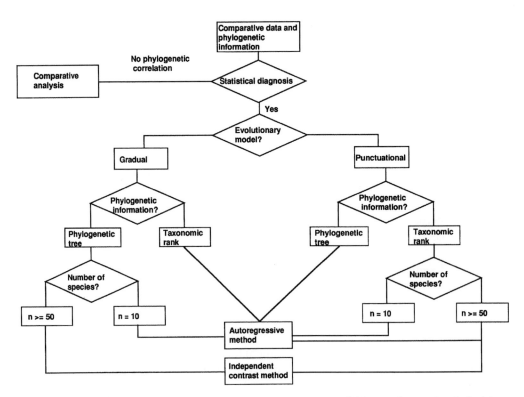

Figure 9 Flow diagram representing various comparative conditions and associated decisions for selecting an appropriate comparative method with either the independent contrasts or autoregressive method.

Few traits are understood well enough to specify a mode of evolutionary rate. Nevertheless, as our simulations and others clearly show, the assumption of evolutionary models is essential for carrying out correct comparative analysis. Perhaps we are now at least understanding the real difficulty in satisfying Haldane's claim for comparative study being a prerequisite to a complete theory of evolution.

ACKNOWLEDGEMENTS

We thank P. Eggleton and R. Vane-Wright for inviting us to the Symposium and soliciting a manuscript contribution. We are grateful to M. Kot, E. Martins, T. Garland, M. Pagel and A. Purvis for assistance with computer programs; to M. Kot, P. Harvey, J. Felsenstein and T. Garland for discussion and encouragement; and to M. Pagel for comments on the manuscript. Computer resources were provided by the College of Liberal Arts, the Department of Zoology, and Science Alliance, all of the University of Tennessee.

REFERENCES

BJÖRKLUND, M., 1990. A phylogenetic interpretation of sexual dimorphism in body size and ornament in relation to mating system in birds. *Journal of Evolutionary Biology, 3*: 171–183.

CHEVERUD, J., DOW, M. & LEUTENEGGER, W., 1985. The quantitative assessment of phylogenetic restraints in comparative analyses: sexual dimorphism in body weight among primates. *Evolution, 36*: 914–933.

CLIFF, A.D. & ORD, J.K., 1981. *Spatial Processes: Models and Applications*. London: Pion.

EDWARDS, S.V. & NAEEM, S., 1993. The phylogenetic component of cooperative breeding in perching birds. *American Naturalist, 141*: 754–789.

ELY, J. & KURLAND, J.A., 1989. Spatial autocorrelation, phylogenetic constraints, and the causes of sexual dimorphism in primates. *International Journal of Primatology, 10*: 151–171.

FELENSTEIN, J., 1985. Phylogenies and the comparative method. *American Naturalist, 125*: 1–15.

FELSENSTEIN, J., 1988. Phylogenies and quantitative characters. *Annual Review of Ecology and Systematics, 19*: 445–471.

GITTLEMAN, J.L., 1991. Carnivore olfactory bulb size: allometry, phylogeny, and ecology. *Journal of Zoology, 225*: 253–272.

GITTLEMAN, J.L., 1992. Carnivore life histories: a reanalysis in light of new models. In N. Dunstone & M. Gorman (Eds), *Mammals as Predators*, 65–86. Oxford: Oxford University Press.

GITTLEMAN, J.L., 1994. Are the pandas successful specialists or evolutionary failures? *Bio Science, 44*: 756–764.

GITTLEMAN, J.L. & KOT, M., 1990. Adaptation: statistics and a null model for estimating phylogenetic effects. *Systematic Zoology, 39*: 227–241.

GITTLEMAN, J.L. & LUH, H.-K., 1992. On comparing comparative methods. *Annual Review of Ecology and Systematics, 23*: 383–404.

HALDANE, J.B.S., 1938. The nature of interspecific differences. In G.R. de Beer (Ed.),

Evolution: Essays on Aspects of Evolutionary Biology Presented to Professor E.S. Goodrich on his Seventieth Birthday, 79–94. Oxford: Oxford University Press.

HARVEY, P.J. & PAGEL, M.D., 1991. *The Comparative Method in Evolutionary Biology.* Oxford: Oxford University Press.

HARVEY, P.H. & PURVIS, A., 1992. Comparative methods for explaining adaptations. *Nature, 351*: 619–624.

LOSOS, J.B., 1990. Ecomorphology, performance capability, and scaling of West Indian *Anolis* lizards: an evolutionary analysis. *Ecological Monographs, 60*: 369–388.

MADDISON, W.P. & SLATKIN, M., 1991. Null models for the number of evolutionary steps in a character on a phylogenetic tree. *Evolution, 45*: 1184–1197.

MARTINS, E.P. & GARLAND Jr, T., 1991. Phylogenetic analysis of the correlated evolution of continuous characters: a simulation study. *Evolution, 45*: 534–557.

MØLLER, A.P., 1991. Sperm competition, sperm depletion, paternal care, and relative testis size in birds. *American Naturalist, 137*: 882–906.

PAGEL, M.D., 1992. A method for the analysis of comparative data. *Journal of Theoretical Biology, 156*: 431–442.

PURVIS, A., 1991. Comparative analysis by independent contrasts (C.A.I.C.). A statistical package for the Apple Macintosh. Version TEST. Oxford: A Purvis. Department of Zoology, University of Oxford.

PURVIS, A., GITTLEMAN, J.L. & LUH, H.-K., 1994. Truth or consequences: accuracy of phylogenetic accuracy on two comparative methods. *Journal Theoretical Biology, 167*: 293–300.

TREVELYAN, R., HARVEY, P.H. & PAGEL, M.D., 1990. Metabolic rates and life histories of birds. *Functional Ecology, 4*: 135–141.

UPTON, G.J.G. & FINGLETON, G., 1985. *Spatial Data Analysis by Example.* Chichester: Wiley.

Investigating the origins of performance advantage: adaptation, exaptation and lineage effects

E.N. ARNOLD

CONTENTS

Phylogenetics and Ecology
ISBN 0-12-232990-2

Key words: Constraint – facilitation – phylogeny – adaptation – exaptation – functional analogue.

Abstract

Performance advantage may arise through the development of either adaptations or exaptations, and both phenomena are influenced by previous events in the history of the taxa concerned. Differences between exaptations and adaptations are discussed, as well as the problems of distinguishing them by phylogenetic tests and the limitations of these. Exaptations are often more positively identifiable than adaptations. However, it is most appropriate to regard derived states, that arise concurrently with changes to selective regimes in which they confer advantage, as provisional adaptations. The origins of exaptations are dealt with as well as their various categories: first-use, addition and transfer. Examples of the investigation of exaptations using lizards and hyenas are given, and the general evolutionary origins of exaptations considered.

Because the word constraint has been used in so many ways, influences on adaptive capacity arising from historical events on the lineages of the taxa concerned are termed lineage effects. The events concerned include alterations in progenitor traits, and changes in selective and developmental influences that may effect these. They may occur and act at various times: before changes in selective regime which produce a problem, after this but before an initial solution is developed, or after such an initial solution. In the latter case, they produce lineage effects with respect to possible adaptive modification of the initial solution.

Like exaptations and the origin of traits that give rise to them, the events causing lineage effects are often identifiable if detailed phylogenetic information is available. They can be investigated most easily by comparing functional analogues, that is independent solutions to a particular problem posed by the selective regime. A protocol is provided for doing this and applied to six independent cases of surface-dwelling lizards that avoid predators by diving into aeolian sand in a variety of different ways. In these examples, likely adaptations and exaptations, modification of precursors by transient or permanent changes in selective regime, and lineage effects arising after an initial solution, can all be identified.

INTRODUCTION

It has long been appreciated that organisms typically possess a range of traits that confer performance advantages in the environments in which they usually exist. In a particular case, a different trait will often not confer the advantage bestowed by the one present and this itself may not be beneficial in a different situation. The frequent happy match of organism to environment is usually explained by the action of evolutionary mechanisms, especially adaptation: the action of natural selection in which traits are promoted by the environment in which they at least initially exist. However, although evidence for the widespread operation of this agency is very strong, there are clear indications that some traits have not been produced in the environmental context in which they confer a particular performance advantage, but arose earlier in the history of the taxon concerned, and are thus exaptations (Gould & Vrba, 1982; see below). Furthermore, different organisms vary in their capacity to produce particular features that confer performance advantage. This is suggested by the way members of different clades may solve the same environmental problems differently. For instance, although basic requirements for aerial flight are essentially the same, insect and vertebrate wings are quite different in construction and, even within vertebrates, the wings of birds, bats and pterosaurs vary greatly in form.

Much work has been devoted to detecting and quantifying such differences in evolutionary capacity between taxa, especially in the field of development, but far less effort has gone into investigating how they have arisen (Lauder & Liem, 1989). Inevitably, like features that become exaptations, such differences are the result of events in the different individual histories of the taxa concerned, that is the later parts of their lineages that are not shared with each other. Knowing how differences in adaptive capacity arose, rather than merely noting their existence, is important, especially because a particular difference may arise in more than one way. For instance, suppose taxon A has the capacity to produce a particular trait but this capacity is lacking in taxon B. This difference may have arisen because ability to produce the trait was primitive but was lost in the lineage of B, or because ability was initially absent but was gained in A. Understanding historical influences on adaptation depends substantially on identifying such events and their causes.

It used to be very difficult to investigate whether traits were likely to have been produced by adaptation and, if so, whether this was to the situation in which they now occur, nor was it easy to find out how differences in adaptive capacity arose. This was largely because the genealogical relationships of the species concerned were unknown and techniques for integrating such phylogenies with other kinds of biological information were unexplored. Consequently, it was not possible to reconstruct the history of the exposure of ancestral forms to particular survival problems, their different evolutionary responses to these and the historical factors that caused those responses, except sometimes in very broad terms. However, in recent years, explicit methodologies of phylogenetic inferrence have been developed (see for instance Hennig, 1966; Felsenstein, 1973, 1981; Arnold, 1981; Wiley, 1981; Sharkey, 1989), and the information base available for this operation has been greatly extended, for instance through the development of DNA sequencing. Because of these developments, robust hypotheses of phylogenetic relationship are often now available. Furthermore, some of the problems of combining them with other aspects of the organisms concerned have been addressed (for example, Greene, 1986; Coddington, 1988; Carpenter, 1989; Arnold, 1990; Maddison, 1990; Brooks & McLennan, 1991; Harvey & Pagel, 1991).

This chapter discusses the use of such methods in investigating the historical aspects of the acquisition of performance advantage. It begins by considering the concept of exaptation and the problems of recognizing and distinguishing possible exaptations and adaptations. The ultimate origins of exaptations and their various categories and evolutionary significance are then explored. This is followed by a discussion of lineage effects – historical influences on the production of adaptations – and their points of origin. Finally, a method is presented of investigating the origins of lineage effects, and illustrated using sand-diving in dune lizards as an example.

EXAPTATIONS

An *exaptation* is a trait of a population, or larger taxonomic unit, that confers performance advantage in a particular way at a specific time but was not produced by natural selection directly for that use. Instead, it arose in some other way, and only subsequently acquired the performance advantage and use under consideration, usually as a result of change in the selective regime (as defined by Baum & Larson, 1991). In contrast, an *adaptation* was produced or at least promoted by natural selection for the use that it has at a specified time. In these definitions ecophenotypic traits, which are responses to environmental factors that develop within the lifetime of an individual, are excluded.

Exaptations were initially defined and distinguished from adaptations by Gould & Vrba (1982), although these authors used the terms initially with reference to current use. Sober (1984) questioned the appropriateness of this restriction when defining adaptations, since a trait could arise by natural selection but later environmental changes might mean that it ceased to be advantageous; this also applies to exaptations, hence the modified definitions used here.

Gould & Vrba (1982) usefully referred to a pre-existing character being *co-opted* to a new use when it becomes an exaptation. They also introduced the inclusive term *aptation* to cover any trait giving a performance advantage, irrespective of how it arose. Aptation consequently covers both exaptation and adaptation in the above senses and replaces a broader use of adaptation that included any trait that confers a performance advantage without considering its origin (e.g. Bock, 1979; Clutton-Brock & Harvey, 1979, 1984; Fisher, 1985). The latter use of adaptation was reasonable at a time when it was difficult to find out how advantageous traits had arisen, unless this had been observed in recent populations. However, phylogeny reconstruction now often allows many individual exaptations to be recognized with some certainty (Greene, 1986; Baum & Larson, 1991), and make distinction of exaptive and adaptive origin of performance advantage appropriate.

The term adaptation has sometimes been confined to the production of complex, polished features, presumably the result of selection acting on more than one aspect of the organism concerned (e.g. Williams, 1966; Stearns, 1986), but to restrict the word in this way leaves simpler products of natural selection, that grade with more complex ones, without a name. Presumably when natural selection promotes single alleles, as in some cases of industrial melanism of moths, or when smut fungi develop the ability to attack new cultivars of cereals, the process is essentially the same as that which ultimately results in the production of complex features, namely the spread of favourable mutations. Certainly, there does not seem to be a clear distinction between the two kinds of event. In a similar way, exaptation may be appropriately used for both simple and complex features.

The use of the term exaptation has been criticized. For instance, Brown (1982) correctly pointed out that subsequent development of functions by pre-existing traits is a well-known and widespread phenomenon for which the long established word preadaptation is already available. However, preadaptation (or preaptation as Gould & Vrba modify this term) and exaptation do not have the same meaning. A trait that acquires a particular function some time after its origin is a preaptation before this happens but an exaptation only afterwards, when it confers the performance advantage concerned.

Rief (1984) also took issue with the varied neologisms introduced by Gould & Vrba (1982), again emphasizing the long-established understanding of 'the phenomena of preadaptation and change of function'. It is easy to sympathize with objections to yet more new evolutionary terms and usages, for they undoubtably can confuse and deter those unfamiliar with them. On the other hand, such new coinings may attract attention to noteworthy phenomena and useful distinctions, as in the present case; if they turn out to be superfluous, they usually cease to be used. In addition, Rief felt that the sharpness of the distinction between co-option and adaptation resulted in a very schematic and fragmented view of characters, and prevented their being treated as entities. Obviously, it is important to consider characters as coherent units and follow the history of their changing functions, but named concepts like exaptation can encourage this to be done in a usefully analytical way.

Coddington (1988) likewise thought exaptation was an unnecessary term arguing that, if a feature developed by some means other than adaptation to its present role, it would only be an exaptation if it exhibited no subsequent adaptive change that fitted it more closely to its new use. If it did alter, the original feature would merely be the plesiomorphic state of the final adapted condition. Certainly, exaptations exist where no subsequent modifications of the trait concerned have taken place. For instance, Gould & Vrba (1982) cite the case of the African Black heron (*Egretta ardesiaca*) which shades the water surface with its wings when hunting and so increases visibility. This behaviour is clearly an exaptive use of the wings which had evolved in a quite different context from shading, but it involves no changes in form of the wings. However, even when subsequent modification occurs, this does not preclude a trait from being an exaptation. By definition, an exaptation involves a feature being co-opted for a use for which it was not produced by natural selection. If natural selection, related to the new use, changes the trait entirely, then clearly it is not an exaptation. But, if such natural selection merely causes modifications, so that the trait continues to be recognizable as essentially the same thing, it is an exaptation still. Only those aspects of the newly co-opted trait that are subsequently modified are plesiomorphic relative to the modifications. Inevitably, there will be border-line cases where the term exaptation is hardly worth employing because the original trait has changed so much that it is scarcely recognizable. However, this does not detract from the general utility of the concept of exaptation. As in the case of names for adjacent colours in the spectrum, many useful words cannot be differentiated from each other with complete precision.

Some exaptations have been dismissed as adaptations considered at the wrong level in the character hierarchy; that is, the traits concerned are not considered at the level at which they arose. Alternatively, it has been suggested that the concept of exaptation, by giving undue attention to pre-existing and therefore plesiomorphic traits, directs attention away from the real evolutionary events, apomorphic changes.

Both these criticisms stem from putting inappropriate emphasis on the methodology of cladistics. While in cladistic analysis and some other approaches to phylogeny reconstruction it is important to consider characters separately and concentrate attention on apomorphic events, no such restrictions are necessary or desirable when dealing with the broader topic of how evolution has occurred. Here a totally atomistic approach to characters is often counterproductive, and the interaction of plesiomorphic and apomorphic traits is a valid area of concern.

The concepts of adaptation and exaptation may be intimately connected in a number of ways, for instance co-opted traits often arise as adaptations. Again, a trait may remain an adaptation at one level but, at a less general one, change from an adaptation to an exaptation. For instance, the hypodermic proboscis of hemipteran bugs is an adaptation to sucking liquids and, initially, was an adaptation to sap-sucking in particular, but in some forms it later became an exaptation at this level to imbibing animal body fluids. Again, an adaptation may involve the co-option of a pre-existing trait to a new use. Thus, like some other lizards, advanced members of the lacertid genus *Meroles* evade predators by diving into loose sand. Sand-diving is a new behaviour in this clade and may well be an adaptation, but it is achieved by redirecting the movements normally used in running on the surface for use when moving into the sand (see below).

Most cases of co-option probably involve evolution in the lineage concerned, that is change in gene frequencies. Behavioural changes frequently associated with alteration in use of a trait are often likely to have a genetic basis, as are any modifications of the trait. On the other hand, it is possible to envisage cases where changes in gene frequencies in the organism in which co-option occurs is minimal or non-existent; for example, when an animal exploits a new food-source very similar to that on which it subsisted before. Such instances are none the less still worth noting since, like exaptations involving genetic change, they may have considerable effects on the success of the organism itself and on other species with which it interacts (see below). Without proper analysis, they can be mistaken for adaptations themselves. Indeed, there are cases where what is essentially the same aptation is likely to be an adaptation in one case and but was acquired by co-option in others. For instance, the various lizards that regularly evade predators by diving into sand do not ventilate their lungs exclusively by lateral movements of the thorax as most other lizards do. Such movements when buried would result in sand falling into the gaps at the sides of the thorax that are created at exhalation, so that inhalation would become impossible. Instead, respiratory movements in submerged lizards occur on the ventral surface of the thorax where space for inhalation can be maintained (Pough, 1969). However, ventral breathing developed only in association with sand-diving in *Meroles*, where it is consequently likely to be an adaptation, while it evolved long before entry into sandy habitats in the phrynosomatid *Uma*, and the agamid *Phrynocephalus* (Arnold, 1994a).

One of the main reasons for trying to recognize exaptations is precisely because they are so easily mistaken for adaptations. If the two kinds of aptation are not differentiated, we risk the possibility of exaggerating the undoubted importance of adaptation in fitting organisms to their environments and of ignoring a phenomenon which, like advantageous mutation, is one of the main sources of beneficial accident in the evolutionary process.

DISTINGUISHING POSSIBLE EXAPTATIONS AND ADAPTATIONS

Possible exaptations and adaptations can be distinguished, at least in principle, if a historical perspective is available. In exaptation the trait (or traits) giving a particular performance advantage arises before the advantage is conferred, which occurs after a change in selective regime. In contrast, the trait constituting an adaptation arises after the selective regime in which it first gives advantage.

It is sometimes possible to observe relevant changes in traits and selective regimes directly in populations of organisms with short generation times, such as bacteria, and they may also be recognizable in organisms that have changed within recorded history. Thus many characteristics of domestic animals that make them desirable to man and so promote their survival appear to be largely the result of direct selection. However, most exaptations and adaptations arose at far earlier times and other means of recognizing them are necessary. Following events in a continuous fossil series would be a direct approach to investigating them but is not usually feasible. The relevant fossils are rarely available, it is necessary to establish ancestor-descendent sequences but difficult to do so (Engelmann & Wiley, 1977; Szalay, 1977), and adequate information about the selective regimes in which fossil forms existed is hard to obtain.

The only alternative is to use phylogenies of holophyletic groups. The employment of phylogenies established by cladistic means in testing hypotheses of adaptation was initially considered by Greene (1986) and later by Coddington (1988). Recently, Baum & Larson (1991) have presented a more extended account of methodology. The phylogenies can involve discrete recent or fossil taxa, or both. Recent taxa have the advantage that information about their usual environment and behaviour is easily available and can be used to infer ancestral states for these attributes.

It may be possible to reconstruct a provisional history of a trait and of that part of its selective regime where it confers a particular performance advantage, to see which originated first. To do this, events on a lineage (a continuous sequence of increasingly more recent internal branches of the phylogeny) are estimated on the basis of conditions on successive external branches. By estimating the most parsimonious conditions at branching points (internal nodes) it is possible to determine on which internal branches (internodes) a trait or an aspect of the selective regime is likely to have originated. However, there is no access to the full continuous sequence of events, especially as some critical phases may not be represented by a node – as, for example, when no speciation event has taken place during the phase concerned. Even when one has, change within the relevant external branch may prevent a faithful reconstruction of the condition at the node so produced. Where many taxa are involved and trait distribution is irregular, optimization techniques, such as that of Farris (1970), may be useful in estimating likely conditions at nodes (Swofford & Maddison, 1988; Carpenter, 1989), as may the use of interactive computer programs, such as PAUP (Swofford, 1990) and MacClade 3 (Maddison & Maddison, 1992).

A possible exaptation may be recognized by the sequence of three node types shown in Fig. 1a, and it might be expected that adaptation could be inferred from the sequence in Fig. 1b. In both cases the basal and final node types are the same. The critical difference lies in the intermediate type between them, which shows

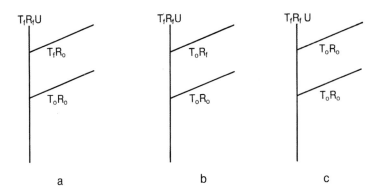

Figure 1 Lineages showing evidence of exaptation (a); apparent adaptation (b); and a situation where both are possible (c). In exaptation, the trait appears before the changed selective environment in which it acquires a new use and performance advantage. In adaptation, the trait appears only after the changed selective environment in which it is advantageous. T = trait: T_o trait undeveloped, T_f trait developed. R = selective regime: R_o original regime, R_f changed regime. U = use.

whether the trait under consideration changes before the selective regime or after it. Phylogenetic sequences indicative of exaptation are actually quite often encountered, which is not surprising as the distinctive intermediate combined state (T_fR_o) is likely to be inherently stable. It will only change to the terminal combined state when the relevant alteration in the selective regime takes place, and this and the trait origin are liable to be independent events (although it is possible to envisage situations in which the evolution of a trait may make the occupation of a selective regime where it confers a performance advantage more probable). Given such independence, the intermediate combined state is quite likely to persist through one or more speciation events, and so be recorded on one or more external branches. The more consecutive branches bearing this, the more parsimonious it is to interpret such phylogenetic sequences as indicating exaptation.

In contrast, the phylogenetic sequence that seems appropriate for recognizing a possible adaptation is met with relatively infrequently. This is because the relevant intermediate state is intrinsically likely to change and consequently often does not persist. There are at least two reasons for this. Firstly, as selection is a dynamic process, it is quite likely, if the necessary variation is present, to occur quickly relative to cladogenesis. This means that the critical intermediate state may never be recorded on an external branch, because no speciation event occurred in the relatively brief interval between the change from R_o to R_f and that from T_o to T_f.

Secondly, even if the critical intermediate combination were initially represented on an external branch, there are reasons for thinking it may often not persist. At point A on the phylogeny shown in Fig. 2, species 4, 5 and 6 have a common ancestor, so they are all derived from the same genotype and phenotype and were subjected to the same new selective regime, R_f. In this situation, it would be expected that, if R_f selected for the change from T_o to T_f in the common ancestor of species 5 and 6 (point B), it would also do so in the ancestor of species 4 (point C). So that, even if the branch originally had T_o in combination with R_f, this association may well be lost in parallel with the common ancestor of species 5 and 6.

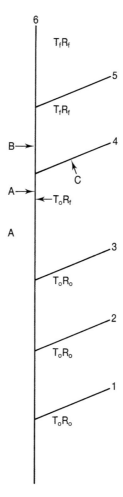

Figure 2 For further explanation, see text. T = trait: T_o trait undeveloped, T_f trait developed. R = selective regime: R_o original regime, R_f changed regime.

Alternatively, the form with the T_oR_f combination may be more likely to become extinct if it does not adapt to its new selective regime, R_f, by developing T_f.

In fact, it is likely that, if T_oR_f is present on a branch, straightforward adaptation is not represented, for something must have prevented the changed selective regime, R_f, acting to produce the trait, T_f from T_o. Various possibilities exist. Relevant mutations may not initially exist or there may have been inherent constraint on variation, preventing selection acting on the common ancestor of species 4, 5 and 6, which only broke down subsequently in the common ancestor of species 5 and 6. Alternatively, some other stronger and conflicting selective factor may have favoured T_o, preventing R_f from acting until after the speciation event that separated species 4 from the common ancestor of species 5 and 6. Such phenomena (lineage effects) are discussed further below. Another less parsimonious possibility is that the change from T_o to T_f occurred in the ancestor of species 4, 5 and 6, but a conflicting selective factor caused reversal, from T_f to T_o, in the ancestor of species 1. In these situations, inherent constraint on variation or a conflicting selective factor may be detectable in species 4.

In spite of reasons for the instability of intermediate combinations of trait and selective regime in adaptive sequences, cases where adaptation does not immediately follow the exposure of a lineage to a particular factor in the regime do occur on occasion. For instance, cichlid fish in rivers in East Africa are found in the same environment as water snails but do not usually eat them or develop specialized tooth structure associated with such a diet. Yet when such forms have invaded large lakes, they have sometimes undergone explosive speciation, and some of the descendent forms have become almost entirely snail feeders with well marked morphological aptations for dealing with such prey (P.H. Greenwood, personal communication). Possibly, in this case, the speciation process facilitates the development of variation, or the large number of species present in lakes results in narrower trophic niches, so that the relevant morphological specialisations are more likely to arise.

Adaptation is thus likely to appear in phylogenetic sequences without a definitive intermediate stage, so that a derived trait and the selective regime in which it confers a performance advantage appear concurrently, as in Fig. 1c. While this condensed sequence is consonant with adaptation it can easily arise in other ways. Thus it could result from exaptation should the definitive intermediate stage be missing, either because no speciation event resulted in its occurrence on one or more external branches of the phylogeny, or because any such branches were subsequently lost by extinction or underwent change. We are consequently left in a rather disconcerting situation where, in individual cases, exaptation can often be recognized with some certainty but adaptation cannot.

This is not to say that hypotheses of adaptation cannot be tested to some extent and tests are similar to those applicable to hypotheses of exaptation. Both kinds make predictions about distribution of traits and selective regimes on phylogenies, and predict particular performance advantages in changed regimes compared with previous ones. For adaptation, a hypothesis is refuted if the new trait develops before the relevant selective regime. If this test is passed, it is possible to check whether the new trait really confers an advantage in the new regime that its plesiomorphic state does not. This applies to both cases where changes in regime and trait are concurrent on the phylogeny and to ones in which there is a delay between these events. A hypothesis of exaptation is refuted if the trait clearly develops after the change in regime and, if this test is passed, it is possible to test whether the trait really confers a new performance advantage in the new regime that it did not perform in the previous one.

The difference between hypotheses of adaptation and exaptation is not in their inherent testability but that the latter can often be strongly corroborated from a phylogeny and information on performance advantage, whereas the former cannot. This is partly because the combination of regime and trait indicative of adaptation is unlikely to appear on phylogenies except in special circumstances. Also, whereas hypotheses of exaptation are about acquisition of performance advantage, something that may be discernible when phylogenies are used, hypotheses of adaptation specify mode of trait origin. There is no way to be sure that a long-standing trait arose by natural selection. It would, however, be a reasonable supposition if the hypothesis passed the tests given above, and if there was cause to believe adaptation by natural selection is the most usual source of new traits.

Interpretation of individual cases where derived traits and selective regimes first appear concurrently

The last point raises the question of how situations like that in Fig. 1c, in which a derived trait and a regime in which it gives a performance advantage first appear concurrently on the same node on a lineage, should be treated? While admitting their essential ambiguity, should they be regarded in practice as provisional evidence of adaptation rather than exaptation, as is often done (Greene, 1986; Coddington, 1988; Baum & Larsen, 1991)? In other words, will we be considerably more often right than wrong to regard individual cases of concurrence as being due to adaptation? One important factor in deciding this is the actual frequency of such concurrences on phylogenies.

As already observed, many of the the traits organisms possess confer performance advantages in the regimes they usually occupy. Recent access to phylogenetic information indicates that such correlations between a trait and a regime in which it is advantageous seem very often to arise in the form of concurrent first appearance of trait and regime on the same node on the lineage of the taxon concerned. This at least approximate temporal correlation is clearly more suggestive of a causal relationship than mere correlation within present species. Many more clades need to be examined to confirm the general nature of this regularity but it is apparent in at least some relatively recent holophyletic groups that have evolved along ecological continua consisting of a series of coercive environments that appear to require numerous aptations. For instance, in the lacertid lizard genus, *Meroles*, first appearance of the 70% of 61 apomorphies for which a performance advantage can be discerned is concurrent with occupation of the environmental situation in which this is so (Arnold, 1990).

Is there any way that exaptation could be responsible for a high proportion of such concurrences? It might be possible if exaptation itself were extremely common. However, as seems very likely, if the origin of a preaptation and that of the regime in which it confers a performance advantage through exaptation are often independent events, exaptations are liable to be distributed more or less randomly relative to the origin of preaptations. Consequently, if many concurrences between first appearance of traits and appropriate regimes represented clandestine exaptations, that is with the preaptation and the relevant regime arising relatively close to each other in time, we would expect numerous cases where the events were widely separated enough for exaptation to be overt. Yet, in general, overt exaptations do not seem to be abundant, compared with concurrences. Thus, in the case of *Meroles*, none of the traits for which initial performance advantages can be discerned shows clear evidence of later co-option to a new use. This sample is inevitably smaller than that showing concurrence between trait and regime, for at least one branching point after origin is necessary to demonstrate exaptation, and this is not always present.

If overt exaptations are generally at a low frequency compared with concurrences, exaptation could only be responsible for a substantial proportion of the latter if the assumption of independence of the appearance of a preaptation from its subsequent co-option were erroneous. If so, there would have to be a mechanism that frequently matched an existing trait to an appropriate regime, yet no such mechanism appears to exist. While it has been suggested that newly arisen traits could be matched to suitable environments by the organisms concerned moving into these

(Dover, 1984), it is highly unlikely that matches could frequently be made. If generation of traits were random relative to the situations in which they ultimately conferred performance advantage, it is difficult to understand why some specialized organisms have such a high proportion of traits that fit them to their specific environment.

On the other hand, adaptation by natural selection provides a plausible mechanism for producing traits giving performance appropriate for a given regime, and there is evidence for its existence. Thus an appropriate genetic mechanism exists which produces variants and enables them to be inherited. Change by selection has not been demonstrated in all groups of organisms, but it is found in such a wide range that the most parsimonious explanation is that it is a feature of living things. Given that concurrences are so common compared with overt exaptation and that adaptation provides the only plausible means of producing such concurrence, it is appropriate to interpret them provisionally as evidence of adaptation.

The above argument can be weakened by an appeal to unremarked extinction (a constant spectre, that may frustrate our best laid plans to reconstruct evolution without direct knowledge). Clearly, if there has been very frequent speciation on a lineage and little or no extinction, the multiplicity of sequential branches will probably allow all exaptations to be discerned. In contrast, if there is no speciation, or alternatively extinction of all the branches that develop, then no exaptations can be recognized. Consequently, the proportion of overt exaptations discernible depends on available tree topology. We would expect exaptations to be most apparent in recent sections of lineage and for an increasing proportion to have been converted to concurrences in older ones.

Other suggested requirements for testing hypotheses about the origin of single adaptations

It has been stated that, to avoid circularity, the basic phylogeny used to test a hypothesis of adaptation must be constructed from features other than those involved in the hypothesis (Coddington, 1988). Presumably this stricture would be applied to hypotheses of exaptation too, but its rationale is unclear. Indeed, it can be argued that excluding a character from the estimation of a phylogeny, merely because the possible adaptive origin of one of its states is being considered, is inappropriate. The reconstruction of relationships should take account of all possible evidence and exclusion of characters may well alter the final result or prevent a result being obtained at all. For instance, suppose the available evidence for the relationships of three taxa, A, B and C consists of characters 1, 2 and 3. Characters 1 and 2 indicate that taxa A and B are more closely related to each other than to C, while character 3 suggests that A and C are more closely related to each other than to B. Parsimony requires that the evidence of characters 1 and 2 should be accepted. If, however, character 1 were to be excluded from the analysis because the possible adaptive origin of its derived state was being investigated, the phylogeny would not be resolvable. Excluding a trait on the unfounded supposition that circularity is thereby avoided is less important with well substantiated phylogenies but, even here, this procedure can create problems if hypotheses of adaptation for a number of traits are being considered simultaneously, since numerous characters would have to be set aside.

The suggestion has also been made that, if more than one successive branch of a phylogeny bears a derived trait, it must have the same function in all of them if a hypothesis of its adaptive origin is to be maintained (Coddington, 1988). This too is not a requirement. As already noted, traits may lose their initial use (Sober, 1984), and many cases of exaptation involve traits that possess one function when they evolve through adaptation but subsequently acquire another. The latter use cannot negate the adaptive origin of the first. The only problem in such cases, where different uses exist on successive branches of a phylogeny, is determining which one arose first. Two successive basal branches bearing a particular use are all that is necessary to establish precedence under the rules of parsimony.

Limitations of phylogenetic methods for testing hypotheses of adaptation involving single origins of derived states

Lauder, Leroi & Rose (1993) have emphasized the limitations of using phylogenies involving single traits to investigate whether adaptation has in fact occurred. These authors point out that appropriately correlated origins of a new aspect of the selective regime and a new trait does not necessarily indicate that the former elicited the latter, and then go on to give examples where such an assumption would be misleading. For instance, a derived trait conferring a performance advantage may not have been the result of direct selection; instead it may have been incidentally promoted when selection acted on a genetically associated trait (as could occur if the traits were pleiotropic effects of the same gene). Again, it is possible that a trait confers advantage in additional unappreciated ways, and it is these clandestine advantages that are actually selected for.

Such postulated cases are indeed possibilities, but they represent alternative hypotheses of adaptive origin to that where the origin of a particular trait is directly elicited by a particular change in the selective regime. The phylogenetic method can only test a given hypothesis to the extent of seeing whether origin of trait and the new aspect of the selective regime may have occurred in the appropriate order. Alternative hypotheses can be tested in the same way and to the same extent, but actually choosing between available hypotheses is a separate operation, requiring in the cases cited above information about any common genetic basis of different traits and about other ways in which they may possibly confer performance advantage.

Using multiple occurrences of traits to recognize the action of natural selection

So far in this paper hypotheses of adaptation have been considered which involve single occurrences of a derived trait. But it is also possible to examine situations where a trait has evolved independently a number of times. In such cases it is feasible to test statistically whether the separate origins of the trait correlate with a change in some aspect of the selective regime. If the correlation is greater than that expected by chance (Maddison, 1990) an explanation for the association is required, one possibility being that the trait arose as a result of natural selection arising from the changed aspect of the selective regime. The case for this is much stronger if the trait can actually be shown to confer a performance advantage in the modified selective regime, or seems likely to do so.

The multiple occurrence approach to recognizing adaptation is not a rival to tests involving the examination of single origins of a derived trait. Indeed, the two kinds of test address different hypotheses: one that an individual occurrence of a trait might arise by adaptation, the other that multiple cases may generally be attributable to it. If possible, the former kind of test should be carried out for each of the separate cases that constitute the sample used in the multiple occurrence approach. Any overt cases of exaptation thus detected should be eliminated before testing for statistical evidence of adaptation. Even when this is done, and an overall positive correlation is found, it cannot be assumed that individual cases where correlation exists represent adaptation, for they could still be clandestine exaptations. However, the chances that they are each the result of adaptation are increased.

It has been emphasized that the statistical approach inevitably concentrates on homoplasious traits. For instance "the result is at least ironic: the worse the fit of a character to the cladogram, the better chance it has to be accepted as an adaptation" (Coddington, 1990). However, the fact that a trait gives little support to a phylogeny says little about its general importance or interest.

Another criticism of the multiple occurrence approach is that it is prone to imprecision (Coddington, 1988, 1990). It is suggested that, in the interest of increasing sample size, there is a temptation to include traits that are not really similar, so the sample becomes heterogeneous. Certainly such samples are inevitably non-homologous, and we do not have the clear evidence of appropriate comparison that is present when comparing a derived trait with the plesiomorphic state it originated from. However, in addition to degree of similarity, which is inevitably somewhat arbitrary, another more objective criterion should be used: in hypotheses of adaptation there should ideally be evidence that the derived states concerned provide a particular performance advantage. This is something that can be checked by experiment but observation can also be informative.

For example, many lizards have a ridge on each side of the snout, a condition that has arisen many times. Initial observations suggest it is often present in forms that burrow rapidly forwards in loose sand to avoid predators. In this situation the ridges appear to confer a performance advantage by facilitating lateral changes in direction of the head that occur during locomotion, something that is inherently testable experimentally. The hypothesis can therefore be put forward that snout ridges have generally evolved by adaptation for this use, and then be exposed to refutation by testing for a statistical correlation between origin of ridges and shift into sandy habitats. In most cases the ridges are formed from the supralabial scales that form the upper lip, but in the tropidurid *Leiolaemus* they arise from the infralabial scales along the lower jaw, so they are essentially quite different. However, in all cases the ridges are of a form which will facilitate motion through the sand and it is thus appropriate to include them all in the test, even though they are quite different in basic anatomy.

OTHER ASPECTS OF EXAPTATION

Origins of traits that subsequently become exaptations

As exaptation is defined in a negative way, by the absence of natural selection for the use concerned at the time of origin of the trait, it is not surprisingly a heterogeneous concept and there are substantial differences in the way the features involved arise. A trait that ultimately becomes an exaptation may have developed through natural selection for some other use, a process that could have involved the promotion of a single allele, or of alleles at several loci, either simultaneously or sequentially. On the other hand, the trait might originally have been the by-product of some other evolutionary event and had no initial advantage or use. For instance, in the case of a single-gene trait, an allele may have been promoted by natural selection because one phenotypic effect was advantageous. So long as other effects of the allele were not harmful, or at least did not outweigh the advantages of the promoted effect, they too would become fixed and thus available for exaptation, if changed selective regimes later made them useful.

When dealing with multi-locus or otherwise more complexly generated features, it is less likely that they would often arise incidentally and be maintained for long periods if they were not advantageous. Certainly, it is difficult to think of many convincing examples. Unless firmly linked to some advantageous feature, such traits would have to be more or less selectively neutral and have, at most, very small costs, both in their ontogeny and their subsequent maintenance. A possible case is repetitive DNA, which may be the ultimate source of new gene loci (Ohno, 1970). Although repetitive DNA might be unadvantageous until it is converted into a new gene, its costs may well be low, as it constitutes such a small part of the fabric of an organism.

One kind of trait that may not be actively maintained by selection, but which seems particularly likely to persist for a long time as it involves no obvious costs, is absence of a feature. For instance, snakes have entirely lost the pectoral girdle and lost or greatly reduced the pelvic one. This initially occurred in connexion with disappearance of the limbs as a concomitant of serpentine locomotion, since without them the girdles are mechanically inessential. Without the restriction of the girdles, snakes could subsequently evolve methods of ingesting and processing very large prey. Absence of features in the earlier stages of ontogeny may also be co-opted to some use. Thus, as Darwin (1859: 197, quoted by Gould & Vrba, 1982) pointed out, the lack of rigid contact between the bones of the developing skull in vertebrates appears to serve no positive function in most groups, but it acquired a use in eutherian mammals where it allows the skull to distort as the young animal passes through the birth passage.

Features of organisms that become exaptations are often quite complex and have not arisen from single evolutionary events. Not infrequently, they are the result of a series of likely adaptations often widely separated in time, some of which have undergone exaptation at least once before. For instance, this is true of sand-diving behaviour in the phrynosomatid lizard, *Uma* (see below).

Categories of exaptation

A number of different kinds of exaptations can be distinguished and, like exaptations in general, they can often be recognized from phylogenies (Fig. 3). In *first-use exaptations* a trait has no performance advantage at its origin, gaining one only when subsequently co-opted to the use in question (use being defined here as the actual way a trait confers performance advantage). In contrast, there are exaptations that already confer a performance advantage in one way before they are co-opted to

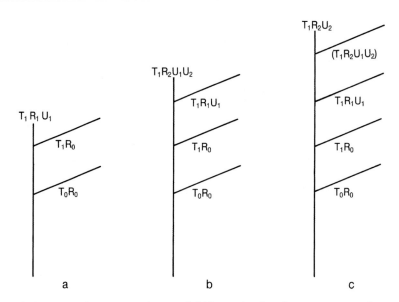

Figure 3 Phylogenies showing evidence of different kinds of exaptation. Each stage may be represented by more than one external branch. If T and R are not polarized by outgroup comparison, three branches bearing T_0 and R_0 are necessary to establish that these conditions are most parsimoniously regarded as primitive. (a), First-use exaptation; (b) addition exaptation; (c) transfer adaptation (here the bracketed intermediate addition phase may not always occur). T = trait: T_0 trait undeveloped, T_1 trait developed. R = selective regime: R_0 original regime; R_1, R_2 successive changed regimes. U = use: U_1, U_2 successive new uses.

a further use. The previous use may have been present from the origin of the trait and provided the selective pressure for its evolution. Alternatively, it may itself be a result of previous exaptation, of either first-use or extra-use kind. Such exaptations can be divided into two groups: *addition exaptations* where the new use is merely added to the first, and *transfer adaptations* involving a shift to a new use with loss of the original one. In fact, although it may not be true in all cases, transfer adaptations often pass through an interim phase as addition exaptations where the two uses occur together. Presumably, when the ancestors of carnivorous hemipteran bugs transferred from sucking sap to feeding on the internal fluids of animals, there was at least a brief phase when they did both. So their hypodermic proboscis was initially an addition exaptation to exploiting animal fluids, but later became a transfer exaptation.

In addition exaptations any modification of the co-opted trait that increases efficiency in its new use is likely to be restricted by the fact that the trait must still perform its original function. No such limitation applies to transfer exaptations, so modification can be greater, although some of them actually exhibit very little change (see below).

Investigations of the origins of aptations

Initial indications of possible exaptation

Because situations in which a trait may be co-opted to a new use arise independently with respect to the origin of the trait itself, the probability of exaptation will increase with time, and long-standing features will consequently be more likely to have been co-opted to new uses than more recently evolved ones. Exaptation is also more likely for traits that have been subjected to a wide range of selective regimes, for instance through adaptive radiation. It is, consequently, not surprising that a basic chordate feature like a postanal tail that has been exposed to many selective regimes should have acquired numerous novel uses, such as for grasping and predator evasion (for example, its employment in lizards for clubbing enemies, stopping burrows, and breaking away from pursuers and distracting them). Such cases are often very obvious, since the new performance advantages clearly develop much later than the trait, as is also true of the secondary use of tongues of gekkonid lizards for intraspecific signalling (*Phelsuma*; Gardner, 1984) and of the claws of crabs for the same purpose (e.g. *Uca*).

The possibility of addition exaptation is suggested if a trait has more than one function within a species, but confirmation of its occurrence and determination of which use came first can only come from detailed phylogenetic information. The extremely flattened head, body and tail of the aberrant arboreal tropical African lacertid lizard, *Holaspis guentheri*, allows it to hunt and hide in narrow crevices beneath bark, and also constitutes an aerofoil which enables it to glide from tree to tree. Phylogenetic analysis (Arnold, 1989a) shows that the flattening first developed in the context of crevice use and was only later co-opted to gliding.

There are no indications within a single species that a transfer exaptation may exist, but its presence may be suggested by the occurrence of alternative uses of a trait in different members of a clade. Sometimes an alternative and earlier use only subsequently becomes apparent, for instance when fossils are investigated. Thus, it turns out that many of the anatomical features now associated with flight in birds arose initially among dinosaurs in the context of prey capture by a mantis-like projection of the forelimbs (Gauthier & Padian, 1985).

Accommodating the eyes in crevice-using lizards (Fig. 4)

Many lizards living on rock exposures utilize crevices as refuges, either when fleeing predators or as nocturnal refuges. This behaviour involves some striking aptations, often including the evolution of very marked dorso-ventral flattening, allowing entrance into narrow fissures as in *Holaspis*. The vertical dimensions of the head are consequently reduced, but the eyes do not usually become correspondingly smaller and, in normal activity, bulge upwards above the skull surface. However, when a lizard flees into a narrow crevice the eyes must be accommodated within the depth of the flattened head. They are most usually pushed downwards by the ceiling of the crevice as the lizard moves deeper into it, so their upper margins are flush with the skull roof and their lower sections bulge through the palate into the buccal cavity (Arnold, 1970). In scincids and lacertids each eye bulges vertically through the corresponding suborbital foramen, a mechanism that has arisen several times. As the occurrence of the foramen on the phylogeny of the forms concerned precedes occupation of crevices, its use for accommodating the eye within the reduced depth of the skull is an exaptation. However the foramen is initially small and triangular, allowing only limited projection of the eye into the buccal cavity, and there is subsequent, presumably adaptive, modification improving the initial exaptation, the foramen becoming larger and more rounded.

Crevice-dwelling cordylid lizards, which entered rocky habitats independently, do

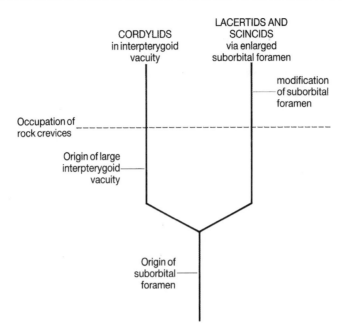

Figure 4   Alternative methods used by lizards for storing part of the eye in the buccal cavity that have both been produced by exaptation. For further explanation, see text.

things differently: although the suborbital foramen is present, the eye shifts medially, so that it partly occupies the interpterygoid vacuity – which is very extensive in cordylids. This again turns out to be an exaptation, for examination of the phylogeny of cordylids (see Lang, 1991) shows that expansion of the interpterygoid vacuity evolved before crevice use, although after the origin of the suborbital foramen. The vacuity may have been utilized instead of the foramen because, being large, it provided immediate housing for a large portion of the eye, whereas the foramen would only have been able to provide this after some modification, as in lacertids and scincids. This case demonstrates that exaptation can produce quite different solutions to what is essentially the same environmental problem, and that an available exaptive solution that functions elsewhere may not be utilized because a more immediately effective alternative arises.

Maintenance of position in crevice-using lizards (Fig. 5)

It is advantageous for rock-dwelling lizards to be able to maintain their position when predators try to extract them from crevices. In nearly all cases they cling with their claws and may push their dorsum hard against the crevice roof, but they also have more specialized methods. The Chuckwalla (*Sauromalus*), a North American iguanid, wedges itelf in place by hyperinflating the lungs, using the buccal cavity as a pump to do so (Smith, 1946). In contrast, cordylid, lacertid and scincid lizards use areas of mobility within the skull, especially the frontoparietal suture, to alter the geometry of the head (Arnold, 1970). This cranial kinesis first allows the head to be flattened and pushed into narrow areas of rock crevices, after which contraction of the adductor muscles of the jaws reverses the alteration and the skull expands dorso-ventrally (Arnold, 1970). The expansion produced results in the upper and

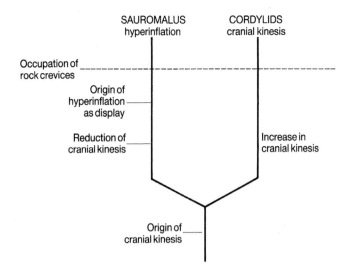

Figure 5 Alternative methods used by lizards for retaining their position in rock crevices. For further explanation, see text.

lower surfaces of the head being pressed against the floor and ceiling of the crevice and holding the lizard in place. The action is analogous to the hand-jamming technique of rock climbers who hold themselves in place by putting a hand in a crevice and flexing it, so the finger tips and heel of the palm press down and the knuckles upwards.

Sauromalus not only inflates in crevices but also during threat behaviour, both when confronted by a predator in the open and in intraspecific display. It can consequently be asked which use of hyperinflation came first, crevice-wedging or threat? When behavioural data is integrated with a phylogeny of all the iguanids (De Queiroz, 1987, as modified by Norrel & De Queiroz, 1990), it is apparent that hyperinflation in threat situations is virtually universal, its occurrence including the forms which are basal to *Sauromalus* on the phylogeny. In contrast, crevice-wedging is confined to that genus. In spite of generally having been regarded as an adaptation, hyperinflation in crevices is consequently an extra-use addition exaptation.

Cranial kinesis within crevices by extant lizards also involves development of an additional use. A mobile fronto-parietal suture and other areas of movement within the skull are primitive and widespread in lizards, and are believed to enhance capture and manipulation of prey (e.g., Frazzetta, 1962; Iordansky, 1966; Condon, 1987). So the use of cranial kinesis in crevices looks initially as if it were another extra-use addition exaptation. Certainly, forms basal to extant crevice-users in phylogenies exhibit kinesis during feeding but do not enter crevices. However, there is circumstantial evidence (Arnold, in preparation) that the mobile frontoparietal suture may actually have first arisen by adaptation in the context of crevice use and only then was rapidly co-opted to feeding. If so, the employment of cranial kinesis by extant rock lizards is not a simple exaptation after all but a reversion to original adaptive use. This case underlines the importance of tracing traits conferring performance advantages back to their origins if the kind of aptive process involved is to be determined. Hyperinflation may be used in *Sauromalus* instead of cranial kinesis which evolved earlier, because that latter is substantially secondarily reduced in iguanids. So, as with accommodating the eyes in crevices, an immediately effective exaptation is used instead of a partial or potential solution that needs modification

Hyena genitalia revisited

The above cases show how phylogenies can be used to recognize exaptation where its occurrence has not previously been suspected and, where competing potential solutions to a problem exist, to see why one may take precedence over another. Here a striking phenomenon that has been regarded as a likely first-use exaptation and widely discussed (e.g. Gould & Vrba, 1982; Hamilton & Tilson, 1984; Frank, 1986) is reassessed.

Although such instances are rare, Gould & Vrba (1982) discuss a case where a complex feature is interpreted as having been been produced as an incidental product of natural selection that only later gained a use. This is the ostensible male form of the female external genitalia in the spotted hyena (*Crocuta crocuta*). In this species, females are larger than the males and dominant to them. Spotted hyenas are often solitary for part of the time but also hunt cooperatively in packs. Kruuk (1972) put forward the hypothesis that the masculinized female external genitalia, and those of the males themselves, have a signalling role in reintegrating solitary wanderers back into their clan during behaviour, which he calls the meeting ceremony, when spotted hyenas approach each other, expose their genitalia and investigate those of the other animal.

Gould & Vrba (1982) suggest that the detailed external resemblance of the female genitalia to those of the male might originally have been an incidental result of selection for larger female size, rather than of direct selection for masculine appearance. Size increase would have been mediated by the increased levels of the hormone, androgen, which are in fact found in female spotted hyenas (Racey & Skinner, 1979). As high androgen levels are responsible for the development of the characteristic features of the external genitalia of males, greater concentrations of the hormone in females could have had the incidental effect of masculinizing the external form of their genitalia. These structures would then have later acquired their role as a signal in the meeting ceremony, or had this enhanced.

How likely is this hypothesis to be correct? It could be partly tested by manipulating the high androgen levels that exist in developing female hyenas (Frank, Glickman & Licht, 1991) to see whether concentration of the hormone affects body size and genital form simultaneously in modern populations. If so, this would corroborate the hypothesis that large size and masculinized external genitalia in females arose together, but the experiment has yet to be made. Examination of the distribution of traits on the phylogeny of the spotted hyena and its relatives would in principle form another kind of test, but the available topology is uninformative. The extant sister-group of the spotted hyena comprises the striped hyena (*Hyaena vulgaris*) and the brown hyena (*H. brunnea*), but these and more distantly related forms lack all the distinctive traits discussed here (Mills, 1978; Racey & Skinner, 1979). There is consequently no indication as to whether large size and masculinized female genitalia are likely to have arisen separately or not. Similarly, no evidence exists that there was a delay indicative of exaptation in the use of female genitalia in the greeting ceremony after they were masculinized, or alternatively, that this use may have appeared simultaneously with masculinization, in which case both may have been promoted by natural selection. However, whatever happened in the particular case of female genitalia there is at least one clear exaptation in spotted hyena social behaviour supported by phylogenetic information. This is the co-option of the form of the male genitalia, presumably originally evolved in the context of copulation, to signalling in the meeting ceremony.

Gould & Vrba (1982) not only use the masculinized external genitalia of female spotted hyenas as an illustration of a possible first-use exaptation, but also to demonstrate how complex traits might evolve as incidental concomitants of the appearance of other features without there being any direct selection for them. However, although the *change* in female external genitalia is incidental to another event and would not involve direct selection if the above interpretation were correct, this may not always have applied to the detailed form of the masculinized genitalia. Their external architecture is not being created for the first time. All that is happening is that the male external morphology is being produced in a different sex during ontogeny by exactly the same mechanism, high androgen levels, that elicits it in the one where it originally evolved. A pre-existing structure, which may well have been originally produced by natural selection, is merely being switched on, a relatively simple event.

Some other phenomena, where there is complex change in form in one sex or growth stage that may well be phylogenetically sudden, also involve the appearance of an already existing morphology, that could have originally developed by natural selection, and is made available in other situations by simple shifts in growth pattern, often mediated by hormones. This applies to masculinization of female genitalia in spider and squirrel monkeys (*Ateles* and *Saimiri*) and the development of mimetic oestrus swelling in male baboons (Racey & Skinner, 1979); the larval morphology of adults in neotonous taxa of salamanders is an analogous case. Such substantial changes may be viable because the adopted morphology has already developed in a selective regime and one not too different from that where it later exists. Radical changes in form where this is not so may be much less likely to lead to evolutionary success.

Evolutionary significance of exaptation

Some aspects of the importance of exaptation in evolution have long been appreciated, although only recently under that name. Darwin (1872) used the idea of different original functions for the simpler early stages of some complex features to counter Mivart's (1871) criticism that natural selection for present use would not have caused the initial development of many anatomical structures. Exaptations thus sometimes allow the development of survival strategies that cannot evolve easily by direct natural selection because the preliminary stages are not advantageous in the context concerned. Presumably, the early stages of the development of wings and feathers would have been of little use for flight and would not have occurred unless they had different uses at the time, such as prey capture and insulation respectively. The likely importance of exaptation in allowing otherwise unattainable solutions to be reached is not confined to extra-use exaptations. Thus, Gould & Vrba (1982) point out that repetitive DNA may have been an effective source of new genes because it was able to develop numerous random changes during its pre-aptive phase.

However, although the evolutionary role of first-use exaptations has been empha-sized (Gould & Vrba, 1982), these appear to be relatively rare and perhaps essentially confined to traits where development and maintenance costs are low, such as loss of features (as already noted). Extra-use exaptations, on the other hand, seem to be far more common. Where they occur, they may be a more immediately effective source of aptations in many situations than adaptation, and not only in cases where natural selection for initial development is inherently unlikely. In adaptation, a particular allele, newly produced by mutation, often only produces part of the final trait and at first is found in only very few members of a species. Consequently, unless promoted by founder effects (Mayr, 1954) or genetic drift (Wright, 1931), it has to provide tangible and more or less continuous benefits from the start, so that natural selection will promote it until it is widespread. Early stages of this process are likely to be especially hazardous: if the environment occupied by the species concerned is heterogeneous, the allele may only be advantageous if it arises in particular areas, and random sampling effects, or short-term environmental changes, can eliminate it.

In contrast, a trait that becomes an extra-use exaptation may well be substantially complete and effective at the time of co-option and also widely distributed, so there would be none of the potential problems of initial spread that characterize muta-tions. Like these, in this process features that could potentially be exaptations develop randomly relative to the use to which some of them are eventually put. However, unlike mutations, this does not mean they are likely to be eliminated if that use is not immediately possible. Since many extra-use exaptations begin as addition exaptations even if they later become transfer ones, the trait is able to persist for a long time before exaptation takes place, because its original use confers advantage and is consequently selected for. This also means the trait will be retained after exaptation, even if the initial benefits of the exaptive use are small or fluctuating. There will consequently be time for the trait to be improved by natural selection, if it is not already optimal for the use to which it was exapted.

Although extra-use exaptations often involve the organism taking on an additional task, they can sometimes involve a radical shift in the way pre-existing ones are handled. For instance, the shell of chelonians presumably developed initially as external protection, but it later took over the main role in skeletal support of the trunk from the vertebral column.

As in the case of limblessness in snakes, exaptations can enable new resources to be exploited. Sometimes the functional change in use involved in exaptation may be slight, for instance when a predator changes to a new kind of prey which can be handled and processed similarly to that previously eaten. Nevertheless, such cases are often of considerable historical and ecological significance. Unspectacular trans-fer exaptations of this kind often occur when predators encounter new prey species through the range extension of one of them. For example, the snake *Lycodon aulicus capucinus* of south and south-east Asia was introduced to Réunion in the south-west Indian Ocean during the 1830s. Because it already had aptations for catching and eating Asian lizards, it was able to exploit and sometimes exterminate endemic forms, such as the skink, *Gongylomorphus bojeri borbonica*. The snake was equipped from the start to utilize a previously unencountered prey, while the Réunion lizards were confronted by methods of predation they had not met, at least during their recent evolutionary history, and to which they had no time to evolve counter-strategies.

Many cases of supposed co-adaptation, where one or both of two interacting species have aptations that benefit them in the relationship and which are supposed to have evolved in response to the other form, may in fact involve transfer exaptations rather than adaptations. Without historical knowledge, the *Lycodon* example could be misinterpreted in this way. As with exaptations in general, such cases can often be clarified by considering phylogenetic information.

LINEAGE EFFECTS: INFLUENCES ON THE PRODUCTION OF ADAPTATIONS ARISING FROM EVENTS IN THE HISTORY OF ORGANISMS

In the biological literature, there has been a marked tendency to emphasize limitations on change in different organisms, such restrictions often being referred to as constraints. Constraint is a very popular term: from 1985 to mid-1992 it was used at least 1342 times in biological titles or as a key word (Biosis, 1992). This frequency partly results from constraint being used in several different ways. Universal constraints (Maynard Smith *et al.*, 1985) act on all living things. Included here are the results of factors that would apply to any possible organisms at any time (ahistorical universal constraints). They include the laws of physics and chemistry, and explain for instance why elephants cannot leap with the insouciance of mice. In addition to these ahistorical factors, there are universal constraints that arise from events in the early history of life (historical universal constraints). For instance, the ubiquity of DNA as a genetic mechanism suggests that organisms possessing it are very unlikely to transfer to some other system, although it would be difficult to prove that such a transfer is completely impossible. The term phylogenetic constraint is often employed for restrictions on the apparent capacity of particular taxa to evolve in particular directions, and has the same meaning as local constraint (Maynard Smith *et al.*, 1985); phylogenetic constraints arise from events on the lineages of the organisms concerned. While often separately recognized, most historical universal constraints are likely to be phylogenetic constraints which have become omnipresent through the extinction of early lineages which did not possess them. Constraint is also sometimes applied to any kind of bias, whether limiting or enhancing particular potential alterations (for instance by Maynard Smith *et al.*, 1985). Again, it may be used not only for effects on evolution but also for the traits of the organism causing these. Indeed it has even been applied to all previously evolved traits of organisms (Brooks & McLennan, 1991), presumably on the basis that all traits may eventually produce some kind of effect on future evolution.

Constraint has thus been employed in biology in a wide variety of ways but usually, as in its everyday sense, to indicate restrictions on evolution. There is consequently a case for using a different, more specific term for historical influences of any kind on adaptation that arise from specific events on the lineages of the taxa concerned. In this chapter, they will be referred to simply as *lineage effects* on adaptation. Where appropriate it can be specified whether they facilitate or constrain a particular evolutionary course, constrain and constraint only being used in this sense. Note that lineage effect applies specifically to historical influences on future evolution rather than to the apomorphic traits that produce them, or to the earlier evolutionary events from which these arise. This is because the effect of particular traits and consequently of the changes leading to them may vary, and they may both facilitate and constrain. For instance, by making one kind of later change more likely, an acquired trait must inevitably make other kinds less so. Again, an earlier change may first facilitate alteration of one feature but later constrain the alteration of another.

Events that produce lineage effects on adaptive capacity

Lineage effects on adaptive capacity can be usefully discussed in the context of the potential production of a particular hypothetical adaptation that would solve a specific problem posed by a changed selective regime (which by definition includes contributions both from the environment and the organism itself; Baum & Larson, 1991). Historical events that affect adaptive solutions are discused below and their possible points of origin are summarized in Fig. 6.

Change in precursor traits: gain, loss and modification

The production of a specific well-defined adaptation is likely to be dependent on the presence of one or more appropriate precursor traits. If a precursor is initially absent, its evolution will facilitate the adaptation. This might be by alteration of pre-existing traits or development of new ones from such events as duplication (for

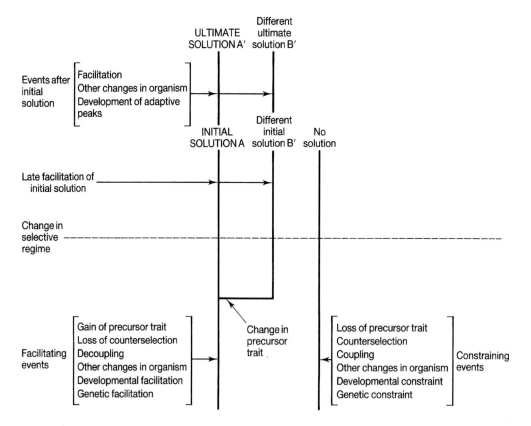

Figure 6 Events that may determine whether a specified adaptive solution (A') to a problem posed by a changed selective regime will evolve.

instance, of genes; Ohno, 1970; or of morphology; Lauder & Liem, 1989). On the other hand the specified adaptation will be prevented by loss of the precursor trait if it already exists. For example, evolutionary loss of limbs in snakes prevents any adaptations involving legs, and suppression of the larval stage in salamanders prevents its later employment as an adult morphology in neotenous forms.

Rather than be totally lost, the precursor may be modified, for example by selection causing change in a direction different from that of the specified adaptation. Such selection may be transient, or it may persist but be largely, although not completely, overridden with the onset of selection pressure for a solution to the problem under consideration. In such cases, the modification of the precursor may result in any solution produced being different, at least in detail from the one initially specified. Evolution of a labial ridge in the phrynosomatid lizard *Uma* provides an example of the effects of such transient selection (see p. 161).

Changes in selection influencing whether adaptation may occur

Even if suitable precursor traits exist, the selective regime may not be able to alter them in the appropriate direction to produce the adaptive solution under consideration. The factors responsible for such inflexibility influence not only changes necessary to convert precursors to the adaptation, but also earlier changes necessary to produce the precursors themselves if they are not already present. Included here is counter-selection: for instance if a precursor trait becomes subject to selection pressure which perhaps maintains its present form but is unconnected with the problem under consideration, later selection for the specified adaptation may not override this. Phenotypes in which the precursor trait tends towards the condition necessary for the specified adaptation may be at an overall disadvantage relative to ones with the initial state. In this situation, loss of the counter-selection will facilitate the specified adaptation while its initial appearance will constrain it. Lamellibranch gills provide a example where selective pressures are likely to conflict, restricting further adaptive change. The gills in these molluscs are employed both for respiration and feeding and the requirements of each of these uses probably limit alterations that would be beneficial in the context of the other.

Counter-selection may not only act directly on a precursor but through one or more other traits to which this is functionally coupled, either directly or through an intermediate feature. For example, the hyolingual system in primitive salamanders is involved in two activities: in tongue projection and in the action of a buccal pump that ventilates the lungs. Because lungs and tongue have their own conflicting hyolingual requirements, evolution of each is restricted by the other. The development of coupling may restrict a possible adaptive solution; conversely, loss of coupling (decoupling) may facilitate its development. For instance the functional coupling of tongue projection and lung ventilation is disrupted in lungless salamanders (Plethodontidae), which no longer have pulmonary ventilation. Here, tongue morphology has been 'released' and subsequently two elaborate projectile feeding patterns, in which the tongue may be rapidly and accurately extruded over long distances, have each evolved three times (Lombard & Wake, 1977; Wake & Larson, 1987; Roth & Wake, 1989). Loss of pulmonary respiration has consequently facilitated evolution of changes in the tongue although it has not directly caused them. In the previous example, coupling has relatively symmetrical effects, both of the linked organs suffering substantial constraints on their evolution. But asymmetrical restriction is also possible, where one structure may be limited by the function of another without the reverse being true.

Some changes in a precursor trait may alter the likelihood of later selective alteration. The changes may themselves be the result of direct selection or they may arise incidentally from adaptive change in another feature. For instance, it has been suggested that spread of the caudifemoralis muscle along the tail base in lizards, possibly selected for in connection with rapid locomotion, reduces the amount of tail that can be shed by autotomy, so the latter antipredation mechanism becomes less effective and is consequently more prone to suppression (Russell & Bauer, 1992; but see also Arnold, 1994b).

Changes in developmental capacity influencing whether adaptation may occur

As well as events producing changes in selective control of a precursor, there are ones that alter developmental capacity. These include things like ontogenetic and genetic events that prevent the generation of deviant phenotypes necessary for the evolution of a specified adaptation or facilitate their appearance. This topic has attracted great attention (see for example Maynard Smith *et al.*, 1985). There are resemblances between effects on adaptive capacity at the gross phenotypic and at the developmental and genetic levels. Thus asymmetric coupling is paralleled by the development of some features being dependent on the presence of others, but not vice versa (part of Riedl's concept of burden; Riedl, 1978), and presence or absence of precursor traits is analogous to the presence or absence of the mutations necessary to produce them. Coupling in the phenotype is echoed by linkage at the genetic level. Changes in organisms that produce alterations in developmental capacity appear likely to interact with those leading to other lineage effects in a complex way. For instance, a precursor trait may be constrained by selection, developmental effects or both. Similarly, it is possible to envisage coupling where one of the traits involved is constrained by developmental rather than selective effects.

Times of origin and action of lineage effects

Lineage effects on adaptation often originate from events that occur before the selective regime poses the problem concerned, but they may also originate afterwards. For instance, an event facilitating development of an adaptation may occur only sometime after the selective regime has changed, resulting in a delay between the advent of a problem and a solution.

Even if an initial solution has developed (either by adaptation or co-option), events on lineages may act subsequently, influencing if and how any modification of the intial solution occurs. Although occurring after an adaptative event, such influences are still lineage effects with respect to the possibility of an ultimate solution different from the initial one and produced by further adaptation.

The initial solution might be improved by further adaptation made possible by subsequent facilitation of traits that increase its effectiveness; for instance this might arise from the occurrence of new mutations. Such facilitation after the initial solution may occur even though the problem remains constant, and should be distinguished from cases where the the solution becomes more sophisticated in response to the problem becoming more severe.

Alternatively, selective forces unrelated to the problem under consideration may alter the organism in such a way that the original solution becomes less effective, and there is then selection pressure for its modification to restore efficiency. For example, sand-diving techniques in the horned lizard (*Phrynosoma*) lineage appear to be altered in response to changes in body form (see p. 161).

Finally, although late facilitation or subsequent changes in other parts of the organisms may alter the initial solution, other events may act to prevent modification. In particular, the initial solution itself, or a later variation of it, may represent an adaptive peak and constrain change. Selection for alteration towards a more effective solution may not take place because phenotypes tending in this direction are at a disadvantage compared with ones possessing the original solution. A possible case involves the transfer of part of the eye into the buccal cavity when rock lizards enter crevices and the eye has to be stored within the restricted confines of the flattened head (page 139). Two distinct routes of movement are used by different species, and one of these may possibly be superior to the other, but change between them is unlikely because the routes are separated by bone and intermediates could not transfer the eye at all. It is of course possible for an organism to move between such peaks, but only if the adaptive landscape changes temporarily, for instance if the initial problem disappears for a time.

Wider lineage effects

So far, lineage effects have been discussed in relation to a solution to a specific problem in one taxon, but they may also be considered in a wider context since they can act on a broad range of descendent species, often producing a variety of different results. When events on a lineage, affecting a particular organismal aspect, facilitate the evolution of greater diversity in that and related aspects among a range of descendent forms, they are called *key innovations* (in the sense of Lauder & Liem, 1989). These may may include such phenomena as decoupling, duplication, and the development of new features that increase the capacity for associated features to change; conversely, other events may constrain future diversity.

USING FUNCTIONAL ANALOGUES TO INVESTIGATE HISTORICAL INFLUENCES ON ADAPTATION AND THEIR ORIGINS

As with exaptations, the origins of lineage effects on adaptive capacity and on any solutions produced, can in principle be investigated by plotting traits and other relevant information on the phylogeny of the taxon concerned and its relatives. This may include the gain, loss and change of the relevant precursor traits of the potential adaptation, including behavioural and functional ones, changes in developmental capacity, and alterations in environmental parameters and other aspects of the selective regime. Once this is done, the relative order of origin of these factors can often be established and suppositions about their causal relationships made, which may then be independently testable.

One way of using this approach is to take a particular event on a lineage, predict its likely effect on descendent forms, and then see if this has in fact occurred. Thus, Lauder & Liem (1989) suggest a procedure for testing hypotheses of key innovation or constraint acting in a descendent clade. A causal model is set up predicting how a particular evolved trait will lead to subsequent diversification or result in restriction of diversity. This hypothesis is then tested by comparing perceived events in the clade originating from the internode in which the trait evolves with those in its immediate outgroup, and perhaps in others. Although this procedure can corroborate the hypothesis proposed, it cannot easily falsify it, for failure to find greater diversity, whether in the descendent clade where it is predicted or its outgroup, could easily be due to extinction reducing this. The possibility also exists that other unconsidered traits may be involved in producing perceived events.

This proposed method can also be applied to cases involving possible effects in a single descendant. It appears procedurally analogous to experimental investigation in which results are compared for cases which differ in their initial conditions. However, not infrequently it disguises the fact that results are known beforehand. For instance, the search for a key innovation or constraint is often prompted by differences already noted between the compared groups. The massive diversity in trophic mechanisms of cichlid fish was appreciated long before the possible role of the pharyngeal jaws as a key innovation promoting this was considered (Liem, 1973).

In fact, when dealing with events that already lie entirely in the past, there is no reason why investigation should inevitably proceed from potential cause to its possible effect. This approach is forced on experimentalists because they need to modify initial conditions and the results of this process inevitably lie out of immediate reach in the future. In contrast, when dealing with events that are totally historical, causes and their results are in principle equally accessible throughout the investigation. Provided effects are not wildly non-linear, it is just as feasible to predict the existence of a cause from later events and then look for it. In fact this technique is more appropriate for investigating the effects of history on adaptation, for the questions asked usually stem from final situations rather than the earlier events from which they arise.

Working from cause to effect does have the advantage that the area of phylogeny to be investigated is generally quite limited, consisting of just the descendent clade and one or more immediate outgroups. In contrast, asking what historical events have determined why a taxon exhibits particular traits may involve occurrences back to the origin of life. This lack of restriction can be avoided by making such questions comparative, inquiring for instance why does taxon A differ from taxon B in a particular respect? If the difference is due to history, the causal event or events will lie in a limited phylogenetic area: the individual lineages of the two taxa as far back as their first common ancestor.

One problem of such an approach is ensuring that differences between taxa are really products of history rather than stemming from differences in their current environments. A way in which this can be done is to consider functional analogues, taxa which have independently solved a specific near-identical problem posed by their selective regime. If the problem is really essentially the same, any differences in its solution or in whether it is solved at all will be due to to the effects of historical events in the individual lineages of the species concerned.

It is usually not possible to make detailed predictions about the historical events that may have resulted in differences in solution, or whether solutions have arisen at all. All that can be hypothesised initially is that such events did occur and that they have a clear causal connexion with the differences encountered. Corroboration can be attempted by taking taxa in pairs and considering all known changes, both intrinsic to the organism and external, on their individual lineages as far back as their last common ancestor. Of course, many changes will be unknown, often including those that are difficult to collect information about, such as alterations in developmental pattern or precise changes in selective regime, and also those changes that are undetectable because taxa incorporating evidence of their existence that would otherwise appear on side branches of the considered lineage have become extinct. This potential incompleteness of data means that, as with postulating the later effects of earlier events, hypotheses proposing that differences in solution result from historical events may sometimes be corroborated, but cannot be refuted.

Given this uncertainty, it might not seem worth carrying out such investigations, but in practice events that seem likely to be largely responsible for different outcomes in different taxa are often encountered, and it is frequently possible to determine if they acted before or after the onset of a problem and its solution if any. Furthermore, once a possible causal event has been identified, the case for its role may be open to test and possible refutation. Thus, on the lineage of the agamid lizard genus *Phrynocephalus*, a very blunt snout arose before the evolution of sand-diving and seems likely to have prevented the development of the forward submergence that occurred in most other sand-divers. In this case, experiments can determine whether the shape of the head really impedes movement into sand and the causal hypothesis is also open to refutation from other sources: a revision of the phylogeny may indicate that head shape did not change before sand-diving occurred; alternatively, investigation of other apomorphies might show that other factors may have been more important.

Although the method using functional analogues can be applied to just a pair of taxa, there are advantages in simultaneously considering several, especially where varied outcomes have been produced, and subjecting them to pairwise comparison. In this situation one or more of the taxa are likely to have evolved a solution to the problem, which helps confirm that it actually exists and is inherently solvable, something that may not be certain in a single case. The range of outcomes also indicates that a variety of historical influences are involved. Comparison is likely to be most effective when dealing with relatively similar forms; if there are many differences between them, it may be difficult to decide which ones are likely to be critical in producing different outcomes. Cases where outcome is essentially the same should also be compared as this may have been achieved in different ways, for instance there may have been convergence of traits from rather different precursors, or the same solution may have been attained by exaptation in one case and adaptation in another.

The success of recognizing lineage effects and their origins depends partly on selecting an appropriate problem posed by the selective regime. This should be relatively simple as it is then easier to define clearly and more easily recognized as essentially the same in different cases. There should also be as many instances as possible where different taxa have been independently exposed to the problem.

Ideally, the taxa concerned should belong to clades with robust phylogenies and it is advantageous if lineages have numerous side-branches. This allows situations at many time points on the lineage to be reconstructed and often enables more changes in more features to be recognized more certainly (Arnold, 1990). The genealogical relationships of such local phylogenies to each other should also be known. Ecological analogues are one of the main potential sources of such fruitfully investigable problems, but many other examples undoubtably exist where different organisms have been exposed to similar problems posed by their selective regimes.

A protocol for investigating the origins of historical influences on adaptation using functional analogues

A procedure for investigating historical influences on adaptation is set out below. It may not always be possible or appropriate to complete all the sections, especially if the requirements just given are not substantially fulfilled. Nor do the sections always need to be carried out in precise sequence, with all taxa being examined before proceeding further.

1. Select a suitable problem. To keep the investigation to a manageable size, it is usually necessary to restrict consideration to a particular taxonomic group and specify the problem closely.
2. Identify cases where different taxa have apparently separately encountered the problem. This can often be achieved without a full phylogeny of all the forms involved. For instance, where the organisms concerned belong to different clades, they can be established as of separate origin if they are not basal within these.
3. If possible, incorporate the taxa concerned into a single continuous phylogeny, so that the sections of their histories that should be used in pairwise comparisons can be determined.
4. Consider cases where a solution to the problem is actually produced. If different solutions occur, confirm that the taxa concerned have really been exposed to essentially the same problem, rather than different ones. For instance, lizards that habitually hide in rock crevices may be very flattened and have smooth skins, or be more robust with large spiny scales supported by osteoderms. These morphotypes could be interpreted as different solutions to the same environmental problem, but they turn out to be associated with crevices of different width relative to the size of the animal: the very flattened, smooth species use narrow fissures while the robust spiny ones occur in large cracks.
5. In each individual case, determine the position on the lineage of the taxon concerned at which any initial solution produced arises. It may then be possible to distinguish events that may produce lineage effects on the initial solution from those resulting in its subsequent modification or restriction. The latter events can only be recognized as distinct if there are one or more internal nodes on the phylogeny later than the initial solution.
6. Separate possible adaptations from clearly supported exaptations.
7. Consider provisional adaptations. Conduct pairwise comparisons of independent cases, whether the solutions differ or not. In each instance check individual lineages of the two taxa concerned as far back as their first common ancestor for events likely to have facilitated or constrained evolution of initial solutions.

This should be done even if the adaptive solutions are essentially the same, as even here, there may be differences in origin. The hypothesized effect on solutions of traits produced by such events can sometimes be tested experimentally.

8. Examine cases of exaptation. These may sometimes arise from more than one preaptation. Locate origins of each preaptation and determine whether it arose in the individual lineage of a case, or in a common ancestor of two or more cases. If the latter, check if the preaptation was co-opted in all descendent cases. Where this has not occurred, various sometimes overlapping alternatives exist including the following: no solution; an adaptive alternative; or co-option of a different precursor which may have been derived from the original one or not. Finally, investigate historical origins of such differences and of complex exaptations. These may turn out, for example, to comprise two or more sometimes independently co-opted traits with or without adaptive modifications.

9. For both apparently adaptive and exaptive initial solutions, consider any lineage effects occurring after the initial solution. Included here are late facilitations, alterations in the solution resulting from other changes in the organism after the initial solution, and constraint produced because the initial solution represents an adaptive peak.

10. Look for possible parallel causes. The case for a particular kind of event producing a specified effect on a solution, either prior or subsequent to its origin, may be enhanced if it is associated with the same effect in different independent cases.

11. Where there are two or more solutions, it may be possible to compare their relative efficiencies. If these are different, the question arises as to why a less effective solution does not evolve towards one that is more so. The possible role of the various lineage effects already discussed in preventing the evolution of greater efficiency should be considered. Studies of relative efficiency may also show if direct adaptation always produces the most effective solution to a problem, or whether this may sometimes arise by a more circuitous evolutionary route.

12. Examine cases in which no solution is apparent. In each, check if the organism has really been exposed to the studied problem. It may, for instance, turn out to occupy a microhabitat where the problem does not exist or this may be avoided in some other way. Other possibilities are that the problem was solved and the solution abandoned or, if failure to solve the problem is often non-lethal, that the relevant selective regime was encountered but no solution evolved. Such cases should be contrasted with forms that have developed a solution and possible causes of the different outcomes looked for. Any other cases of failure to find a solution should also be compared to see if causes are the same.

AN EXAMPLE: PREDATOR AVOIDANCE BY SAND-DIVING IN LIZARDS

Lizards living in open aeolian sand environments are often exposed to predators in situations where refuges such as burrows or vegetation are unavailable or distant.

One widely adopted strategy for evading pursuit in these circumstances is to submerge rapidly in the loose sand. Lizards that have adopted this behaviour had to solve the problem of entering a dense and only semifluid medium as quickly as possible and moving some way within it. The problem of effective sand-diving is a suitable one for investigating historical influences on adaptation as it is relatively simple and well defined. Loose aeolian sand of the sorts customarily utilized by diving lizards is quite consistent in its range of physical properties (Bagnold, 1941), so animals entering it will meet with similar difficulties. Sand-diving has evolved at least nine times in surface-dwelling lizards assigned to six different families, and there are also instances of lizards living in situations where such behaviour appears appropriate but does not take place.

Lineages which have evolved along an ecological continuum into stringent and demanding environments often have robust phylogenies based on morphology, and have many side branches, and this gives, as already discussed, a good basis for reconstructing the history of the taxa concerned. This is true of the groups considered here where lineages have progressed from mesic to arid aeolian sand habitats. Recent studies of higher lizard relationships (Estes, De Quieroz & Gauthier, 1988; Etheridge & De Queiroz, 1988; Arnold, 1989b) make it possible to incorporate the groups into a general phylogeny. Comparisons of different sand-diving methods and an analysis of their origins are given in detail elsewhere (Arnold, in press) and are consequently only summarized here.

Methods of sand-diving in different groups

The methods of sand-diving adopted by different lizard groups are highly varied. In the advanced members of the south-west African lacertid genus *Meroles*, lizards dive head first into the sand using all their limbs, and flexing the head and body laterally into serpentine waves, the tail being buried by broad whip-like sweeps (Fig. 7). In contrast, the agamid *Phrynocephalus*, of central and south-west Asia, descends vertically. The body is flattened and oscillated laterally by alternate extension and contraction of the limbs on each side, and is tilted downwards in the direction of each sideways thrust so that sand is pushed onto the back. The head and tail are subsequently buried separately by lateral movements (Fig. 8).

Like the advanced *Meroles*, the phrynosomatid *Uma*, from California and adjoining Mexico, dives into the sand head first, but the way of doing so is different. Initially the head is twisted rapidly about its sagittal axis like an awl and forward motion is largely produced by the hindlimbs, the forelimbs being quickly laid back along the flanks. Lateral movements are confined to low-amplitude waves passing from the head to the fore-body, which contribute to forward progress by moving the shoulders backwards and forwards alternately, so areas of large posteriorly-directed scales that they bear act like a pair of ratchets. Finally, the distal tail is concealed by a very rapid, small-amplitude vibration (Fig. 9). Advanced members of another phrynosomatid genus, *Phrynosoma*, in North America, such as *Phrynosoma modestum*, also bury themselves rapidly in aeolian sand. Here the head is swung from side to side and then pushed under the surface, the forelimbs again being soon laid along the flanks as the lizard moves forwards powered by the hind legs. Once the foreparts are buried, the hind quarters are rapidly oscillated from side to side, so that they sink vertically (Fig. 10).

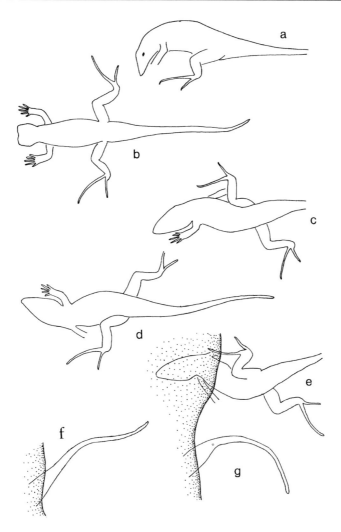

Figure 7 Sand-diving in the lacertid *Meroles cuneirostris*: (a, b) initial position – limbs turned forwards to some extent and the snout turned steeply downwards; (c, d) use of both pairs of limbs; (e) forelimbs folded back as they enter the sand; (f, g) successive positions of the tail in the final stages of burial.

In the skinks of the genus *Scincus*, of Arabia and North Africa, animals again enter the sand head first, rapidly laying both pairs of hind legs along the flanks. After this they progress by rapid high-amplitude sinusoidal movement of the body, and swim through the sand rather like a fish in water (Fig. 11). Sand-diving using this method has been independently acquired by a number of other scincids, including the *Chalcides-Sphenops* clade in the north African region, and in south-west Africa, similar behaviour also occurs in the cordylid, *Angolosaurus skoogi*. Some lacertids, like advanced *Acanthodactylus* of North Africa and south-west Asia live alongside other diurnal, surface-dwelling aeolian dune taxa that habitually dive into the sand, such as *Phrynocephalus* and *Scincus*, but not do so themselves.

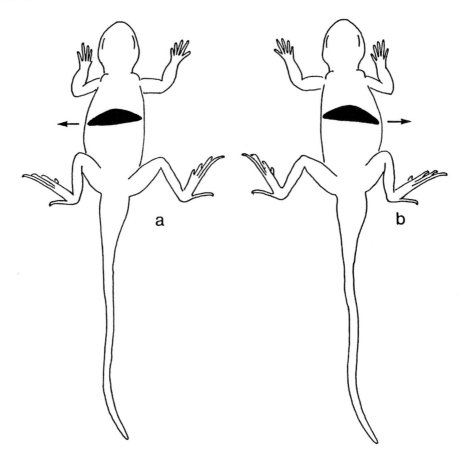

Figure 8 Sand-burial in the agamid *Phrynocephalus*: (a) the body is tilted and extension of one set of limbs and retraction of the other pushes one sharp lateral edge into the sand, so some of this is moved on to the back; (b) this movement is repeated in the other direction and the reciprocating action rapidly repeated, so that the body quickly sinks beneath the surface. Subsequent rapid lateral movements bury the head and then the tail.

Historical origins of differences in sand-diving technique

Phylogenetic information confirms that all the occurrences of sand-diving just described arose independently. The genealogical relationships of the taxa concerned and so the areas of the phylogeny that need to be compared to investigate the origins of their different approaches to sand-diving, are shown in Fig. 12. In no instance is there evidence that the different methods employed reflect different mechanical problems. Sand-diving appears to have developed in association with entry into aeolian dune habitats in advanced *Meroles* and *Phrynocephalus*, so direct adaptation may have been involved in these cases. In the other four instances, there is clear evidence of exaptation, involving two quite different precursors.

The phylogeny of the species of *Meroles* and their relatives is shown in Fig. 13, together with information on habitat, usual means of predator evasion and refuges used when inactive. Advanced forms use their method of sand burial not only to rapidly evade predators by day but also to conceal themselves when inactive,

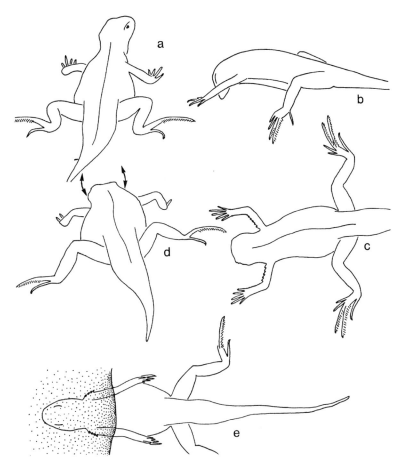

Figure 9 Rapid sand-burial in the phrynosomatid *Uma scoparia*: (a) before diving; (b, c) head lowered and limbs forward at beginning of dive; (d) initial penetration of sand with head being twisted round its sagittal axis; (e) foreparts buried by a combination of straightening of body, low-amplitude lateral movements of head and of shoulders and upper forelimbs with their coarse projecting scales, and action of hindlimbs; final rapid low-amplitude vibration of tail.

especially at the end of their daily activity. However, as it arises first on the lineage of advanced *Meroles* as a means of fast predator avoidance, the burial mechanism is likely to have evolved in this context. This is supported by the fact that the movements used in sand-diving are a modification of normal running motion used in flight. So here sand-diving may be a direct adaptation to predator avoidance, even though the actual method is substantially co-opted from running movements. Presumably the use of sand-diving for burial when inactive is also an exaptation.

At least two other lacertids, *Ophisops elegans* and *Psammodromus hispanicus*, have independently evolved burial behaviour, but in *Ophisops* at least this is only used at the end of activity periods in places where it is concealed by vegetation. The method employed is quite different from advanced *Meroles*: burial is interrupted by long pauses, the forelimbs are soon laid back along the flanks and body flexure is slight. Such burial takes place in relatively dense substrates, often incorporating roots, and it may well be an adaptation to such conditions. Slow burial may be appropriate in such an energy intensive activity where speed is not essential. Laying

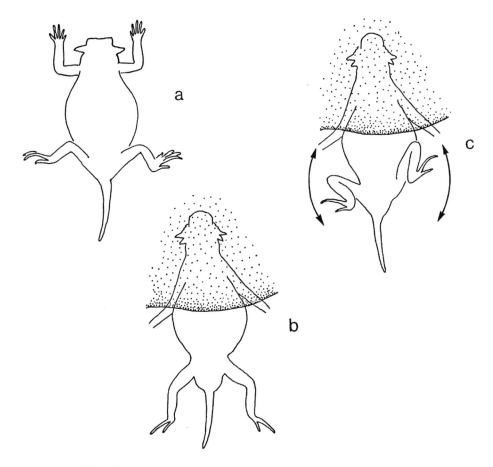

Figure 10 Sand-burial in *Phrynosoma modestum.*

the forelimbs back reduces cross sectional area of the lizard which, with small-amplitude body motions, reduces the amount of substrate that has to be displaced.

Sand-diving in *Phrynocephalus* can also be provisionally regarded as an adaptation to prey avoidance although, unlike *Meroles*, it is not certain that diving arose before nocturnal burial, and the method used is very different. When apomorphies on the separate lineages of sand-diving lizards are examined, there is one discernible event that may have been responsible for the distinctive outcome in *Phrynocephalus*. This is the development of a very blunt snout on its lineage which evolved before entry into aeolian sand habitats. Such a head shape is likely to make forward progress into sand very difficult and so may have necessitated the evolution of a submergence technique that did not require penetration using the head. Thus, the apparent historical cause of the aberrant diving pattern of *Phrynocephalus* is loss of a necessary precursor to the forward diving found in most lizards adopting this behaviour, namely a more or less streamlined snout. Other lineage effects that may reinforce this functional restriction on forward sand-diving are as yet uninvestigated. The reason for the evolution of a very blunt snout in the *Phrynocephalus* lineage is unknown, as is whether it is maintained by selection pressure or by developmental

Figure 11   Sub-sand locomotion in the scincid, *Scincus mitranus.* As the foreparts enter the sand, the pairs of limbs are successively laid back along the body, which is thrown into strong lateral sinusoidal waves that carry the lizard forward until it is completely buried.

restrictions that would counter change towards a more streamlined form. In species that dive forwards into the sand, the initial streamlining of the head becomes more marked and continuous lateral ridges develop on the upper labial scales, making movement of the head through the sand easier. As might be expected, this head configuration has not developed in *Phrynocephalus.*

The phylogeny of the phrynosomatid clade including the aeolian sand-diving *Uma* and advanced *Phrynosoma* is shown in Fig. 14. Members of nearly all the main branches are capable of burial in the substrate, but not those constituting the most basal one (*Petrosaurus*), which suggests burial arose in the internode leading to the rest of the clade. In the first branch where it occurs (*Uta–Urosaurus–Sceloporus*) burial is generally similar to that of the lacertid *Ophisops*. It is carried out clandestinely in relatively firm substrates before a period of inactivity, rather than in the context of immediate predator evasion. Burial is intermittent, the forelimbs are soon held to the sides and body flexion is relatively slight. The only major difference from *Ophisops* is that the distal tail is submerged by being rapidly vibrated. As with that genus, this burial pattern may well have arisen adaptively for the use it has in the *Uta–Urosaurus–Sceloporus* clade but many of its characteristics, including tail vibration, have later been co-opted into the mode of sand-diving used by *Uma.*

Sand-diving in *Uma* has other distinctive features not found in the foregoing burial mode, including sagittal twisting of the head. Behaviour of species constituting the *Callisaurus–Cophosaurus–Holbrookia* clade helps elucidate their origin. These animals also bury at night and possess all the features of the burial pattern found in the *Uta–Urosaurus–Sceloporus* clade, but have additionally developed the ability to bury

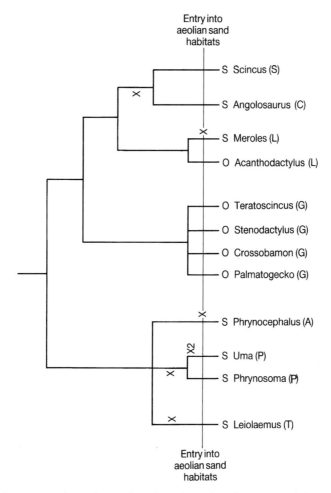

Figure 12 Phylogenetic relationships of aeolian dune lizards. S = sand-diving; O = not sand-diving; X = Approximate position where basic pattern of burial first developed, X2 subsequent modification of basic pattern in *Uma* lineage; A = Agamidae; C = Cordylidae; G = Gekkonidae; L = Lacertidae; P = Phrynosomatidae; S = Scincidae; T = Tropiduridae.

very rapidly in substrates which are more compact than aeolian sand. They exhibit the rapid head twisting of *Uma* and the upper labial scales are obliquely arranged with ridges forming a series diagonal of cutting edges. When the head is twisted the snout acts like a reciprocating drill bit enabling the lizard to enter quite compact substrates rapidly. It is likely that the pattern of sand-diving in *Uma* is directly derived from this behaviour, although there are further morphological modifications that make it more effective in loose sand, such as the development of large ratchet-like scales on the shoulders and a continuous labial ridge.

Apart from the distributions of relevant traits on the phylogeny, there are functional reasons for believing the *Uma* lineage passed through a phase where rapid penetration of firm substrates occurred. The head twisting that occurs in this genus is not an essential part of aeolian sand-diving, for none of the other lizards with this escape behaviour uses it. *Uma* is also unique among aeolian sand divers in

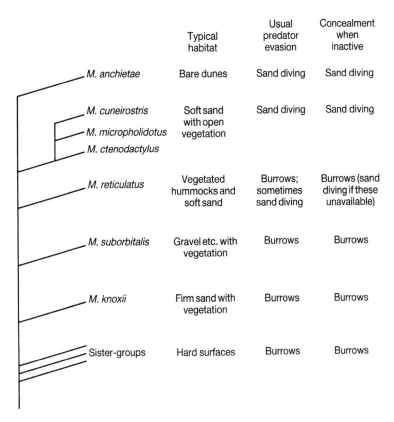

	Typical habitat	Usual predator evasion	Concealment when inactive
M. anchietae	Bare dunes	Sand diving	Sand diving
M. cuneirostris	Soft sand with open vegetation	Sand diving	Sand diving
M. micropholidotus			
M. ctenodactylus			
M. reticulatus	Vegetated hummocks and soft sand	Burrows; sometimes sand diving	Burrows (sand diving if these unavailable)
M. suborbitalis	Gravel etc. with vegetation	Burrows	Burrows
M. knoxii	Firm sand with vegetation	Burrows	Burrows
Sister-groups	Hard surfaces	Burrows	Burrows

Figure 13   Phylogeny of *Meroles* (Lacertidae), showing changes in habitat, usual predator evasion and concealment when inactive.

possessing a lateral ridge on the snout made up of diagonally elongated upper labial scales, rather than roughly rectangular ones. Although having no functional advantage in aeolian sand, these two features form an important part of the drilling mechanism used in firm substrates: diagonal elongation of the upper labial scales increases the length of the cutting ridges and, as already noted, head twisting operates them. This being so, it is much more likely that these features were selected for in this context rather than aeolian sand. The unique head movements and structure of the lateral ridge of *Uma* among aeolian sand-divers provide an example of how a transient period of selection may result in modified solutions to problems.

The phrynosomatid phylogeny suggests that sand-diving in advanced *Phrynosoma* also arose from the burial pattern adopted before periods of inactivity by members of the *Uta–Urosaurus–Sceloporus* clade. Many elements of this are still present, such as putting the arms rapidly to the sides and small-amplitude lateral movements. However, the failure to achieve complete forward burial by this method and the incorporation of rapid lateral oscillation of the hindquarters so that they sink vertically appear to result from post-solution lineage effects acting on the original pattern, and resulting from later changes of habitus in *Phrynosoma* including the development of a blunt snout and very plump body (the latter probably associated with adoption of myrmicophagy).

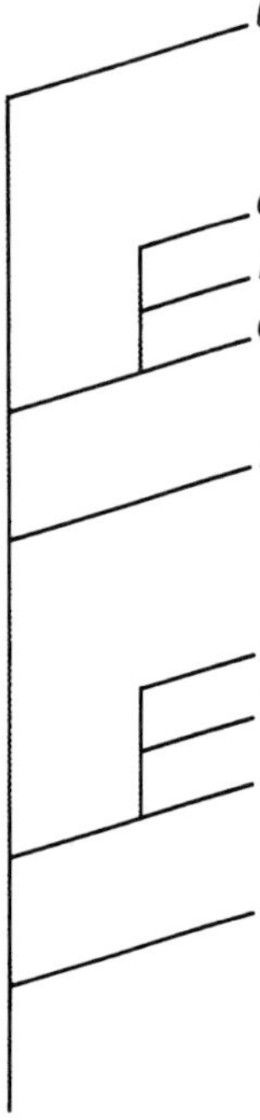

	Typical habitat	Flight behaviour includes sand diving	Night retreat includes submergence in substratum	Burial components
Uma	Ground-dwelling on aeolian dunes	yes	yes	shimmy, head rocking, tail vibration
Cophosaurus Holbrookia Callisaurus	Ground-dwelling on firmer, often sandy substrates	yes	yes	shimmy, head rocking, tail swinging and vibration
Phrynosoma	Ground-dwelling on sandy substrates	sometimes	yes	lateral oscillation
Sceloporus Urosaurus Uta	Ground-dwelling or climbing on firm substrates, vegetation	not usually	often	shimmy, tail vibration
Petrosaurus	Climbing on rocky surfaces	no	no	–

Figure 14 Phylogeny of *Uma, Phrynosoma* (Phrynosomatidae) and their relatives, showing changes in habitat, escape behaviour and nocturnal concealment.

The various sand-diving skink groups and the cordylid *Angolosaurus* exhibit the same method of forward burial, but this pattern of locomotion with both pairs of limbs laid back and high-amplitude body movements is widespread in the two families as a whole, often being used for locomotion on the surface, and for concealment in periods of inactivity among litter or in the substrate. This distribution suggests that the pattern arose in a common ancestor of both families and was later co-opted with little if any modification for use in loose sand habitats (although the forms concerned developed convergent morphological features, such as stream-lined snouts with lateral upper labial ridges).

As noted, some species of the lacertid genus *Acanthodactylus* occur in the same aeolian dune habitats as regular sand-divers like *Scincus* and *Phrynocephalus*, yet do not exhibit diving behaviour themselves. One possibility is that they use an alternative technique of predator avoidance that has made the evolution of sand-diving unnecessary. In fact, *Acanthodactylus* usually respond to pursuit by running into ready-made burrows at the base of often small shrubs, but possession of this strategy does not preclude sand-diving, for another lacertid, *Meroles reticulatus* uses both techniques. Another possibility is that, in spite of occurring in the same environments as sand-diving species, advanced *Acanthodactylus* occupy a structural microhabitat where this behaviour is unnecessary. This may well be the case.

Among sparsely vegetated sand dunes near Sharjah, in the United Arab Emirates the mean distance of animals when first sighted from shrubs (and therefore burrows) was only 0.97m in *Acanthodactylus schmidti*, compared with 6.21m in *Scincus mitranus* and 4.67m in *Phrynocephalus arabicus* (data from Arnold, 1984). As many habitual sand-divers only use this technique when hard pressed or far from cover, the usual closeness of *Acanthodactylus schmidti* to its burrows may make such behaviour unnecessary. It is of course possible that there are lineage effects that are responsible for the lack of sand-diving in *Acanthodactylus* compared with its close relative, *Meroles*, and with other sand-diving forms. However, when the individual lineage of *Acanthodactylus* is compared with that of each of the sand-diving forms, multiple spread into really open sandy habitats by the latter are the only discernible events that provide an explanation as to why they submerge in loose sand and *Acanthodactylus* does not.

Summary of historical influences on sand-diving in lizards

Thus, among sand-diving lizards, a range of historical influences can be discerned that can be traced provisionally to specific events. In only advanced *Meroles* is sand-diving developed straightforwardly as an apparent adaptation to evading pursuing diurnal predators, and even here the actual method of burial adopted represents an exaptation, albeit with modification of a pre-existing behaviour pattern. In *Phryno-cephalus* simple forward burial is prevented by loss of an appropriate precursor, necessitating an alternative method. In the four other independent cases examined, sand-diving is an exaptation of burial during periods of inactivity, ultimately derived from only two preaptations which, however, are very different. In skinks and cordylids, the method concerned is co-opted with little modification, but in phryno-somatids events have been much more complex. Thus sand-diving in *Uma* is derived ultimately from a primitive phrynosomatid pattern associated with inactivity, but this is first co-opted and modified to rapid penetration of firm substrates during predator avoidance before being co-opted and modified once more in the context of aeolian sand. Some of the behavioural and structural modifications produced by transient selection for rapid penetration of firm substrates influence the final diving pattern and associated morphology in *Uma*. Although sand-diving in *Phrynosoma* has the same initial origins as that of *Uma*, it resembles *Phrynocephalus* in having a strong vertical component functionally associated with head shape. However, this similarity develops in quite different ways. In *Phrynocephalus* change in head shape restricted the initial adaptive solution, whereas in *Phrynosoma* head shape modified the solution only after it developed.

CONCLUDING REMARKS

Although developmental constraints on adaptation tend to be emphasized, many other kinds of lineage effects exist. Such historical influences have long been recognized in a general way, especially when considering differences between old and disparate groups such as phyla, but the recent availability of detailed phylo-genies and methods for integrating these with other kinds of data, means that comparative approaches can be used to investigate lineage effects among clades that are much more alike. However, as with other uses of phylogenies to reconstruct history, the recognition of lineage effects and the events leading to them is not always straightforward. The approach depends on the availability of phylogenies that are robust and in which, at least, those component hypotheses of relationship critical to the matter under investigation are well supported. Furthermore, even when confidence can be placed in a phylogeny, its topology influences what can be said about trait distribution on it and with what degree of certainty: lineages with many side branches are likely to be more informative than others. Regrettably, there is evidence that phylogenies based on morphological data that are robust and exhibit usefully pectinate topologies may often have had rather special ecological histories. If these are concentrated on, only a subset of evolutionary patterns may be being considered (Arnold, 1990).

A further cause of difficulty is changes on side branches that cause inappropriate conditions to be reconstructed on lineages, if the original condition is no longer represented or only inadequately so. Finally, extinction of taxa constituting side branches of a lineage will obscure the history of changes in traits and their relationship to alterations of other features. Problems with the robustness of a phylogeny and the effects of its topology at least have the advantage that they can be evaluated, and such assessments can be used to decide how much confidence can be placed on conclusions drawn. In contrast, extinction leaves no certain traces, yet it is widespread and can change interpretations radically, as is easily demonstrated by artificially removing species or subclades from a phylogeny. One way in which robust phylogenies of clades arise is by extinction of some branches, so remaining relationships are supported by more synapomorphies. Thus, frustratingly, the robust phylogenies needed for historical investigations may often be paid for in terms of extinct branches, which can result in misleading interpretations.

For the reasons just given, it is unlikely that there will ever be anything approaching a total picture of the effects of history on the evolution of current organisms. But investigations to date suggest that there are many parts of the tree of life where robust phylogenetic results can be obtained and these are very much worth investigating. Use of DNA sequences may increase the range of well-substantiated phylogenies available, and may circumvent the problem of morphological data producing much less strongly supported phylogenies in some ecological situations than others.

Although it does not make it certain, concurrent appearance of many apomorphic features at a node at least raises the possibility that branches may have been lost prior to it. Also, while extinction can cause information to be lost and misinterpretations to be made, trait distributions on phylogenies where it may have had this effect are often recognizable. Conversely some situations cannot have easily been produced by extinction remoulding apparent trait distribution. It is also possible to avoid compounding the problem of extinction by omitting known taxa the inclusion of which might effect conclusions drawn (this is a caution that applies to phylogeny reconstruction itself: in homoplasious data sets, omission of taxa sometimes reduces overt conflict but increases erroneous relationships; Arnold, 1981). Overall, it seems likely that, as in sand-diving lizards, there will be many situations where lineage effects and their origins can be usefully investigated. Certainly enough analysable phylogenies exist to get a much better idea as to how historical factors operate.

The widespread occurrence of lineage effects at low taxonomic levels beginning to be discerned underlines the complexity of the evolutionary process, and it is apparent that not only do historical influences often result in different outcomes in similar selective situations, but they may also sometimes produce similar results in quite different ways. Also, once related lineages have accumulated some differences, for instance as a result of adapting to disparate environments, these may differentially influence response to later situations, even ones that are very similar. Historical influences consequently make it more likely that divergent lineages will continue to diverge, something which often helps to make phylogeny estimation from non-molecular data a feasibility. The recognition of widespread lineage effects does not conflict with the idea that adaptation by natural selection is the primary promoter of evolutionary novelties that confer performance advantages, but it indicates that the way in which selection proceeds is likely to be heavily dictated by its own earlier effects.

REFERENCES

ARNOLD, E.N., 1970. 'Functional aspects of taxonomic characters of lacertid lizards'. Unpublished D.Phil. thesis, University of Oxford.

ARNOLD, E.N., 1981. Estimating phylogenies at low taxonomic levels. *Zeitschrift für zoologische Systematik und Evolutionsforschung, 9*: 1–35.

ARNOLD, E.N., 1984. Ecology of lowland lizards in the eastern United Arab Emirates. *Journal of Zoology, 204*: 329–354.

ARNOLD, E.N., 1989a. Systematics and adaptive radiation of equatorial African lizards assigned to the genera *Adolfus, Bedriagaia, Gastropholis, Holaspis* and *Lacerta* (Reptilia, Lacertidae). *Journal of Natural History, 23*: 525–555.

ARNOLD, E.N., 1989b. Towards a phylogeny and bigeography of the Lacertidae: relationships within an old-world family of lizards derived from morphology. *Bulletin of the British Museum (Natural History)* Zoology, *55*: 209–257

ARNOLD, E.N., 1990. Why do morphological phylogenies vary in quality? An investigation based on the comparative history of lizard clades. *Proceedings of the Royal Society of London, Series B, 240*: 135–172.

ARNOLD, E.N., 1994a. Do ecological analogues assemble their common features in the same order? An investigation of regularity in evolution, using sand-dwelling lizards as examples. *Philosophical Transactions of the Royal Society, B, 344*: 277–290.

ARNOLD, E.N., 1994b. Investigating the evolutionary effects of one feature on another: does muscle spread suppress caudal autotomy in lizards? *Journal of Zoology, London, 232*: 505–523.

ARNOLD, E.N., in press. Identifying the effects of history on aptation: origins of different sand-diving techniques in lizards. *Journal of Zoology, London.*

BAGNOLD, R.A., 1941. *The Physics of Blown Sand and Desert Dunes.* London: Methuen.

BAUM, D.A. & LARSON, A., 1991. Adaptation reviewed: a phylogenetic methodology for studying character macroevolution. *Systematic Zoology, 40*: 1–18.

BIOSIS, 1992. BIOSIS Preview, Life Sciences; Biomedia file 55 (1985 to August 1992). BIOSIS.

BOCK, W.J., 1979. The scientific explanation of macroevolutionary change – a reductionistic approach. *Bulletin of the Carnegie Museum of Natural History, 13*: 20–69.

BROOKS, D.R. & McLENNAN, D.A., 1991. *Phylogeny, Ecology and Behaviour: a Research Programme in Comparative Biology.* Chicago: University of Chicago Press.

BROWN, J.L., 1982. The adaptationist program. *Science, 217*: 884, 886.

CARPENTER, J.M., 1989. Testing scenarios: wasp social behaviour. *Cladistics, 5*: 131–144.

CLUTTON-BROCK, T.H. & HARVEY, P.H., 1979. Comparison and adaptation. *Proceedings of the Royal Society of London, Series B, 205*: 547–565.

CLUTTON-BROCK, T.H. & HARVEY, P.H., 1984. Comparative approaches in investigating adaptation. In J.R. Krebs & N.B. Davies (Eds), *Behavioural Ecology, an Evolutionary Approach*, 2nd edition, 7–29. Oxford: Blackwell Scientific Publications.

CODDINGTON, J.A., 1988. Cladistic tests of adaptational hypotheses. *Cladistics, 4*: 3–22.

CODDINGTON, J.A., 1990. Bridges between evolutionary pattern and process. *Cladistics, 6*: 379–386.

CONDON, K., 1987. A kinematic analysis of mesokinesis in the Nile Monitor (*Varanus niloticus*). *Experimental Biology, 47*: 73–87.

DARWIN, C., 1859. *On the Origin of Species*, 1st edition. London: John Murray.

DARWIN, C., 1872. *On the Origin of Species*, 6th edition. London: John Murray.

De QUEIROZ, K., 1987. Phylogenetic systematics of iguanine lizards, a comparative osteological study. *University of California Publications in Zoology, 188*: 1–203.

DOVER, G.A., 1984. Improbable adaptations and Maynard Smith's dilemma. Public lecture, Oxford, 1984.

ENGELMANN, G.F. & WILEY, E.O., 1977. The place of ancestor–descendant relationships in phylogeny reconstruction. *Systematic Zoology, 26*: 1–11.

ESTES, R., De QUEIROZ, K. & GAUTHIER, J., 1988. Phylogenetic relationships within Squamata. In R. Estes & G. Pregill (Eds), *Phylogenetic Relationships of the Lizard Families*, 283–367. Stanford, CA: Stanford University Press.

ETHERIDGE, R. & De QUEIROZ, K., 1988. A phylogeny of the Iguanidae. In In R. Estes & G. Pregill (Eds), *Phylogenetic Relationships of the Lizard Families*, 118–281. Stanford, CA: Stanford University Press.

FARRIS, J.S., 1970. Methods for computing Wagner trees. *Systematic Zoology, 19*: 83–92.

FISHER, D.C., 1985. Evolutionary morphology: beyond the analogous, the anecdotal, and the ad hoc. *Paleobiology, 11*: 120–138.

FELSENSTEIN, J., 1973. Maximum-likelihood estimation of evolutionary trees from continuous characters. *American Journal of Human Genetics, 25*: 471–492.

FELSENSTEIN, J., 1981. A likelihood approach to character weighting and what it tells us about parsimony and compatibility. *Biological Journal of the Linnean Society, 16*: 183–196.

FRANK, L.G., 1986. Social organisation of the spotted hyaena, *Crocuta crocuta*. II. Dominance and reproduction. *Animal Behaviour, 34*: 1510–1527.

FRANK, L.G., GLICKMAN, S.E. & LICHT, P., 1991. Fatal sibling aggression, precocial development and androgens in neonatal spotted hyenas. *Science, 252*: 702–704.

FRAZZETTA, T.H., 1962. A functional consideration of cranial kinesis in lizards. *Journal of Morphology, 111*: 287–319.

GARDNER, A.S., 1984. The evolutionary ecology and population systematics of day geckoes (*Phelsuma*) in the Seychelles. Unpublished Ph.D. thesis, University of Aberdeen.

GAUTHIER, J. & PADIAN, K. 1985. Phylogenetic, functional and aerodynamic analyses of birds and their flight. In M.K. Hecht, J.H. Ostrom, G. Viohl & P. Wellnhofer (Eds), *The Beginnings of Birds: Proceedings of the International Archaeopteryx Conference, Eichstatt*. Eichstatt: Bronner & Daentler.

GOULD, S.J. & VRBA, E.S., 1982. Exaptation – a missing term in the science of form. *Paleobiology, 8*: 4–15.

GREENE, H.W., 1986. Diet and arborality in the Emerald Monitor, *Varanus prasinus*, with comments on the study of adaptation. Fieldiana, *Zoology, N.S., 31*: i–iii + 1–12.

HAMILTON III, W.H. & TILSON, R.L., 1984. Social dominance and feeding patterns of Spotted hyaenas. *Animal Behaviour, 32*: 715–724.

HARVEY, P.H. & PAGEL, M., 1991. *The Comparative Method in Evolutionary Biology.* Oxford: Oxford University Press.

HENNIG, W., 1966. *Phylogenetic Systematics.* Urbana: University of Illinois Press.

IORDANSKY, N., 1966. Cranial kinesis in lizards: contribution to the problem of the adaptive significance of skull kinesis. *Zoologichesteii Zhurnal, 45*: 1398–1410 (translated through Smithsonian Herpetological Information Services).

KRUUK, H., 1972. *The Spotted Hyena, a Study of Predation and Social Behaviour.* Chicago: University of Chicago Press.

LANG, M., 1991. Generic relationships within the Cordyliformes (Reptilia: Squamata). *Bulletin de l'Institut Royal des Sciences Naturelles de Belgique, 61*: 121–188.

LAUDER, G.V & LIEM, K.F., 1989. The role of historical factors in the evolution of complex organismal functions. In D.B. Wake & G. Roth (Eds), *Complex Organismal Functions: Integration and Evolution in Vertebrates*, 63–78. Chichester: John Wiley.

LAUDER, G.V., LEROI, A.M. & ROSE, M.R., 1993. Adaptations and history. *Trends in Evolution and Ecology, 8*: 294–297.

LIEM, K.F., 1973. Evolutionary stategies and morphological innovations: cichlid pharyngeal jaws. *Systematic Zoology, 22*: 425–441.

LOMBARD, R.E. & WAKE, D.B., 1977. Tongue evolution in lungless salamanders, family Plethodontidae. II. Function and evolutionary diversity, *Journal of Morphology, 153*: 39–80.

MADDISON, W.P., 1990. A method for testing the correlated evolution of two binary characters: are gains and losses concentrated on certain branches of a phylogenetic tree? *Evolution, 44*: 539–557.

MADDISON, W.P. & MADDISON, D.R., 1992. *MacClade version 3. Analysis of phylogeny and character evolution.* Sunderland, MA: Sinauer.

MAYNARD SMITH, J., BURIAN, R., KAUFFMAN, S., ALBERCH, P., CAMBELL, J., GOODWIN, B., LANDE, R., RAUP, D. & WOLPERT, L., 1985. Developmental constraints and evolution: a perspective from the Mountain Lake Conference on development and evolution. *Quarterly Review of Biology, 60*: 265–287.

MAYR, E., 1954. Change of genetic environment and evolution. In J. Huxley, A.C. Hardy & E.B. Ford (Eds), *Evolution as a Process*, 157–180. London: Allen & Unwin.

MILLS, M.G.L., 1978. The comparative socio-ecology of the Hyaenidae. *Carnivore, 1*: 1–7.

MIVART, St. G.J., 1871. *The Genesis of Species.* London: Macmillan.

NORREL, M.A. & De QUEIROZ, K., 1991. The earliest iguanine lizard (Reptilia: Squamata) and its bearing on iguanine phylogeny. *American Museum Novitates*, (2997): 1–16.

OHNO, S., 1970. *Evolution by Gene Duplication.* New York: Springer.

POUGH, F.H., 1969. The morphology of undersand respiration in reptiles. *Herpetologica, 25*: 223–227.

RACEY, P.A. & SKINNER, J.C., 1979. Endocrine aspects of sexual mimicry in spotted hyenas *Crocuta crocuta. Journal of Zoology, London, 187*: 315–326.

RIEDL, R., 1978. *Order in Living Organisms.* New York: John Wiley.

RIEF, W.-E., 1984. Preadaptation and the change of function – a discussion. *Neues Jahrbuch für Geologie und Palaeontologie, 1984*: 90–94.

ROTH, G. & WAKE, D.B., 1989. Conservatism and innovation in the evolution of feeding. In D.B. Wake & G. Roth (Eds), *Complex Organismal Functions: Integration and Evolution in Vertebrates*, 7–20. Chichester: John Wiley.

RUSSELL, A.P. & BAUER, A.M., 1992. The *m. caudifemoralis longus* and its relationship to caudal autotomy and locomotion in lizards (Reptilia: Sauria). *Journal of Zoology, London, 227*: 127–143.

SHARKEY, 1989. A hypothesis-independent method of character weighting for cladistic analysis. *Cladistics, 5*: 63–86.

SMITH, H.M., 1946. *Handbook of Lizards.* Ithaca, NY: Comstock.

SOBER, E., 1984. *The Nature of Selection*. Cambridge, MS: MIT Press.

STEARNS, S.C., 1986. Natural selection and fitness, adaptation and constraint. In D.M. Raup & D. Jablonski (Eds), *Patterns and Processes in the History of Life*, 23–44. Heidelberg: Springer-Verlag.

SWOFFORD, D.L., 1990. PAUP: Phylogenetic analysis using parsimony, version 3.0. Available from the author: Illinois State Natural History Survey, 172 Natural Resources Bldg., 607 E. Peabody, Champaign, IL 61820, U.S.A.

SWOFFORD, D.L. & MADDISON, W.P., 1988. Reconstructing ancestral character states under Wagner parsimony. *Mathematical Bioscience, 87*: 199–229.

SZALAY, F.S., 1977. Ancestors, descendants, sister groups and testing of phylogenetic hypotheses. *Systematic Zoology, 26*: 12–18.

WAKE, D.B. & LARSON, A., 1987. Multidimensional analysis of an evolving lineage. *Science, NY, 238*: 42–48.

WILEY, E.O., 1981. *Phylogenetics. The Theory and Practice of Phylogenetic Systematics*. New York: John Wiley.

WILLIAMS, G.C., 1966. *Adaptation and Natural Selection, a Critique of Some Current Evolutionary Thought*. Princeton, NJ: Princeton University Press.

WRIGHT, S., 1931. Evolution in Mendelian populations. *Genetics, 16*: 97–159.

CHAPTER

7

Distinguishing phylogenetic effects in multivariate models relating *Eucalyptus* convergent morphology to environment

DANIEL P. FAITH & L. BELBIN

CONTENTS ──────────────────────

Keywords: Phylogeny – ecology – multivariate analysis – convergent evolution – eucalypts – morphology – environmental gradients.

Abstract

Recent methodological developments in comparative biology largely have focused on comparisons of one variable (e.g. morphological) to another variable (e.g. environmental). Here, we describe recent development and application of a multivariate approach, based on a variant of multidimensional scaling, that relates a suite of morphological variables to one or more environmental variables. This procedure has some properties typical of comparative methods.

1. Observations over diverse taxa are interpreted as providing evidence for association between morphology and environment; patterns attributable only to common ancestry must be distinguished from those attributable to convergence.

Phylogenetics and Ecology
ISBN 0-12-232990-2

2. Convergently derived character states are most informative about associations between morphology and environment.

The multivariate approach also has distinctive properties. It identifies overall patterns of correspondence between morphology and environment in which the pattern for any one morphological variable may be poor; it is only assumed that, on average, a shared derived morphological state will imply shared environment. Further, not one but several different environmental variables may be associated with the resulting scaling pattern of taxa. One limitation of this multivariate approach has been the absence of a procedure that evaluates the degree to which common ancestry rather than convergence is responsible for the resulting association pattern. Here, we describe several tests that help to determine whether the multivariate pattern is attributable to convergence over the suite of characters. In an application to highly convergent morphological data for species of *Eucalyptus* in south-east Australia, the tests show that apparent patterns of association of morphology with environment may reflect only common ancestry.

INTRODUCTION

Other papers in this volume (e.g. Coddington, Pagel, Gittleman & Luh) highlight recent developments in comparative methods that generally have focused on determining whether one variable/character is related to another variable (see also Harvey & Purvis, 1991). A key question addressed by these methods is whether the relationship found between the two variables exceeds that which could be expected merely through common-ancestry. 'Homoplasy approaches' (Coddington, Chapter 3) address this issue by using cases of convergence as the basis for independent assessments of relationship. While the procedure described and extended in this chapter also is a homoplasy approach (Faith, 1989), it departs from the usual paired-univariate framework in that it attempts to relate, not one, but a suite of convergent morphological characters to one or more environmental/ecological variables.

Our multivariate approach may also be distinctive in emphasizing, at least initially, hypothesis generation more than hypothesis testing. Comparative methods often start with an hypothesis that two particular characters are related, and then search for cases of convergence as a means of obtaining valid assessments (tests) of relationship (for review see Harvey & Purvis, 1991). The approach discussed in this chapter reverses this process. It begins with observed convergence, possibly over many characters, and the initial analysis of these data then has the goal of identifying environmental variables that can be hypothesized to explain the patterns of convergence.

Because of the exploratory nature of this multivariate procedure, not only in searching for common pattern among different convergent characters but also in examining alternative environmental variables, there has been little focus on hypothesis testing (Faith, 1989). As a homoplasy method, a test is needed that evaluates whether the correspondence between environment and morphology is a product of observed multivariate convergence, or simply reflects common-ancestry ('phylogenetic effects'). In this paper, a simple procedure for carrying out such a test is presented.

The first section of this chapter will review multivariate analysis of convergence

(Faith, 1989). The subsequent sections will explore an application of this approach to analysis of convergent morphology of species of *Eucalyptus* in south-east Australia. This will illustrate properties of the method and highlight the importance of testing for phylogenetic effects.

ANALYSIS OF CONVERGENCE

Methods in comparative biology have paralleled cladistic methods in focusing on derived character states as most informative regarding evolutionary patterns and processes (e.g. Coddington, 1988). In cladistics, shared-derived characters are assumed to indicate common-ancestry patterns; in comparative biology shared-derived characters may indicate common adaptation and/or common environment. As in cladistics, the latter interpretation of shared-derived characters can be placed in a multivariate framework. Here, the corresponding pattern summarizing covariation among characters is not a tree topology but an ordination of the taxa (Faith, 1989). While pattern-similarity of taxa within the ordination primarily reflects similarity in shared-derived characters, the ordination analysis is nevertheless complementary to cladistic analysis. In the ordination, taxa are to be linked as a result of convergently shared-derived characters; consequently, these same taxa generally will not be closely related on the phylogenetic tree.

Figure 1 outlines the initial steps in the multivariate analysis of convergence. The approach begins with a given phylogenetic hypothesis for a set of taxa (Fig. 1a). This tree is used to infer patterns of character state change for a given morphological data set for these taxa. These data may or may not be the same data used to infer the tree (Faith, 1989). We agree with Brooks & McLennan (Chapter 1) that the problem of circularity (using a phylogenetic tree to learn about the same data used to derive the tree) is of less concern than is the degree of corroboration of the tree used to make inferences about character change.

In Fig. 1a, the hypothetical cladogram of ten taxa shows the diagnosis of characters A, B, C and D. Character state optimization (see e.g. Swofford, 1992) implies that characters A and D show convergence while B and C do not. A new data set is formed (Fig. 1b) using the convergent character states only; the derived state in each case is coded as '1' and all taxa not having that state are coded as '0'. This data set is used to produce an ordination of the ten taxa (Fig. 1c). The ordination method is a particular form of multidimensional scaling (Faith, Minchin & Belbin, 1987; Belbin, 1991a,b) which arranges the taxa such that those taxa sharing convergently-derived states (sharing 1-states rather than 0-states) will tend to be adjacent in the space. A key element in such an analysis is the choice of the dissimilarity measure used to summarize pairwise relationships among the taxa, for input into a multidimensional scaling programme. If the taxa having 1-states (for a given character) are assumed to cluster together in any part of the environmental space (a general 'unimodal response pattern' in that the 1-states form a 'mode' surrounded by 0s; see Faith *et al.*, 1987), then Bray-Curtis dissimilarities are appropriate as they have been found to provide a robust analysis under this assumption (Faith, Minchin & Belbin, 1987).

For the example of Fig. 1, the solid loops enclose those taxa having the derived state for character D (left loop) and character A (right loop). In practice, the resulting

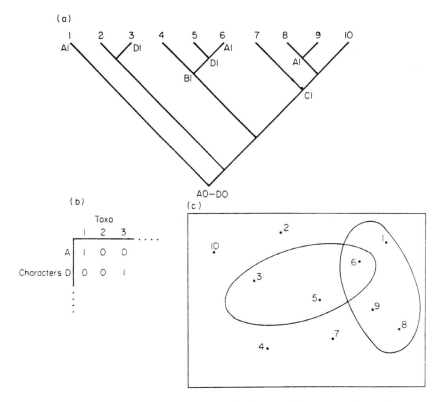

Figure 1 A reproduction of fig. 2 from Faith (1989) showing the steps in the multivariate analysis of convergence. (a) A hypothetical cladogram of ten taxa showing the diagnosis of characters A–D. A and D show convergence while B and C do not. (b) A new data set is formed using the convergent characters only; the derived state in each case is coded as '1'. (c) An ordination of the ten taxa in which taxa sharing convergently-derived states tend to cluster in the space. The solid loops enclose taxa having the derived state for character D (left) and character A (right).

pattern represents a compromise over the set of all convergent characters. As in cladistic analysis, a single pattern accounting for all shared-derived characters may not be found; this means that the taxa having some character-1-states will not be adjacent in the ordination space (Faith, 1989).

The ordination pattern, in using a small number of dimensions, should reveal patterns of variation among the taxa that in some way explain the convergence over many of the characters. One or more environmental variables then may show significant (product-moment) correlation along some projection through the ordination space. Significance can be judged initially by comparing the magnitude of the observed correlation to that found for permutations of the given environmental variable (Faith & Norris, 1989). The corresponding significance level is easily adjusted for multiple tests (over many environmental variables).

A limitation of this approach, as originally applied by Faith (1989), is that the test described above is not adequate on its own to assess the importance of convergence in revealing environmental correlates. The recovery of an environmental factor may be due to 'residual' common-ancestry information remaining in the new convergence data set. For example, in Fig. 1a, the convergently derived state for character D unites two members of the same monophyletic group (taxa 5 and 6). Consequently, the ordination pattern, in bringing taxa with the derived state for character D close together, inevitably also reflects common-ancestry information. This common-ancestry information on its own might have been sufficient to reveal the underlying environmental factor. If this is the case, then the factor may be judged to have no necessary relationship with the derived character states. In contrast, if the recovery of environment is found to be dependent on convergence, then these individual cases of convergence over different characters have provided support for some non-trivial connection between the derived character states and the environmental factor (e.g. the environmental factor or its correlates represents a selective force on the morphological character(s)). To distinguish between these two possibilities, a method for separating out the phylogenetic effects (of common ancestry) is described in the next sections of this paper.

CONVERGENT MORPHOLOGY IN *EUCALYPTUS*

The gradual progress made in *Eucalyptus* systematics over recent decades has highlighted the importance of morphological convergence in this group (e.g. Johnson, 1972; Ladiges & Humphries, 1983; Chippendale, 1988). The possible role of environment in explaining morphological variation among species has been emphasised: "A question for the future will be to determine whether there are general adaptive tendencies in the several sections and subgenera which are linked to their present distributions and the climates and conditions of their geographic origins" (Johnson, 1972:28). As part of a larger research programme at CSIRO, the goal of which is to understand the distribution and abundance of eucalypt species and communities, we have begun to explore the links between morphology and environment, based on a number of existing eucalypt data bases.

The distribution data for this study are derived from an extensive geographic data set (Austin, unpublished) for the south-east corner (approximately 40 000 km2) of New South Wales, Australia. A map of the study area can be found in Austin, Nicholls & Margules (1990). This reference also provides background information on the distribution data and associated environmental measurements at survey sites. Other data sources are described below.

Environmental, phylogenetic and morphological information

Environmental data
Based on records from the extensive geographic data base for eucalypts (Austin, unpublished; see also Austin *et al.*, 1990), values for different environmental variables measured at each sample site could be used to characterize any individual species, by summarizing those values measured over sites where the species was recorded as present. The forms of summary in this exploratory study consisted of arithmetic mean, median and standard deviation (each summarizing variation for a taxon over sample sites). These summaries resulted in a total of 109 non-invariant variables. Given the exploratory nature of the initial ordination analysis, the full suite of available environmental variables are included here (Appendix 1).

These available environmental data were used to choose a subset of eucalypt taxa for this study. A sample of 35 taxa were chosen, with the requirement that the chosen taxa represented a range of phylogenetic positions (see below), and a range of environments, as indicated by plots of rainfall and temperature values for sites where the taxa were recorded as present. This procedure provided a manageable subsample of taxa for analysis (the 35 taxa are listed on the tree in Fig. 2). The two subspecies of *E. piperita*, and the two closely related species, *E. radiata* ssp. *radiata* and *E. robertsonii*, were not distinguished in the environmental data set; consequently the assigned environmental values were the same for the two taxa in each case.

Phylogenetic and morphological information

An estimate of the phylogenetic relationships for the subset of taxa used in this study was obtained from a series of publications on different eucalypt groups (Ladiges, Humphries & Brooker, 1983, 1987; Ladiges & Humphries, 1986; Ladiges, Newnham & Humphries, 1989). Figure 2 shows these relationships (the literature-derived tree was plotted using Hennig86; Farris, 1988). The taxon names in the

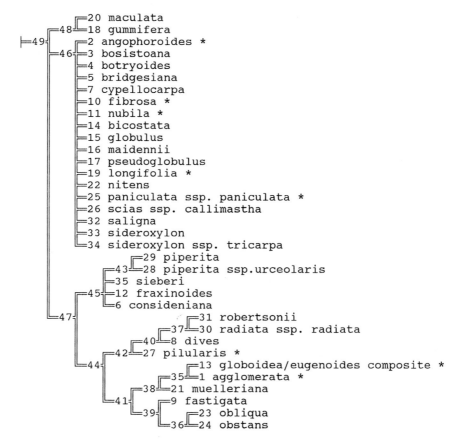

Figure 2 Estimated phylogenetic relationships among the 35 taxa used in this study. Numbers and 8-character labels identify species. The distribution of the convergently-derived states of one of the characters, 'bud-shape fusiform', is shown by the location of the asterisks.

figure are species names, all within the genus *Eucalyptus*. In the analyses that follow, taxa will be identified using the corresponding sequence numbers from the tree.

The morphological characters used in this study (Chippendale, 1988; Belbin, Gill & Chippendale, unpublished) represent a set that is to some degree independent of those characters used in the phylogenetic studies referred to above. The characters in this original data set were optimized on the given tree (Fig. 2), using Hennig86. This procedure is important in providing estimates of ancestral states and of character state changes (see also Wenzel & Carpenter, Chapter 4). The inference of character state changes resulted in a new data set consisting of 49 new characters, all showing convergence on the tree (see Appendix 2a,b).

It should be noted that these new characters do not have a one-to-one relationship with the original characters. The original characters were multistate, unordered characters; for example, the character fruit-shape had six different states (Appendix 2). Any of the states of such a character could produce a new character in the convergence data set in either of two ways:

1. The character state is shown to have arisen two or more times independently (optimization based on parsimony unequivocally excludes a single origin of the state). Common environment is hypothesized to account for these independent events; therefore those taxa that have gained this state are coded as '1' and all other taxa are coded as '0' for this new character.
2. The character state is shown to have been lost two or more times independently (optimization based on parsimony unequivocally excludes a single loss of the state). Common environment is hypothesized to account for these independent events; therefore those taxa that have lost this state are coded as '1' and all other taxa are coded as '0' for this new character.

Note that the same multistate character can produce several new characters (e.g. if both 'fruit shape–conical' and 'fruit shape–urceolate' have been independently derived more than once).

To illustrate a typical case of convergence, the distribution of the convergently-derived states of one of the character states, 'bud-shape fusiform', is shown by the location of the solid terminal branches (Fig. 2). This derived state can be inferred (conservatively) to have arisen three times.

The majority of the new characters in our convergence data set were defined as a result of independent change to the same character state (scenario 1 above). We discuss the implications of also including convergent losses in our analysis in the Discussion section below.

Concern has been expressed about the likelihood for errors in the inference of patterns of character state change, and their impact on comparative methods. One weakness of this study is that the phylogenetic relationships depicted here cannot yet be regarded as well-corroborated; any errors in tree estimation naturally can lead to errors in inferences about character change.

In addition to these problems due to errors in phylogenetic tree estimation, other difficulties will include possible reticulation among eucalypt species, and possible biases resulting from our convention for inferring convergence via character state optimization (Peter Weston, personal communication; Wenzel & Carpenter, Chapter 4). Any biases in the latter case may be accentuated through our use of only a subset

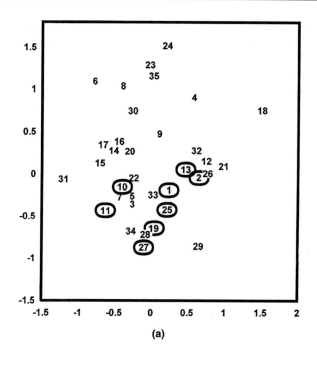

(a)

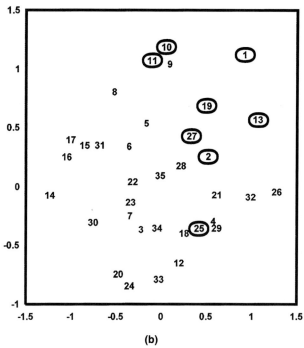

(b)

Figure 3 The first two dimensions of a three-dimensional HMDS solution. Identification numbers correspond to sequence numbers on the cladogram. The circled taxa are those having the convergently derived state for the character 'bud-shape fusiform'. Several labels on the ordination are shifted slightly to avoid overlap. (a) Solution based on Bray-Curtis dissimilarities among taxa. (b) Solution based on Manhattan dissimilarities among taxa.

of taxa on the phylogenetic tree. Nevertheless, the analysis presented here provides a useful first test of morphological/environmental relationships in eucalypts and, more importantly, illustrates the need for tests for phylogenetic effects.

Analyses and tests

Choice of pattern model

The convergence data set, consisting of 49 characters for the 35 taxa, produced a three-dimensional ordination (Fig. 3a; dimensionality tested following procedure in Faith & Norris, 1989). This analysis used Bray-Curtis dissimilarities as input to the multidimensional scaling programme, allowing for possible unimodal relations of morphology to ordination space (see above).

An alternative analysis model used dissimilarities defined by Manhattan distance (Fig. 3b). This variant effectively searches for strictly monotonic, rather than general unimodal relations between morphology and ordination space (for discussion see Faith *et al.*, 1987). For our binary (0,1) data, this relation implies that the 1-states for a given character, in general, will be at one end of the space. To some extent, this difference in character-to-ordination relationship can be seen for the character, 'bud-shape fusiform' (compare Fig. 3a to 3b).

The choice of model for inferring pattern, as a first step in any comparative method, is an important issue, highlighted in other papers presented in this volume (e.g. Pagel; Wenzel & Carpenter – chapters 2 and 4). In this study, the more specific model implied by Manhattan distance (more specific because a linear relationship is a special case of unimodal) will be justified if the character-to-ordination relation tends to be linear/monotonic even when the more 'robust' Bray-Curtis method is used. To evaluate this, the 49 morphological variables were used to search for vectors or projections in the ordination space of maximum correlation (Fig. 4). This plot, comparing correlation values for the two ordinations, shows that most of these linear correlations are high for both ordinations, indicating that the character-to-ordination relation was generally linear/monotonic. Visual examination of plots of character values on the original ordinations supported this conclusion. Further support is found in examining a histogram of Bray-Curtis dissimilarity values; the small number of values equal to the maximum of 1 indicates that character-to-ordination relations were not likely to be unimodal (see Faith *et al.*, 1987).

Figure 5a shows the product-moment correlations of the 109 environmental variables with the Bray-Curtis ('B-C') and Manhattan ('MAN') ordinations. The line of equal correlation is also shown. The magnitude of the correlations is larger in general for the Manhattan-based ordination. For this ordination, high correlations (correlations greater than about 0.5 could be judged significant at 0.05 level after correction for multiple tests) were found for many of the mean (and median) radiation variables for winter months, and also the mean/median radiation for January. Additionally, correlations were high for average (mean or median) values for slope, elevation and mean temperature. We will discuss some of these correlations further below.

Figure 5b shows the same plot for random environmental variables. These data were simulated using the 'uniform random distribution' option in the 'RAND' module of PATN (Belbin, 1991b). The result for random variables contrasts with that for the real data in that there is an approximately even number of points on the two sides

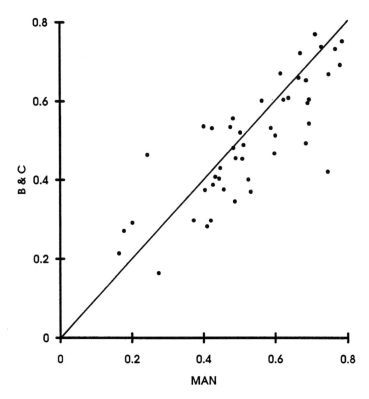

Figure 4 Plot showing the product-moment correlations of 49 morphological variables with the Bray-Curtis ('B-C') and Manhattan ('MAN') ordinations. The line of equal correlation is also shown.

of the equal-correlation line. This result suggests that the larger correlations found for the Manhattan analysis are not simply an artifact relating to ordination method.

These results highlight the problem of choosing among models in the context of comparative methods (Wenzel & Carpenter, Chapter 4). If different models of equal 'status' yield different results relating to morphology-environment relations, then arguably not much confidence is possible in the analysis. Particularly concern has been expressed regarding models for inferring ancestral states; in the present context, the model of interest relates to assumptions about the pattern of convergence (general unimodal versus linear relationships with the environmental space). The more specific Manhattan model arguably has greater status given the initial comparisons described above. This represents the proper basis for distinguishing between the pattern models; the Manhattan-based result could not be chosen simply because it produced better recovery of environmental patterns.

Distinguishing effects of homology and homoplasy

In the general description of convergence analysis, we emphasized that a given set of convergent characters can also contain information relating to common ancestry patterns (as in character D in Fig. 1). Because the resulting ordination pattern in part reflects this information, the finding that a given environmental variable has a high correlation with the space may reflect only common ancestry effects. Taxa in the

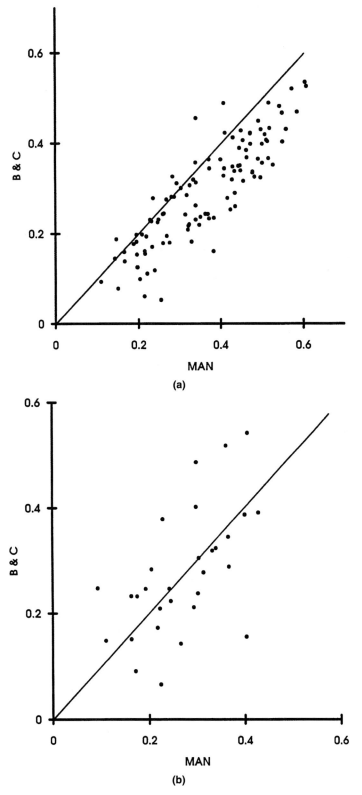

Figure 5 (a) Plot showing the product-moment correlations of 109 environmental variables with the Bray-Curtis ('B-C') and Manhattan ('MAN') ordinations. The line of equal correlation is also shown. (b) Plot showing corresponding results for random environmental variables.

same monophyletic group may share both a given derived state and a given environmental state as a result of common ancestry. Such a scenario implies that there is no strong evidence for a process link between the two (Harvey & Purvis, 1991). On the other hand, if the recovery (high correlation) of the environmental variable depends on the independent occasions of correspondence between morphology and environment, then a hypothesized relationship between the two is better supported.

Examination of the ordination of Fig. 3b, in light of the phylogenetic relationships presented in Fig. 2, suggests that neighbouring taxa in the space may or may not be close relatives. In Fig. 6, identification of members of each of three different monophyletic groups suggests that the configuration of taxa in the space may primarily reflect patterns of convergence.

We propose the following simple test to distinguish common-ancestry and convergence effects. A new data set is produced from the original set of convergent morphological characters by re-coding each independent derivation of a character's 1-state as a separate character. In this way, the new ordination can only reflect the residual common ancestry patterns that were contained in the original data set. Consequently, if an environmental variable now does not produce a high correlation, then the original recovery of the variable would appear to depend on convergence.

The re-coding procedure yielded a new matrix consisting of 78 characters for the 35 taxa (after deletion of characters with only one species having a 1-state). As an example, for the phylogenetic tree (Fig. 2) the character, 'bud-shape fusiform' created two new characters: one having a 1-state only for taxa 2, 10, 11, 19 and 25 (as numbered on the tree), and the other having a 1-state for taxa 1 and 13. The

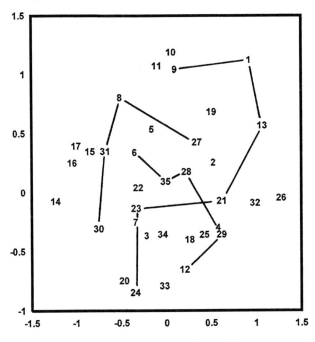

Figure 6 The Manhattan ordination showing taxa in three different monophyletic groups. Taxa within the same group are connected by line segments.

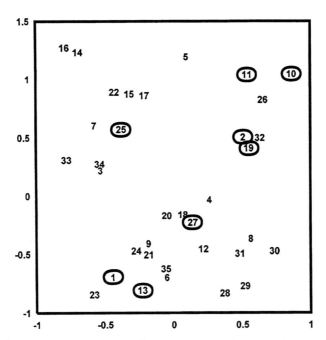

Figure 7 An ordination using Manhattan dissimilarities on the new data set in which derived states from different monophyletic groups are coded as separate characters. The circled taxa are those having the convergently derived state for the character, 'bud-shape fusiform'. Several labels on the ordination are shifted slightly to avoid overlap. Numbering matches that of Fig. 3.

additional independent origin of 'bud-shape fusiform' in taxon 27 does not result in a coded character in the new data set because the single 1-state would be uninformative in the ordination analyses.

The new ordination is shown in Fig. 7. For comparison with the original ordination (Fig. 3b), the taxa having the derived state for character 'bud-shape fusiform' are again circled. As suggested by this character, the new procedure does not in general bring the taxa in different monophyletic groups together in the ordination space. For example, taxa 1 and 13 as sister taxa represent a single derivation of 'bud-shape fusiform' and are close together in the space, but are separated from members of other monophyletic groups having the same character state (Fig. 7).

Are the correlations for environmental variables changed for this ordination? Following recalculation of correlations for each of the 109 environmental variables, the correlation values are compared for the two ordinations (Fig. 8a). There appear to be cases where the correlation for a variable is much reduced in the 'no-convergence' case; on the other hand, the opposite pattern also occurs frequently. It would appear that such a pattern over many variables might result even for random environmental variables. To evaluate this, the correlation analysis for the two ordinations was repeated for 200 random variables, using PATN (Fig. 8b). This analysis suggests that the pattern of contrasts observed for the actual 109 variables indeed is similar to that expected for random data.

A further test may be appropriate for particular environmental variables. For example, the variable 'mean elevation' (Fig. 8a) shows a large reduction in

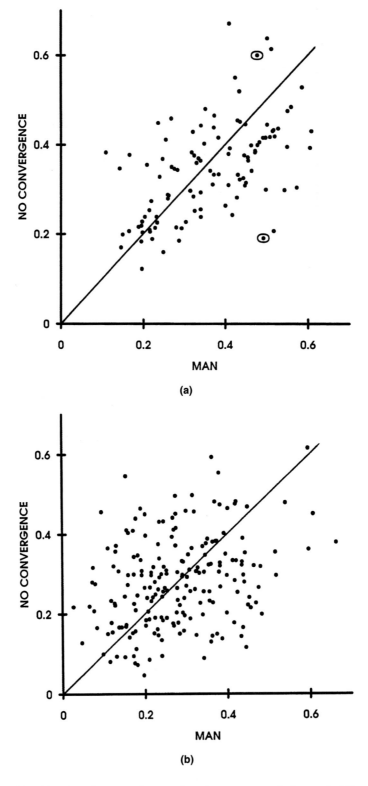

Figure 8 (a–b) (a) Plot showing the product-moment correlations of 109 environmental variables with the Manhattan ordinations based on the convergence (x axis) and no-convergence (y axis) data sets. The line of equal correlation is also shown. The lower circled point is that for the variable, 'mean elevation'; the upper circled point is variable 'mean slope'. (b) Plot showing corresponding results for random environmental variables.

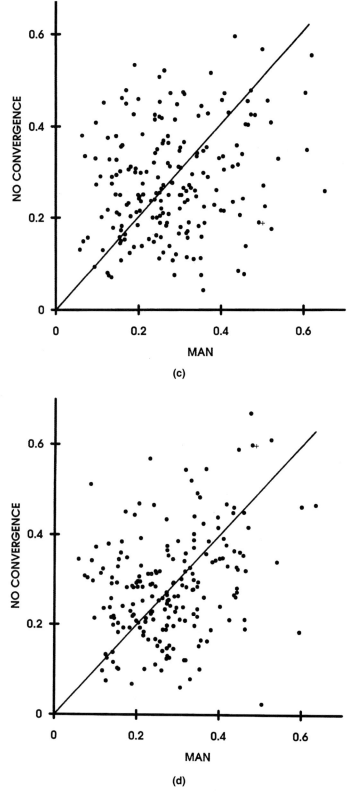

Figure 8 (c–d) (c) Plot showing corresponding results for 200 permutations of one of the original environmental variables, 'mean elevation'. The original point for this variable from Fig. 8a is also plotted and labelled with an adjoining '+' symbol. (d) Plot showing corresponding results for 200 permutations of one of the original environmental variables, 'mean slope'. The original point for this variable from Fig. 8a is also plotted and labelled with an adjoining '+' symbol.

correlation value in the non-convergence case. Two hundred random permutations of this variable's values were produced, and the correlations with each ordination recalculated (Fig. 8c). These results, in showing a wide spread of points away from the line of equal correlation, suggest that the observed contrast in correlation values could arise with reasonable frequency for such a variable. Consequently, we conclude that the high correlation found in the original analysis need not be attributed to the effects of convergence.

The test was used above in a simple graphical manner. The test can be used quantitatively if necessary; however, *a posteriori* tests must be treated carefully. For example, if a variable was chosen for evaluation because its correlation was observed to be, say, greater than 0.50 (along the 'MAN' axis), then only random data providing a correlation at least as large would be part of the randomization test; among these, a random point would match the original if its vertical distance from the equal-correlation line was as large or larger.

It may also be of interest to test correlations with the opposite pattern of contrast. For example, the variable 'mean slope' (Fig. 8a) shows a high correlation under non-convergence and a lower correlation under convergence. A significant difference would suggest that common ancestry implies a strong correspondence between morphology and this variable, but that this relationship changes from group to group. In the present case, the graphical permutation test suggests that this correlation contrast easily could occur by chance (Fig. 8d).

DISCUSSION

Our conclusion regarding eucalypt morphology/environment relationships, following evaluation of many of the observed environmental correlations using the tests developed here, is that there is no evidence from these data for a relationship between environment and morphology that goes beyond what might be expected from phylogenetic effects. Although it is clear that there is extensive morphological convergence among these taxa, and that the patterns of convergence are somewhat consistent from character to character (as summarized in the low-dimensional ordination space), it does not appear that the environmental variables used in this study represent the factors that explain convergence. This example therefore effectively illustrates the utility of these tests in evaluating whether phylogenetic effects can account for apparent correspondence between morphology and environment.

While the main purpose of our eucalypts example was to illustrate the use of our tests, further investigation of eucalypt convergent morphology and environmental factors could be pursued using one or more improvements in data and analysis:

1. Use of a better corroborated, more resolved phylogeny, when available. This would result in more confident assertions about patterns of convergence.
2. Investigation of alternative codings for convergent characters. The present study interpreted some convergence events as arising when the same character state was independently lost one or more times, but the usual cases showing convergent gain of the same state (the predominant pattern) may be more revealing about common environment. While a re-analysis of our data with convergent-

losses excluded produced a similar overall ordination pattern, future work might include a full analysis based on convergent-gains only.

3. Refinement of links between survey distribution data (used to infer environmental values for species) and eucalypt taxonomy. During the course of our work, revisions in eucalypt taxonomy (particularly in the '*globulus*' group) meant that the exact distribution of a given taxon often had to be re-assessed. A more stable taxonomy and more complete survey data will allow for better assessments of the environmental descriptions for different species.

CONCLUSION

Combining tests for phylogenetic effects with multivariate analysis of convergence has produced an overall approach that may be a useful complement to other comparative methods (e.g. Harvey & Purvis, 1991). The multivariate convergence pattern can reveal a significant environmental correlation that may not have been detectable using any one of the morphological variables on its own (Faith, 1989). Exploring these multivariate patterns may lead to a better understanding of complex patterns and processes of adaptation, including aspects relating to evolutionary constraints, co-option of characters for other functions, non-aptations and different character states serving the same function.

ACKNOWLEDGEMENTS

We thank P. Eggleton and R. Vane-Wright for the invitation to speak at the symposium, and P. Cranston and other members of the Canberra 'Coopers and Cladistics Discussion Group' for their continued interest in this topic. We thank M. Austin and M. Gill for making their extensive data sets available for this study, M. Doherty for taxonomic advice and M. Cawsey for technical support. C. Humphries and J. Coddington provided many useful suggestions for improving the chapter.

REFERENCES

AUSTIN, M.P., NICHOLLS, A.O. & MARGULES, C.R., 1990. Measurement of the realized qualitative niche: environmental niches of five Eucalyptus species. *Ecological Monographs, 60:* 161–177.

BELBIN, L., 1991a. Semi-strong hybrid scaling, a new ordination algorithm. *Journal of Vegetation Science, 2:* 491–496.

BELBIN, L., 1991b. *PATN software package.* Canberra: CSIRO Division of Wildlife and Ecology.

BROOKS, D. & McLENNAN, D., 1993. Historical ecology as a research programme: scope and limitations. In P. Eggleton & R.I. Vane-Wright (Eds), *Phylogenetics and Ecology*, 1–27. London: Academic Press.

CHIPPENDALE, G.M., 1988. *Eucalyptus, Angophora* (*Myrtaceae*). In A.S. George (Ed.), *Flora of Australia, 19:* 1–447. Canberra: Government Publishing Service.

CODDINGTON, J.A., 1988. Cladistic tests of adaptational hypotheses. *Cladistics, 4:* 3–22.

CODDINGTON, J.A., 1994. The roles of homology and convergence in studies of adaptation. In P. Eggleton & R.I. Vane-Wright (Eds), *Phylogenetics and Ecology*, 53–78. London: Academic Press.

FAITH, D.P., 1989. Homoplasy as pattern: multivariate analysis of morphological convergence in Anseriformes. *Cladistics, 5:* 235–258.

FAITH, D.P. & NORRIS, R.H., 1989. Correlation of environmental variables with patterns of distribution and abundance of common and rare freshwater macroinvertebrates. *Biological Conservation, 50:* 77–98.

FAITH, D.P., MINCHIN, P.R. & BELBIN L., 1987. Compositional dissimilarity as a robust measure of ecological distance. *Vegetatio, 69:* 57–68.

FARRIS, J.S., 1988. HENNIG86, version 1.5. Available from author, 41 Admiral St, Port Jefferson Station, New York 11776, U.S.A.

GITTLEMAN, J. & LUH, H., 1994. Phylogeny, evolutionary models and comparative methods: a simulation study. In P. Eggleton & R.I. Vane-Wright (Eds), *Phylogenetics and Ecology,* 103–122. London: Academic Press.

HARVEY, P.H. & PURVIS, A., 1991. Comparative methods for explaining adaptations. *Nature, 351:* 619–624.

HILL, K.D., 1991. *Eucalyptus.* In G.J. Harden (Ed.), *Flora of New South Wales,* 76–142. Sydney: Royal Botanic Gardens.

JOHNSON, L.A.S., 1972. Presidential address: evolution and classification in *Eucalyptus, Proceedings of the Linnean Society of New South Wales, 97:* 11–29.

LADIGES, P.Y. & HUMPHRIES, C.J., 1983. A cladistic study of *Arillastrum, Angophora* and *Eucalyptus* (Myrtaceae). *Botanical Journal of the Linnean Society, 87:* 105–134.

LADIGES, P.Y. & HUMPHRIES, C.J., 1986. Relationships in the stringybarks, *Eucalyptus* L'Herit. informal subgenus *Monocalyptus* series Capitellatae and Olsenianae: phylogenetic hypotheses, biogeography and classification. *Australian Journal of Botany, 34:* 603–632.

LADIGES, P.Y., HUMPHRIES, C.J. & BROOKER, M.I.H., 1983. Cladistic relationships and biogeographic patterns in the peppermint group of *Eucalyptus* (informal subseries Amygdalininae, subgenus *Monocalyptus*) and the description of a new species, *E. willisii. Australian Journal of Botany, 31:* 565–584.

LADIGES, P.Y., HUMPHRIES, C.J. & BROOKER, M.I.H., 1987. Cladistic and biogeographic analysis of Western Australian species of *Eucalyptus* L'Herit. informal subgenus *Monocalyptus* Pryor and Johnson. *Australian Journal of Botany, 35:* 251–281.

LADIGES, P.Y., NEWNHAM, M.R. & HUMPHRIES, C.J., 1989. Systematics and biogeography of the Australian 'green ash' eucalypts (*Monocalyptus*). *Cladistics, 5:* 345–364.

PAGEL, M., 1994. The adaptationist wager. In P. Eggleton & R.I. Vane-Wright (Eds), *Phylogenetics and Ecology,* 29–51. London: Academic Press.

SWOFFORD, D.L. 1992. PAUP: Phylogenetic Analysis Using Parsimony, ver 3.0. Available from the author, Illinois State Natural History Survey, 172 Natural Resource Bldg., 607 E. Peabody, Champaign, IL. 61820, U.S.A.

WENZEL, J. & CARPENTER, J., 1994. Comparing methods: adaptive traits and tests of adaptation. In P. Eggleton & R.I. Vane-Wright (Eds), *Phylogenetics and Ecology,* 79–101. London: Academic Press.

APPENDIX 1

List of environmental variables used to describe sites where a given taxon was recorded as present.

slope (degrees); aspect (degrees); elevation; latitude; longitude; mean rainfall (mm.); mean temperature; radiation index; site disturbance rating; soil depth; plotsize; nutrient rating

topography: topo 1 (ridge); topo 2 (slope); topo 3 (lower slope); topo 4 (gully); topo 5 (flat); topo 6 (unknown); topo 7 (other; e.g. dune)

lithology: lith 1 (volcanics); lith 2 (hard sediments); lith 3 (soft sediments); lith 4 (granites); lith 5 (others)

seasonality: seas 1 (summer rainfall); seas 2 (aseasonal); seas 3 (winter rainfall)

average radiation per day in given month: janrad; febrad; marchrad; aprilrad; mayrad; junerad; julyrad; augrad; septrad; octrad; novrad; decrad

average radiation per day over a year

APPENDIX 2

A: list of morphological characters showing convergence on phylogenetic tree of Figure 2.

Bark: smooth; rough; flaky
Leaf shape: narrow; lanceolate; broad; falcate; not falcate
Leaf apex: acuminate; uncinate; rostrate
Leaf base: cuneate;
Leaf colour: green; grey; glaucous;
Leaf texture: coriaceous; thick;
Leaf upper/lower: discolorous; concolorous;
Veins: prominent; not prominent;
Vein angle: very acute; acute;
Intramarginal vein: double
Bud shape: ovoid; clavate; fusiform;
Bud surface: glaucous
Operculum shape: hemispherical; conical; acute; rostrate
Operculum length: less than hypanthium; greater than hypanthium
Operculum width: less than hypanthium; equal to hypanthium; greater than hypanthium
Peduncle shape: terete; narrowly flattened
Fruit shape: globular; hemispherical; conical; pyriform; turbinate; urceolate:
Fruit ornamentation: verrucose
Disc: enclosed
Valves: enclosed; excert

B: data matrix representing convergent states (coded as 1); the rows are the 35 taxa in tree-sequence order (Fig. 2), and the columns correspond to the sequence of states in Appendix 2A.

```
0000011100 0101101000 0000011010 1011000100 000000111
0001000000 0000001110 0000101000 0010000100 000000111
0001000000 0000001001 1000100000 0000110100 000000000
1110111110 1000000111 0000000000 0000000100 000000001
0011000001 0000001001 1000110010 0111000000 010000111
0010000100 0011000000 0010000001 0000000000 001100010
1101000000 0000001001 1001110010 0000000100 000000000
0000111100 0100000001 1000000001 0000000000 011000110
0000000010 1100000101 1000000010 0000000000 011000111
0000010000 0011000001 1001011010 0011000000 011000101
0001000000 0011100001 1000011110 0011000000 010000101
0001000011 1000000000 0000100000 0000000100 000010000
0000011110 1100001100 0000011010 0010000100 000000011
1101000000 0000011001 1001100111 0100000100 000101111
0001000000 0000000001 1000010101 0100000100 000100111
1101000000 0000000001 1001100101 0110011100 010000111
0011000000 0000000001 1000000111 0100000100 000100111
0000000000 0000000110 0010000000 0000000000 000010000
0011000000 0001000000 0000011010 1011000000 000000001
1101001100 0000001000 0000110001 0100000000 000000000
0000010000 0100001110 0000100000 0010000100 000000001
1101000000 0000001001 1000110010 0000000100 010000011
0000110000 1100000001 1001000001 0000000000 000010000
1100110000 0000010000 0010000001 0000001000 000011000
0000000000 0000001111 1000111010 0000110100 011000000
0000010010 1000000110 0010110010 0011011100 010000011
0000000000 0000000001 1000001010 1111000100 000000000
0000000000 0001000001 1000110010 0011000100 000000000
0000000000 0000000000 0010110010 0011000000 000010000
0001100100 0000001000 0000000000 0000110000 001000101
0001000000 0011001001 1000000100 0000000000 001000000
0010000010 1000000110 0000110010 0010000100 001000011
0001001100 0001001001 0000110010 0100110100 000010000
0000000000 0000001001 1000110010 0100110000 000000000
0010000000 1000000000 0000000001 0000000100 011000000
```

CHAPTER

8

Testing ecological and phylogenetic hypotheses in microevolutionary studies

R.S. THORPE, R.P. BROWN, M. DAY, A. MALHOTRA, D.P. McGREGOR & W. WÜSTER

CONTENTS

Keywords: Geographic variation – Mantel tests – DNA phylogenetics – field experiments – parallel selection.

Abstract

The patterns of variation among populations may reflect ecogenetic adaptation by natural selection for current ecological conditions. However, the geographic variation may also reflect phylogenesis when past events, such as vicariance, have had an influence on the population differentiation. The influence of these causative factors can be difficult to disentangle. A range of procedures for determining the causative factors in morphological variation are investigated in relation to geographic variation in reptiles. These include: quantitative tests (including tests of anagenesis in trees, randomization tests of congruence, Mantel tests, partial Mantel tests and canonical correlations); parallel patterns on adjacent islands which have parallel ecological

Phylogenetics and Ecology
ISBN 0-12-232990-2

differentiation but independent geological histories; large-scale field experiments which indicate the role of natural selection in relation to ecological conditions; and the role of molecular phylogenies in helping to determine the cause of geographic variation in morphology.

INTRODUCTION

The interaction between ecology and phylogeny has a particular meaning for studies of geographic variation, because it is at this intraspecific level that these two forces can directly combine to influence allele frequency, and hence evolution. Previous publications (Thorpe, 1991; Thorpe *et al.*, 1991) have expanded on the idea that one can usefully think of geographic variation within a species as being influenced by these two forces, that is ecogenesis (natural selection for current ecological conditions) and phylogenesis (past, historical events, including the effects of genetic drift and selectively neutral mutations).

This conceptualization can be facilitated by examples. Species recently introduced to new areas that have clearly differentiated geographically supply examples of ecogenesis, as historical events can have had little, or no, influence on the geographic pattern of differentiation in their new regime. Examples of this are sparrows in North America (Johnston & Selander, 1971) and New Zealand (Baker, 1980), Mynas in New Zealand (Baker & Mooed, 1979) and *Drosophila buzzatii* in Australia (Barker & Mulley, 1976; Thorpe *et al.*, 1991).

On the other hand, constellations of species pairs, or parapatric races, which have contact zones with coincident direction (rather than exactly coincident position) and a position that does not relate to a present barrier, or ecotone, may be best interpreted in light of past events. For example, a number of species pairs and parapatric races have north-south orientated contact zones aggregating around central Europe. These species include the grass snakes (*Natrix natrix*), hedgehogs (*Erinaceus*), crows (*Corvus*) and water snakes (*N. maura* and *tessellata*) (Thorpe, 1979). One historical factor affecting all these species was the Pleistocene ice caps, which at their maxima would have split the distribution of the precursor species into south-east and south-west refugia. These would have differentiated over time, and on expanding in post-glacial times have met their sister-taxa along contact zones. If one takes the grass snake example, one can find numerous racial differences between forms either side of the contact zone, but these do not coincide with an ecotone and cannot be ascribed to selection for current ecological conditions.

While these examples may represent the two extremes, in most cases these two factors interact in a complex manner, and to understand the relative importance of these forces and their role, one needs to disentangle them. Methods that may elucidate the causes of geographic variation will be discussed under four headings: quantitative tests; natural situations; natural experiments; and molecular data.

QUANTITATIVE TESTS

Some of this area has recently been reviewed elsewhere (Thorpe, 1991; Thorpe *et al.*, 1991) and these aspects will not be treated in depth here. One can look at tests for: (a) pattern congruence; (b) pattern of anagenesis in phylogenetic trees; and (c) tests of association between observed and hypothesized patterns.

Pattern congruence

If a major historical process, such as a vicariance event, influences the racial differentiation of a species, then the entire genome is subject to the same event, and one has the expectation that most major character systems will reflect this process. This predicts high congruence between patterns of geographic variation in different characters and character systems caused by 'phylogenesis'. On the other hand, different characters and character systems may be predominantly adapted to different facets of the environment (e.g. rainfall *versus* vegetation) by natural selection This predicts the possibility of low congruence between patterns of geographic variation caused by 'ecogenesis'.

One may therefore be able to elucidate the cause of the variation by inspecting the levels of congruence. This, however, is complicated by the fact that congruence between character systems is influenced by the number of characters employed. A random re-sampling procedure (Thorpe, 1991, and references therein) enables one to compare the congruence between character sets for a given number of characters. It is predicted that the congruence-against-character-number curve should be higher for phylogenesis than ecogenesis. When tested using models, which on the basis of independent evidence are thought to represent ecogenetic and phylogenetic differentiation (geographic variation in the Palaearctic grass snake, *Natrix natrix*, is the 'phylogenetic' model; Thorpe, 1984; and geographic variation in the Tenerife gecko, *Tarentola delalandii*, is the 'ecogenetic' model; Thorpe, 1991), then, as predicted, the former has higher congruence levels than the latter (Thorpe, 1991).

Pattern of anagenesis in phylogenetic trees

One can (erroneously) subject taxa to a phylogenetic analysis even if their differentiation is not caused by phylogenesis, and even in these cases, one would expect the position of the representative population on the tree to be related to its geographic origin. Consequently, the existence of a geographically interpretable 'tree' does not necessarily indicate a phylogenetic cause. However, the pattern of anagenesis in a tree should indicate causation because only if the differentiation is phylogenetically caused should one expect the extent of anagenesis to be related to the time of divergence, that is the branches arising near the root should be longer than branches arising further from the root. This is what one finds in trees based on largely selectively neutral nucleic acid data (Fig. 1) (Thorpe *et al.*, 1993). Consequently, when morphological trees have a clear pattern of anagenesis related to time of divergence, phylogenesis is implicated (as in the *N. natrix* example; Thorpe, 1984). When no such pattern is evident, as in the Tenerife gecko (Thorpe, 1991), then ecogenesis is implicated. This is discussed in greater depth in Thorpe (1991) and references therein.

Tests of association between observed and hypothesized patterns

Correlating an observed pattern of geographic variation in a character with a pattern derived from a hypothesized cause (e.g. rainfall) can be useful. Of course,

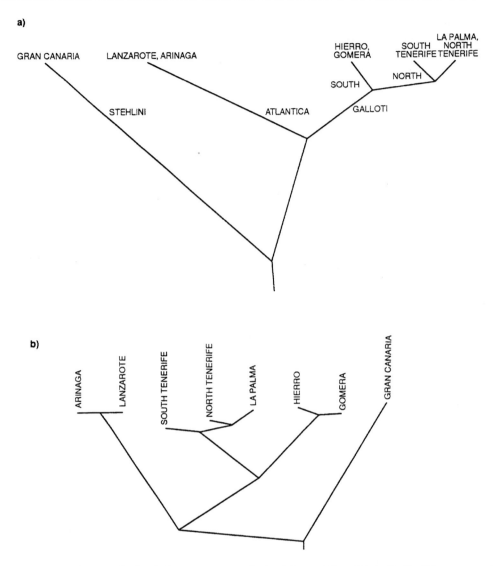

Figure 1 Fitch-Margoliash phylogenetic trees derived from: (a) mtDNA RFLPs (Thorpe *et al.*, 1993); and (b) a 1005 base pair sequence of mtDNA (Thorpe *et al.*, in press) of *Gallotia* populations listed by island name (Arinaga, Lanzarote = *G. atlantica*; Gran Canaria = *G. stehlini*; Tenerife, La Palma, Gomera, Hierro = *G. galloti*). Note that these trees, derived from 'selectively neutral' data, have branch lengths which tend to be related to time (position) of origin, unlike those perturbed by ecogenesis (Thorpe, 1991).

correlation does not prove causation, but the lack of a correlation can enable one to reject a hypothesis.

Partial correlations

A comprehensive range of hypotheses have to be tested for this procedure to be useful, but this raises the problem that the hypothesized patterns may be inter-correlated. This can be overcome by using partial correlation when both the observed and hypothesized patterns are unidimensional. This is the procedure used by Thorpe & Brown (1989a,b) to relate colour pattern in the Tenerife lacertid, *Gallotia galloti*, to several putative causal hypotheses.

Mantel tests

However, some multivariate patterns and hypotheses (e.g. geographic distance between sites) cannot readily be expressed in single dimensions and so matrices are compared by Mantel tests (Mantel, 1967; Dietz, 1983; Manly, 1986, 1991; Schnell, Douglas & Hough, 1986; Smouse, Long & Sokal, 1986; Dow, Chevereud & Friedlaender, 1987; Cheverud, Wagner & Dow, 1989; Brown & Thorpe 1991a,b; Brown, Thorpe & Baez, 1991; Malhotra & Thorpe, 1991a; Sokal, Oden & Wilson, 1991; Thorpe, 1991; Thorpe & Brown, 1991; Thorpe & Baez, 1993). A matrix representing the affinity between localities based on the observed pattern is compared to a matrix representing the affinities between localities based on the hypothesized cause of variation. The units of these matrices are not independent so instead of using tabulated degrees of freedom the probability of association in a Mantel test is derived from randomizing one of the matrices numerous times. The observed coefficient of association (e.g. correlation or regression) is then compared to the distribution of the coefficient derived from the randomizations.

An example of the use of a simple Mantel Test is given by the comparison of generalized ecological conditions to generalized morphology in the Dominican anole, *Anolis oculatus* (Malhotra & Thorpe, 1991a). This showed a significant relationship between generalized morphological divergence and generalized ecological conditions.

Simultaneous (partial) Mantel tests

Mantel tests suffer from the same drawback as ordinary correlations in that several causes may actually, or potentially, influence the pattern of divergence, and the various hypothesised patterns (independent variables) may be inter correlated. Consequently, one needs to partial out the effects of the various effects with a simultaneous (partial) Mantel test. Until recently, software that could consider more than two independent matrices simultaneously was not (readily) available and this was a major limitation. However, one can now use Mantel tests to compare simultaneously a matrix of observed differences among sites to numerous matrices of hypothesized differences between sites. Three example of the use of simultaneous Mantel tests are given here: Philippine cobras, Tenerife lacertids and Dominican anoles.

Philippine cobras Sixteen scalation characters were recorded from 44 specimens of Asiatic cobras (*Naja naja* species complex) from the southern Philippines and Borneo, and the affinities between individuals expressed as a taxonomic distance. Three causative factors, each expressed as a dissimilarity matrix between individuals, were considered. There is a longitudinal climatic gradient across the Philippines with the eastern sector being wetter than the western sector. Since rainfall and humidity can influence geographic variation in reptilian scalation (Thorpe & Baez, 1987), this climatic gradient was considered as a putative causal factor. In contrast to this ecogenetic hypothesis reflecting current conditions, one could hypothesize that the affinities are determined by a historical (phylogenetic) factor, that is their origin on the Pleistocene mega-islands (Fig. 2) that resulted from the lower sea levels during

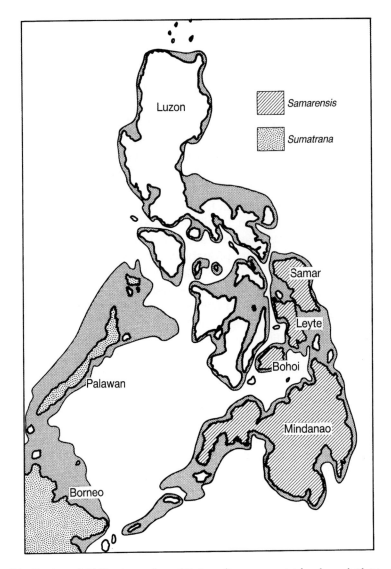

Figure 2 Distribution of Philippine cobras (*Naja* sp.) on current islands and Pleistocene mega-islands (enclosed by shaded area). The scalation variation of cobras (*samarensis* and *sumatrana*), distributed across the southern two mega-islands, was subjected to a partial Mantel test. This shows that past mega-island origin, rather than current climate, influences the geographic variation (see text for explanation).

the Pleistocene glaciations (Heaney, 1985, 1986). Finally, one can test whether the affinities are determined by their current island origin. The results of the simultaneous Mantel test (Table 1) show that even when the current island origin is taken into account the scalation is still significantly influenced by the historical factor of Pleistocene mega-island origin, whilst the ecogenetic factor (climate) has no significant effect.

Tenerife lacertids The above example was based on only three hypotheses, but one may have to test more hypotheses than this. A previous study of *Gallotia galloti*

Table 1 Philippine Cobras. Absolute partial correlations and probabilities from simultaneous Mantel tests of association between observed variation in scalation and three putative causal hypotheses. See text for explanation

Pleistocene mega-islands (phylogenesis)	Climate (ecogenesis)	Current islands
0.13 ($P<0.01$)	0.02 insignificant	0.34 ($P<0.001$)

on Tenerife used partial correlations to test the association between colour pattern (67 locality means) and a wide range of inter-correlated ecogenetic and phylogenetic hypotheses (Thorpe & Brown, 1989a,b, 1991). Here, these hypotheses are tested using simultaneous Mantel tests, with geographic proximity as one of the independent matrices. Briefly, the causative hypotheses that can be tested to see if they account for the colour pattern differentiation within Tenerife are as follows: (1) that the precursor islands of Anaga and Teno had phylogenetically divergent populations on them which met to produce the observed pattern after the eruption of Teide joined the islands; (2) that the cloud layer around Tenerife has separated high and low altitude populations long enough for them to be phylogenetically divergent; (3) that the colour pattern, like the scalation, is ecogenetically adapted to altitude; (4) that the colour pattern is ecogenetically adapted to the distinct climatic/vegetational biotypes which meet in a sharp ecotone, the north being humid with lush vegetation and the south being hot, arid and barren; and (5) that the affinities are determined by geographic proximity, that is, opportunity for gene flow. A fuller explanation of these hypothesized patterns is given in Thorpe & Brown (1989b).

The results (Table 2) clearly show that the multivariate generalized ('total') colour pattern is significantly associated only with the ecotone hypothesis (hypothesis 4). All other hypotheses, both ecogenetic and phylogenetic, can be rejected. However, considerable care must be used when interpreting multivariate patterns when ecogenesis is thought to be the causative factor, because different individual characters may be selected for in different facets of the environment. These

Table 2 Tenerife lacertids. Absolute partial correlations and probabilities from simultaneous Mantel tests of association between individual colour pattern characters plus 'total-multivariate' pattern and numerous causal hypotheses. See text for explanation.

Character	An/Teno (phylo)	Cloud (phylo)	Altitude (ecogen)	Ecotone (ecogen)	Proxim	Sel/GF
1	0.1061	0.0495	0.0166	0.6029***	0.1583	
2	0.0201	0.0753	0.0461	0.0639	0.0182	
3	0.0268	0.0070	0.0296	0.5422***	0.0191	
4	0.0924	0.0289	0.0320	0.6360***	0.1506	
5	0.0045	0.0595	0.0579	0.2462**	0.2141*	
6	0.0273	0.0087	0.0488	0.6240***	0.0181	
Total	0.0608	0.0770	0.0024	0.6400***	0.1517	4.2

* indicates significance beyond $P<0.01$ (Bonferroni correction of 0.05), ** $P<0.0005$, *** $P<0.0001$.

individual adaptations may be obscured by multivariate generalization. Mantel tests on each of the six individual characters (Table 2) indicate that, in this case, multivariate generalization has not obscured adaptation of individual characters to different environmental factors as five of the six characters give the same result as the generalized pattern (although proximity is just significant for character 5) and the sixth character is not significantly related to any hypothesis. A biological interpretation of the geographic variation in the colour pattern, in relation to natural selection for cryptic dorsal coloration *versus* sexual selection for blue lateral coloration, is given in Thorpe & Brown (1989b).

Dominican anoles The mountainous island of Dominica (Lesser Antilles) is subject to prevailing easterly winds which result in pronounced environmental differentiation within the island, both with longitude and altitude. Generalized morphology is significantly related to generalized ecology when subjected to a simple Mantel test (Malhotra & Thorpe, 1991a). When generalized body dimensions (13 characters adjusted for size independence), generalized scalation (10 characters) and generalized colour pattern (22 characters) are tested (using simultaneous Mantel tests) against the potential causative environmental factors broken down into the biotic environment (vegetation), temperature, rainfall, altitude and geographic proximity, one can see that in females shape is significantly associated with the biotic environment, scalation with rainfall and colour pattern with both (Malhotra, 1992) (Table 3). Once again as ecogenesis is the cause of the variation individual characters may be associated with different facets of the environment which would be obscured by multivariate generalization. This can be seen to be the case when the individual scalation characters are tested against these hypotheses. Whilst rainfall is the predominant 'cause' of scalation variation, the variation in the scales between the orbits is also associated with temperature and altitude, and the number of enlarged white scales is also associated with the vegetation type (biotic environment) (Table 4). This emphasizes the need to treat each character separately when dealing with ecogenetically caused variation.

Canonical correlations

Not only can the various components of the character variation be coalesced by multivariate generalization, or treated separately, the various components of the causative factors can also be rather arbitrarily coalesced, or separated. For example, in the Dominican anole case, the various facets of the physical environment may be

Table 3 Dominican anoles. Absolute partial correlations and probabilities from simultaneous Mantel tests of association between character set patterns (females) and numerous causal hypotheses. See text and Malhotra (1992) for explanation.

Character set	Proximity	Veg. type	Temp.	Altitude	Rainfall
Body proportions	0.06	0.14*	0.02	0.05	0.12
Scalation	0.07	0.07	0.11	0.02	0.34**
Colour pattern	0.10	0.34**	0.09	0.04	0.25**

* indicates significance beyond $P < 0.05$, ** $P < 0.01$, *** $P < 0.001$

Table 4 Dominican anoles. Absolute partial correlations and probabilities from simultaneous Mantel tests of association between individual scalation characters plus 'total-multivariate' pattern (females) and numerous causal hypotheses. See text and Malhotra (1992) for explanation.

Character	Proximity	Veg. type	Temp.	Altitude	Rainfall
BSC	0.05	0.13	0.14	0.11	0.34***
LAM	0.11	0.01	0.02	0.10	0.07
VENTSC	0.08	0.04	0.00	0.04	0.14*
SUPLABS	0.10	0.04	0.06	0.05	0.29**
SUBLABS	0.06	0.06	0.01	0.08	0.23*
SCBORB	0.10	0.06	0.18*	0.23**	0.26*
SS	0.10	0.26**	0.10	0.06	0.25*
Total	0.07	0.07	0.11	0.02	0.34**

* indicates significance beyond $P < 0.05$, ** $P < 0.01$, *** $P < 0.001$.

treated separately, or coalesced by multivariate generalization. This underlines the utility of being able to treat each character and causative factor separately *a priori* and then look for constellations of characters that are related to constellations of causative factors. Canonical correlation attempts to do this.

To illustrate this the Dominican anole case study is used. Six ecological factors (three biotic and three physical) were canonically correlated with 46 individual morphological characters (Malhotra, 1992) in males. Table 5 summarizes the results for the second canonical variate which shows that occurrence of littoral woodland is

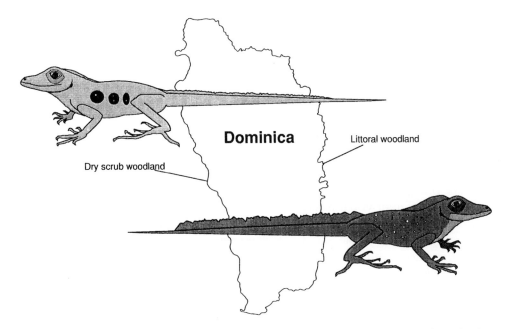

Figure 3 Dominican *Anolis oculatus* from the Caribbean dry scrub woodland (left) and Atlantic wet littoral forest (right). A functional complex of scalation, body proportion and colour characters give a different visual image for each of these anoles living in the different habitats. The anole living in the more open dry scrub woodland are light coloured with black blotches, whilst those living in the denser littoral woodland are more saturated in colour and have enlarged light lateral scales and a profile dominated by large neck crest and tail crest (Malhotra, 1992).

Table 5 Canonical correlation loadings on the first two variates for: (A) ecological variables; and (B) selected morphological variables, of Dominican *Anolis* (after Malhotra, 1992).

A

Ecological factor	CV1	CV2
Temperature	0.699	0.445
Altitude	−0.755	−0.430
Rainfall	−0.880	−0.256
Dry scrub woodland	0.896	−0.420
Rainforest	−0.878	−0.301
Littoral woodland	0.072	0.911

B

Character systems	Morphological character	CV1	CV2
Body shape	Head width	0.485	−0.256
	Lower leg length	−0.469	0.014
	Toe length	−0.524	−0.150
	Tail depth	0.101	0.358
Scalation	Number of scales at mid-body	0.533	0.123
	Number of ventral scales	0.146	0.355
	Enlargement of white scales	−0.186	0.351
Colour	Dorsal colour – magenta	−0.287	0.379
	Dorsal colour – cyan	−0.399	−0.045
	Dewlap colour – cyan	−0.364	−0.265
	Eyeskin colour – cyan	−0.681	−0.120
	Ventral colour – magenta	−0.224	0.519
	Ventral colour – cyan	−0.428	0.313
	Number of black patches	−0.118	−0.533

associated with 'visual effect' characters from each of the character systems, that is size of the tail crest (body dimensions), enlargement of white lateral scales (scalation) and intensity of magenta and absence of black spots (colour pattern). These characters produce an overall functionally coherent effect in adapting the male lizards to the different visual background offered by the littoral woodland compared to the other vegetation types (Fig. 3). While canonical correlations are successful in showing *a posteriori* constellations of characters (and causes) that cut across the subdivisions into 'man-made' character systems, they have two limitations. They are non-probabilistic and are unable to take into account multidimensional patterns and hypotheses represented by matrices, for example geographic proximity in two dimensional space. This problem is addressed by Malhotra & Thorpe (in press).

NATURAL SITUATIONS

In some circumstances the natural situations are such that a cause of geographic variation can be deduced. If, within a region, such as an island, two species show

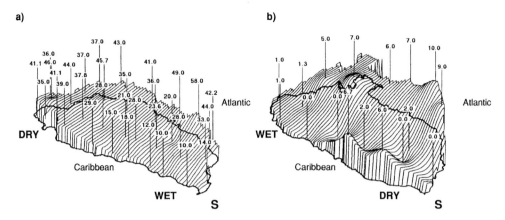

Figure 4 Parallel variation in the Lesser Antillean anoles. Isometric plots of intensity of dorsal magenta in: (a) *Anolis oculatus* on Dominica; and (b) *Anolis marmoratus* on Basse Terre (Guadeloupe) showing parallel variation. The intensity of the magenta component of the dorsal colour is greatest in the Atlantic littoral woodland and least in the wettest part of the Caribbean dry scrub woodland for both Dominica and Basse Terre anoles (Malhotra, 1992).

parallel patterns of variation, this does not allow one to deduce cause, because these species may have both ecogenetic factors and historical factors in common. However, in a situation where two islands have parallel patterns of environmental differentiation within them, but independent histories, then parallel character variation in these allopatric species can be ascribed to ecogenesis as they have only parallel ecological conditions in common.

Two examples of this are given here. To the north of Dominica the adjacent island of Basse Terre (Guadeloupe) is another mountainous island subject to the same weather patterns. Consequently it has parallel environmental differentiation to Dominica, for example wet littoral woodland on the Atlantic coast, hot dry scrub woodland on the Caribbean coast and very wet cooler rainforest at higher altitudes. A series of characters from the scalation, colour pattern and body dimensions show parallel patterns of geographic variation in the Dominican anole, *Anolis oculatus*, and the Guadeloupean anole, *Anolis marmoratus* (Malhotra, 1992). Their parallel variation in the increased intensity of the magenta component of dorsal hue in association with Atlantic littoral woodland is illustrated here (Fig. 4). This parallel variation can be ascribed to ecogenesis rather than phylogenesis for the reasons given above.

The second example comes from Canary island skinks (Brown, Thorpe & Baez, 1991). Gran Canaria and Tenerife are adjacent mountainous islands subject to the same prevailing northerly trade winds which result in a humid lush northern sector and a hot arid barren southern sector on these islands. On Gran Canaria the skink, *Chalcides sexlineatus*, has a northern form with a brown tail and a southern form with a bright blue tail. This is thought to reflect alternative anti-predator strategies as the tail can be autotomized and bright colour will attract a predator's attention away from the vulnerable head to the disposable tail. Parallel variation in the tail colour (Fig. 5) and multivariate generalized colour pattern can be seen in *Chalcides viridanus* on the adjacent island of Tenerife. In both cases simultaneous Mantel tests are used to test colour pattern variation against various alternative causal hypotheses.

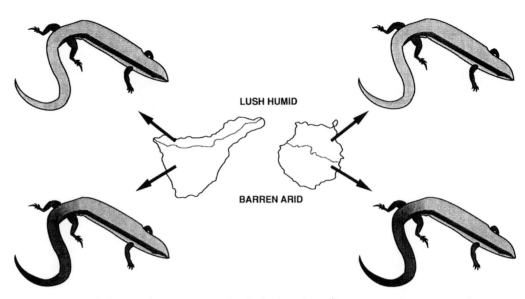

Figure 5 Parallel variation in Canary Island skinks. The colour pattern variation in the Gran Canarian skink (*Chalcides sexlineatus*) (right) is associated with climatic differences within the island. These differences are paralleled by colour pattern variation by a related skink (*Chalcides viridanus*) on the adjacent island of Tenerife (left). In particular, skinks from the dry barren south of both islands tend to have blue tails (Brown, Thorpe & Baez, 1991).

These tests, and the parallel variation, indicated ecogenetic adaption to the two biotic regimes (Brown, Thorpe & Baez, 1991).

NATURAL EXPERIMENTS

For several reasons field experiments on evolution are very difficult and consequently, rarely carried out. The time scale of evolutionary processes generally exceeds that of a study and there can be considerable practical difficulties in obtaining comparable, discrete, experimental entities. However, Endler's outstanding study of the balance between natural selection for crypsis and sexual selection for bright coloration of guppies in Trinidadian stream pools has shown the value of field experiments in investigating causative factors in microevolution (Endler, 1980), as have a few other studies (e.g., Halkka & Raatikainen, 1975; Knights, 1979).

Terrestrial species, such as anoles, may not be naturally separated into convenient experimental units. Consequently, Malhotra & Thorpe separated natural forest into four large experimental enclosures and utilized a lizard-proof fence to segregate populations of the Dominican anole (Malhotra & Thorpe, 1991b, 1993; Malhotra, 1992; Thorpe & Malhotra, 1992). The lizards were cleared from each of the four enclosures and replaced by examples of *Anolis oculatus* from the four ecotypes (including a control). The multivariate morphology of each specimen was recorded before translocation. In as brief a time as two months there was a difference in the morphology of the survivors compared to the non-survivors within an enclosure. For the montane ecotype (which come from the locality with an environment most different to that of the enclosures) the difference was significant for both males and females, and the extent of morphological difference was significantly correlated to the extent of ecological difference between enclosure and locality of origin across the ecotypes (Malhotra & Thorpe, 1991b; Thorpe and Malhotra, 1992). This indicates the importance on natural selection in determining the pattern of geographic variation and the surprising rapidity of its action. Partial repetitive experiments in different ecozones, and allochronic repeats have confirmed this conclusion.

MOLECULAR DATA

The study of evolution has been strongly influenced by investigations of the differences among island populations in archipelagos, but often little thought has been given to the relative importance of the underlying causes. Both phylogenesis (e.g. the pattern of colonization) and ecogenesis (natural selection for different environmental conditions on the different islands) are likely to have contributed to the pattern of inter-island differentiation. Their relative contribution can be very difficult to determine. If one wishes to test to what extent a pattern of morphological divergence among islands is due to ecogenetic adaptation one needs to take into account the phylogenetic history. However, at the intraspecific level, if all one has is the phylogenetic tree based on morphology, one will not be able to do this in case the tree is corrupted by this convergence. A way out of this circular problem is to erect a tree on an independent information system which is least likely to be perturbed by ecogenesis. Mitochondrial DNA is such an information system (Wilson *et al.*, 1985).

The Canary island lizards of the genus *Gallotia* provide a model for this approach. The islands are volcanic in origin and arose from the sea floor. The western islands are separated by deep sea and have not been joined to one another, the islands to the east, or the African mainland, in the time scale pertinent to this study. The ages of origin of the islands can be estimated. The eastern islands are oldest with those to the west being progressively younger (Abdel-Monem, Watkins & Gast, 1971, 1972; Anguita & Hernan, 1975, 1986; Carracedo, 1979; Anchocea *et al.*, 1990; and references in Thorpe, Watt & Baez, 1985 and McGregor, 1992). These western islands are occupied by one extant species, *Gallotia galloti* (except for a small relict population of *G. simonyi* on Hierro). For the reasons given above their distribution can only be explained by dispersal. These islands differ in form and position, and consequently in climate. Any morphological differences among island populations could therefore be due the dispersal history (phylogenesis), or adaptation to the different ecological conditions (ecogenesis).

Numerous morphological differences exist among the island populations of *G. galloti* (Thorpe, 1985a,b; Thorpe, Watt & Baez, 1985), but we can take body size as an illustration (Fig. 6). Adult body size differs among island populations even when raised under constant laboratory conditions. In north Tenerife and south Tenerife the lizards are large whilst in the west the lizards from La Palma are smaller and those from Gomera and Hierro are smallest. Attempts to relate body size to island size have been shown to be flawed by Thorpe (1985b), but reduced body size could either reflect ecogenetic adaptation to similar ecological conditions on far western islands (La Palma, Hierro and Gomera), or phylogenesis (i.e. a clade of smaller animals occupying La Palma, Gomera and Hierro).

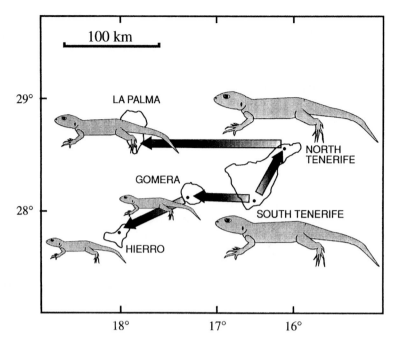

Figure 6 *Gallotia galloti* on the western Canary islands vary in size. Those from Tenerife are largest while on islands to the west they are smaller, particularly in Gomera and Hierro. When the colonization path is deduced from a DNA phylogenetic tree (Thorpe *et al.*, in press) it is apparent that these smaller lizards on La Palma, Hierro and Gomera do not constitute a separate 'small-body' clade, but represent ecogenetic adaptation of the northern and southern lineages that separately colonized westwards.

Phylogenetic trees (Fitch-Margoliash and Wagner trees) were derived from a range of mitochondrial DNA data, that is six-cut restriction fragment length polymorphisms (RFLPs), 4-cut RFLPs, cytochrome b sequence, cytochrome oxidase I sequence and 12s rRNA sequence (Thorpe *et al.*, in press). Data from a novel RAPD (random amplified polymorphic DNA) analysis of nuclear DNA was also used to construct a tree (Thorpe *et al.*, in press). None of these 6 trees indicated that the populations of La Palma, Hierro and Gomera formed a separate clade (Thorpe *et al.*, in press).

When the colonization sequence amongst the islands was deduced from either geography plus tree topology, or topology plus anagenesis, using rules explained in Thorpe *et al.* (in press), then an origin in southern Tenerife is indicated with two independent expansions westward from older to progressively younger islands. Hierro is colonized from south Tenerife via Gomera; and La Palma is colonized from north Tenerife. Various estimates for the time of geological origin of the islands (McGregor, 1992; Thorpe *et al.*, in press) can be compared to the estimates for time of colonization based on molecular clock assumptions, whether taken overall (Thorpe *et al.*, in press), or for each DNA data type separately (Table 6) (McGregor, 1992). These procedures agree in so far as they indicate that in each case the hypothesized time of colonization of an island is only after its estimated geological origin.

One can see that this phylogenetic analysis, independent of the morphology, allows one to reject the hypothesis that the evolution of small size is primarily due

Table 6 Origin (oldest rocks) of western Canary Islands (Anguita & Hernan, 1975) and divergence times for their *Gallotia galloti* populations (in millions of years) based on various mitochondrial DNA data. DNA dates derived from McGregor (1992), but see also Thorpe *et al.* (1993) and Thorpe *et al.* (in press).

Island	Origin	RFLP	Cyt b	Cyt ox I	12s rRNA
Tenerife	15.7	5.5	5.5	8.0	11.0
Gomera	12.0	2.0	2.5	4.0	3.5
La Palma	1.6	0.4	1.1	0.2	–
Hierro	0.8	0.5	0.6	–	–

to 'phylogenesis'. Small size appears to be an ecogenetic adaptation which occurs independently in the two lineages.

This approach has been combined with the use of Mantel tests (Thorpe, in press). The observed matrix of morphological differences (single characters or multivariate generalizations) are the dependent matrix, with independent matrices representing various hypothesized causes. Phylogenesis, for example, is represented by patristic distances on the molecular phylogenetic tree, whilst ecogenesis is represented by several, potentially important ecological variables. This approach is related to the 'comparative method' (Cheverud, Dow & Leutenegger, 1985).

Inter-island variation in the endangered *Iguana delicatissima* provides a further example of how DNA data may aid in the interpretation of morphological variation. This species has a patchy distribution across the Lesser Antilles from Martinique and Dominica in the south to Anguilla in the north. Morphological variation between (and in some cases within) islands is revealed by multivariate analysis of 60 variables from the scalation, colour pattern and body dimensions. Is this geographic variation due to adaptation to the ecological differences between islands, or does it reflect the historical colonization sequence? Analysis of the sequence of the mitochondrial gene cytochrome b in other iguanids reveals substantial inter-population variation (11–14% in *Anolis oculatus*; Malhotra, 1992). There are clear differences in cytochrome b sequence between the two species of iguana in this area (*Iguana iguana* and *I. delicatissima*). However, in *I. delicatissima* there are no cytochrome b sequence differences among the island populations of Desirade, Anguilla and Dominica, even though they are from opposite ends of the species range. This implies that the populations of this species have been dispersed recently, probably by human factors (pre-Columbian Indians) and the observed morphological differences are largely, or solely, due to ecogenetic adaptation to the different ecological conditions on each island.

SUMMARIZED DISCUSSION

It is evident, that whilst it may be difficult to determine the relative importance of 'phylogeny' and 'ecology' in causing a pattern of geographic variation, there are several procedures which may help us do so. Simultaneous Mantel tests appear to be a particularly useful technique, even though there is an arbitrary element in how the observed and hypothesized patterns are considered. Parallel patterns found in independent situations may be useful where they are found, and when possible manipulative field experiments can be very revealing. However, these two procedures will not be of widespread utility. DNA data can now play a very important role in revealing intraspecific phylogenesis and aiding us in interpreting variation in morphological data.

ACKNOWLEDGEMENTS

The senior author (RST) wishes to thank Bryan Manly for access to his Mantel test programs, and the following bodies for financial support: Science and Engineering Research Council (GR/F/19968 to RST and A. Cumming); Natural Environment Research Council (GR3/6943 to RST and A. Cumming); The Leverhulme Trust; The Carnegie Trust for the Universities of Scotland and The Royal Society. We acknowledge additional support from SERC (research studentships to WW, AM, MD) and NERC (research studentship to RB and a postdoctoral fellowship to WW).

REFERENCES

ABDEL-MONEM, A., WATKINS, N.D. & GAST, P.W., 1971. Potassium–Argon ages, volcanic stratigraphy, and geomagnetic polarity history of the canary islands: Lanzarote, Fuerteventura, Gran Canaria, and La Gomera. *American Journal of Science, 271*: 490–521.

ABDEL-MONEM, A., WATKINS, N.D. & GAST, P.W., 1972. Potassium–argon ages, volcanic stratigraphy and magnetic polarity history of the Canary Islands: Tenerife, La Palma and Hierro. *American Journal of Science, 272*: 805–825.

ANCHOCEA, E., FUSTER, J.M., IBARROLA, E., CENDRERO, A., COELLO, J., HERNAN, F., CANTAGREL, J.M. & JAMOND, C., 1990. Volcanic evolution of the island of Tenerife (Canary Islands) in the light of new K-ar data. *Journal Volcanolology Geothermal Research, 44*: 231–249.

ANGUITA, F. & HERNAN, F., 1975. A propagating fracture model versus a hot spot origin for the Canary Islands. *Earth and Planetary Science Letters, 27*: 11–19.

ANGUITA, F. & HERNAN, F., 1986. Geochronology of some Canarian dike swarms: contribution to the volcano-tectonic evolution of the archipelago. *Journal Volcanolology Geothermal Research, 30*: 155–162.

BAKER, A.J., 1980. Morphometric differentiation in New Zealand populations of the house sparrow (*Passer domesticus*). *Evolution, 34*: 638–653.

BAKER, A.J. & MOOED, A., 1979. Evolution in the introduced New Zealand populations of the common myna, *Acridotheres trictis* (Aves: Sturnidae). *Canadian Journal of Zoology, 57*: 570–584.

BARKER, J.S.F. & MULLEY, J.C. 1976. Isozyme variation in natural populations of *Drosophila buzzatii*. *Evolution, 30*: 213–233.

BROWN, R.P. & THORPE, R.S., 1991a. Description of within-island microgeographic variation in body dimensions and scalation of the skink *Chalcides sexlineatus*, with testing of causal hypotheses. *Biological Journal of the Linnean Society, 44*: 47– 64.

BROWN, R.P. & THORPE, R.S., 1991b. Within-island microgeographic variation in the colour pattern of the skink, *Chalcides sexlineatus*: pattern and cause. *Journal of Evolutionary Biology, 4*: 557–574.

BROWN, R.P., THORPE, R.S. & BAEZ, M., 1991. Lizards on neighboring islands show parallel within-island microevolution. *Nature, 352*: 60–62.

CARRACEDO, J.C., 1979. *Paleomagnetismo e Historia Volcanica de Tenerife.* Tenerife: Aula de Cultura.

CHEVERUD, J.M., DOW, M.M. & LEUTENEGGER, W., 1985. The quantitative assessment of phylogenetic constraints in comparative analyses: sexual dimorphism in body weight among primates. *Evolution, 39:* 1335–1351.

CHEVERUD, J.M., WAGNER, G.P. & DOW, M.M., 1989. Methods for the comparative analysis of variation patterns. *Systematic Zoology, 38:* 201–213.

DIETZ, E.J., 1983. Permutation tests for association between two distance matrices. *Systematic Zoology, 32:* 21–26.

DOW, M.M., CHEVERUD, J.M. & FRIEDLAENDER, J., 1987. Partial correlation of distance matrices in studies of population structure. *American Journal of Physical Anthropology, 72:* 343–352.

ENDLER, J.A., 1980. Natural selection on color patterns in *Poecilia reticulata. Evolution, 34:* 76–91.

HALKKA, O. & RAATIKAINEN, M., 1975. Transfer of individuals as a means of investigating natural selection. *Hereditas, 80:* 27–34.

HEANEY, L.R., 1985. Zoogeographic evidence for Middle and Late Pleistocene landbridges to the Philippine Islands. *Modern Quarternary Research in South-East Asia, 19:* 127–143.

HEANEY, L.R., 1986. Biogeography of mammals in SE Asia: estimates of rates of colonization, extinction and speciation. *Biological Journal of the Linnean Society, 28:* 127–165.

JOHNSTON, R.F. & SELANDER, R.K., 1971. Evolution in the house sparrow. II. Adaptive differentiation in North American populations. *Evolution, 25:* 1–28.

KNIGHTS, R.W. 1979. Experimental evidence for natural selection on shell shape in *Cepaea hortensis* (Mull). *Genetica, 50:* 51–60.

McGREGOR, D.P., 1992. 'Mitochondrial DNA Evolution in Canary Island Lizards (genus: *Gallotia)*'. Unpublished Ph.D Thesis, University of Aberdeen.

MALHOTRA, A. 1992. 'What Causes Geographic Variation: a Case Study of *Anolis oculatus*'. Unpublished Ph.D Thesis, University of Aberdeen.

MALHOTRA, A. & THORPE, R.S., 1991a. Microgeographic variation in Anolis oculatus on the island of Dominica, West Indies. *Journal of Evolutionary Biology, 4:* 321–335.

MALHOTRA, A. & THORPE, R.S., 1991b. Experimental detection of rapid evolutionary response in natural lizard populations. *Nature, 353:* 347–348.

MALHOTRA, A. & THORPE, R.S., 1993. An experimental field study of an eurytopic anole, *Anolis oculatus. Journal of Zoology, London, 229:* 163–170.

MALHOTRA, A. & THORPE, R.S., in press. An evaluation of Mantel tests: testiry geographic patterns in the Dominican anole.

MANLY, B.F.J., 1986. Randomization and regression methods for testing associations with geographical, environmental and biological distances between populations. *Researches on Population Ecology, 28:* 201–218.

MANLY, B.F.J., 1991. *Randomization and Monte Carlo methods in biology.* New York: Chapman & Hall.

MANTEL, N., 1967. The detection of disease clustering and a generalized regression approach. *Cancer Research, 27:* 209–220.

SCHNELL, G., DOUGLAS, M. & HOUGH, D.J., 1986. Geographic patterns of variation in offshore spotted dolphins (Stenella attenuata) of the eastern tropical Pacific Ocean. *Marine Mammal Science, 2:* 186–213.

SMOUSE, P.E., LONG, J. & SOKAL, R.R., 1986. Multiple regressions and correlation extensions of the Mantel test of matrix correspondence. *Systematic Zoology, 35:* 627–632.

SOKAL, R.R., ODEN, N.L. & WILSON, C., 1991. Genetic evidence for the spread of agriculture in Europe by demic diffusion. *Nature, 351:* 143–145.

THORPE, R.S., 1979. Multivariate analysis of the population systematics of the ringed snake *Natrix natrix. Proceedings of the Royal Society of Edinburgh, 78:* 1–62.

THORPE, R.S., 1984. Primary and secondary transition zones in speciation and population differentiation: a phylogenetic analysis of range expansion. *Evolution, 38*: 233–243.

THORPE, R.S. 1985a. Relative similarity between subspecies of the western Canary Island lizard, *Gallotia galloti. Bonner Zoologische Beitrage, 36*: 529–532.

THORPE, R.S., 1985b. Body size, island size and variability in the Canary Island lizards of the genus *Gallotia. Bonner Zoologische Beitrage, 36*: 481–487.

THORPE, R.S., 1991. Clines and cause: microgeographic variation in the Tenerife gecko *Tarentola delalandii. Systematic Zoology, 40*: 172–187.

THORPE, R.S., in press. The use of DNA divergence to help determine the correlates of evolution of morphological characters.

THORPE, R.S. & BAEZ, M., 1987. Geographic variation within an island: univariate and multivariate contouring of scalation, size and shape of the lizard *Gallotia galloti. Evolution, 41*: 256–268.

THORPE, R.S. & BAEZ, M., 1993. Geographic variation in scalation of the lizard *Gallotia steblini* within the island of Gran Canaria. *Biological Journal of the Linnean Society, 48*: 75–87.

THORPE, R.S. & BROWN, R.P., 1989a. Testing hypothesized causes of within-island geographic variation in the colour of lizards. *Experientia, 45*: 397–400.

THORPE, R.S. & BROWN, R.P., 1989b. Microgeographic variation in the colour pattern of the lizard *Gallotia galloti* within the island of Tenerife: distribution, pattern and hypothesis testing. *Biological Journal of the Linnean Society, 38*: 303– 322.

THORPE, R.S. & BROWN, R.P., 1991. Microgeographic clines in the size of mature male *Gallotia galloti* (Squamata: Lacertidae) on Tenerife: causal hypotheses. *Herpetologica, 47*: 28–37.

THORPE, R.S. & MALHOTRA, A., 1992. Are *Anolis* lizards evolving? *Nature, 355*: 506.

THORPE, R.S., WATT, K. & BAEZ, M., 1985. Some interrelationships of the Canary Island lizards of the genus *Gallotia. Bonner Zoologische Beitrage, 36*: 577–584.

THORPE, R.S., BROWN, R.P., MALHOTRA, A. & WÜSTER, W., 1991. Geographic variation and population systematics: distinguishing between ecogenetics and phylogenetics. *Bollettini di Zoologia, 58*: 329–335.

THORPE, R.S, McGREGOR, D. & CUMMING, A.M., 1993. Molecular phylogeny of the Canary Island lacertids (*Gallotia*): mitochondrial DNA restriction site divergence in relation to sequence divergence and geological time. *J. Evol. Biol., 6*: 725–735.

THORPE, R.S., McGREGOR, D.P., CUMMING, A.M. & JORDAN, W.C., in press. DNA evolution and colonization sequence of island lizards in relation to geological history: mtDNA RFLP, cytochrome b, cytochrome oxidase, 12s rRNA sequence, and nuclear RAPD analysis. *Evolution.*

WILSON, A.C., CANN, L.C., CARR, S.M., GEORGE, M., GYLLENSTEN, U.B., HELM-BYCHOWSKI, K.M., HIGUCHI, R.G., PALUMBI, S.R., PRAGER, E.M., SAGE, R.D. & STONEKING, M., 1985. Mitochondrial DNA and two perspectives on evolutionary genetics. *Biological Journal of the Linnean Society, 26*: 375–400.

Adaptation and phylogenetic inference

ADRIAN E. FRIDAY

CONTENTS

Keywords: adaptation – Comparative Method – evolutionary trees – maximum likelihood – parsimony – phylogeny.

Abstract

The Comparative Method focuses renewed attention on the distribution of character states among species, and particularly on patterns of homoplasy. It has placed emphasis on the elucidation of adaptive change by examination of the reconstructed pattern of occurrence of derived character states on evolutionary trees. Species that are not related by immediate common ancestry, but that appear to demonstrate separate acquisitions of derived states, may be counted as independent instances when evaluating theories of adaptation. The assumptions underlying phylogenetic estimation are discussed, and it is argued that information about adaptation can be fed into attempts to reconstruct evolutionary trees, as well as emerging from those reconstructions.

INTRODUCTION

It is profitable to examine the relationship between ecological principles and the procedures of phylogenetic analysis through an evaluation of the sorts of assumptions made in the latter.

Phylogenetics and Ecology
ISBN 0-12-232990-2

Fisher's Method of Maximum Likelihood was introduced into the field of phylo-genetic analysis by Edwards & Cavalli-Sforza (1963, 1964; Cavalli-Sforza & Edwards, 1967). Because of conceptual and computational difficulties, found at the time to be inhibiting, Edwards & Cavalli-Sforza also considered the feasibility of more approx-imate approaches to phylogenetic estimation. This led to the proposal of various approximations to full maximum likelihood estimation. These approaches would now be recognized under the broad heading of 'parsimony' methods. Although the term 'parsimony' has, unfortunately, been used in different ways in the literature of evolutionary biology, it is used below exclusively in the context of estimation.

Shortly after the period during which Edwards & Cavalli-Sforza were investigating their dual approach to the estimation of phylogenetic trees, Hennig's cladistic methodology was also becoming more widely known (Hennig, 1966) in combina-tion with the accessory Principle of Parsimony. It is significant that the two approaches were concerned with different types of data. The likelihood approach was employed initially in the analysis of gene frequencies, whereas Hennig's methods were introduced in connection with qualitative data in the form of morphological characters.

Demonstrations, in parallel, by Felsenstein (1973) and Thompson (1975) that the Method of Maximum Likelihood could, indeed, be made to work in the practical estimation of phylogenetic trees brought to the subject a well-defined, statistical framework. Unfortunately, the clear distinction between the different components of phylogenetic inference, which was a feature of these initial studies, has not always been maintained: method of estimation and model of process have frequently been confounded. It is common to encounter references to 'the maximum likelihood method', meaning not simply the method of estimation, but the combination of this with a particular model of evolutionary change (usually a form of Poisson process model). The advantages of keeping as clear a distinction as possible between model, on one hand, and method of estimation, on the other, should be apparent in the issues dealt with below.

It also emerged from these studies that minimum evolution approaches could not necessarily, for the data and models in question, be relied upon to provide good approximations to the full maximum likelihood solutions. There has since been a tendency to consider the two approaches to estimation, likelihood and parsimony, to be somehow in conflict, despite strenuous attempts, described below, to demon-strate the conditions under which they would agree in their respective, best-supported estimates of the pattern of evolutionary history.

LIKELIHOOD JUSTIFICATIONS OF PARSIMONY

There have been several, widely-quoted attempts to characterize, in stochastic terms, the model (or models) underlying parsimony approaches. Farris (1973) achieved an element of synthesis by strongly defending phylogenetic estimation as essentially a statistical concern and suggesting a stochastic model with minimal assumptions about evolutionary processes. Farris's arguments were based on the intuitively appealing suggestion that if a tree were divided into a large number of sections, each representing a suitably short period of time, the probability of change for a character on any given segment would be small. The additional assumptions in his formulation were constructed in such a way that change was allowed to be reversible; that the probabilities of change for different characters, although small, could be (reasonably) different; and, most importantly, that, although individual changes were of low probability over the segments chosen, there was no restriction on the total number of changes for a given character over the whole tree. The last feature was included to negate the criticism that a minimum evolution approach assumed that change was inherently rare over an entire tree and that instances of homoplasy were, correspondingly, even rarer. Farris's assumptions were intended to be fairly realistic and unexceptionable, and yet to lead to the demonstration that, under these conditions, the tree topology of least length, in terms of state changes, was also the solution of highest likelihood.

Felsenstein independently gave much attention to the assumptions under which the most parsimonious solution for tree topology was equivalent to that of highest likelihood. The elements of his approach were established in an early publication (Felsenstein, 1973) and were subsequently developed in more detail (Felsenstein, 1978, 1981). His 1981 formulation again assumed that a (two-state) character may change state over a period of time, represented by the length of a branch on the tree, with a defined probability. The emphasis was then on the limiting value for this probability as the rate of change for the given character becomes very small. Specifically, if the rate of change per unit time for the character concerned is u, and the period of time under consideration is t (that is, t is the time elapsed for a given branch on the tree) then the probability of change of state approaches ut (and the probability of no change is, accordingly, 1−ut). Having previously established a general procedure for the calculation of the likelihood of a tree, in the case of many characters, Felsenstein demonstrated that, given the limiting value for the probability of change, the likelihood takes a simple form: it is essentially the product of the individual ut values for all instances of change on the tree. If the likelihood is expressed as its logarithm, then this log-likelihood is the sum of the relevant values, ln(ut). These log values (made positive) were characterized by Felsenstein as the 'weights' of the changes and he noted that, when all values of the weights are equal, the tree pattern of highest likelihood is the Wagner parsimony solution (Kluge & Farris, 1969; Farris, 1970).

This demonstration of the simple behaviour of the likelihood additionally assumes not only that the values of ut are small, but also that they are similar to each other, which, in turn, requires that the branch lengths (in terms of time) be approximately equal. Felsenstein sets up conditions under which, by limiting the disparity between branch lengths, it is not possible for one change on a short branch to be less probable than two changes on long branches. In other words, a reconstruction involving one change will contribute a very much greater amount to the total likelihood than one involving two changes in the same character on two different branches. Under these conditions, the reconstruction of non-divergent change is minimized. Having established the formulation of the 'weights' in probabilistic terms, Felsenstein goes on to elucidate, in similar fashion, the nature of several other approaches to phylogenetic reconstruction, such as compatibility analysis.

By these arguments, both Farris and Felsenstein attempted to demonstrate conditions under which the maximum likelihood estimate of the tree pattern is the same as that derived using a parsimony approach. It should be emphasized, however, that whereas Farris was attempting to justify the use of parsimony in general, Felsenstein was more concerned to define the conditions under which parsimony approaches would perform poorly. In particular, attention was directed at the failure of these approaches to be statistically consistent in general. Parsimony approaches would, under some conditions, continue to support incorrect phylogenetic estimates even when provided with ever larger quantities of data. Under the same conditions, however, the likelihood approach would exhibit statistical consistency (Felsenstein, 1978).

A difference between the formulations of Farris and Felsenstein concerns the expected behaviour of homoplasy under the respective models. It has already been noted that Farris explicitly set conditions under which the number of independent state changes of a character on a tree is not limited; whereas it follows from Felsenstein's model that such events would be uncommon, because any change on the tree is uncommon. In Farris's treatment the number of short tree-segments is potentially very large and independent occurrences of change in the same character are thus permitted to be frequent, despite the low probability of each change within its own segment and the fact that the tree might be a simple one with only several branches. Felsenstein, however, assigns low probabilities of change on a per branch basis: separate occurrences of change on different branches would be correspondingly more unlikely. Despite this formal difference, the tactic of searching for the tree pattern that requires the minimum total number of reconstructed changes remains the correct one in both cases. In other words, in both cases a tree pattern that involves the fewest instances of homoplasy will be favoured, even though that number of instances might be considerable. It is evident that for Felsenstein, finding frequently that the minimum number of instances of homoplasy is, in fact, still substantial should lead to doubts about whether the model is admissible. Those who are sceptical about the feasibility of applying a stochastic formulation at all might well hold that the model should, indeed, be rejected on these grounds.

Farris's model appears not to have this problem, but from its inception his formulation was strongly attacked by Felsenstein on the grounds that it lacked statistical consistency. The exchange of arguments can be followed chronologically in a parallel series of publications (Farris, 1973, 1983; Felsenstein, 1973, 1978, 1981, 1985). The issues have also recently been re-examined in detail, and new insights gained, by Goldman (1990) whose account may be consulted for a more complete review of the relevant literature and for a critical discussion of the important statistical theoretic issue of just what the two approaches really are attempting to estimate. In summary, Goldman underwrites the main conclusion of both Farris and Felsenstein that the 'best' tree topologies obtained by using likelihood and parsimony as methods of estimation will coincide, in the case of a stochastic model of evolutionary change, when the probabilities of state change involved are small. Effectively, this means that the models under which each approach is operating are brought into equivalence by the enforcement of a limiting condition, but the methodology of parsimony is not, of course, making direct use of the probability values. Without the limiting condition, the tree topologies obtained by the two approaches will not necessarily be the same.

The formulations described above make the assumption of statistical independence between characters and also of independence of change in a given character on different parts of the tree. Such assumptions might be regarded as biologically quite unrealistic in view of the manifest, experimentally verifiable, functional relations between morphological characters. The same criticism applies, incidentally, to the stochastic models currently in use with molecular data. In mitigation, however, it is currently difficult to see how to drop the assumptions of independence and still formulate any sort of stochastic model that would be mathematically tractable in the context of phylogenetic estimation.

THE PRINCIPLE OF DIVERGENCE

It is reasonable to question whether the assumptions made explicit in these probabilistic formulations really do accord with those being made in approaches that customarily employ the Principle of Parsimony. Perhaps approaches that use parsimony might better be characterized as being based on deterministic models (as suggested in Bishop & Friday, 1985). Certainly, in practice, maximum likelihood estimation is rarely used with morphological characters, largely because the particular models of evolutionary change currently available are generally felt to be unconvincing for these data. One perceived source of difficulty, for example, is that the stochastic models do not include any apparent recognition of the role of natural selection, which is, in any case, sometimes characterized as a deterministic force and contrasted with other components of the evolutionary process that quite clearly can be characterized as stochastic. It is not, however, necessarily the case that selection is being ignored in stochastic models of evolutionary change, even though no explicit mention is made of it. Many processes that contain a multiplicity of complex interacting components can, nevertheless, be modelled with a reasonable degree of success by quite simple stochastic models: the net behaviour of a complex system need not appear complex, and a model stands or falls by how consistently relevant it is found to be in conjunction with data, not by how literally it mirrors the real, individual components of the system in question.

Most recent phylogenetic studies of comparative morphological data have used some version of Hennigian ('conventional') cladistic methodology. This implies a conviction that evolutionary processes, in all their complexity, nevertheless lead to a tell-tale patterning of the distribution of character states among species: a patterning simple and strong enough to retain evidence of evolutionary relationships in the embedded hierarchic structure marked out by sequential evolutionary innovations. Regardless of what probabilities of change might have been operative, the application of the Principle of Parsimony ensures that the solution favoured is that which involves the smallest number of non-divergent events. This is consistent with the point of view, occasionally put forward, that preferring the most parsimonious cladogram is equivalent to minimizing the number of *ad hoc* hypotheses needed to 'explain' instances of homoplasy or non-congruence. In this context, Felsenstein (1985) has criticized Farris (1983) for retreating from his former, statistical, attitude to phylogenetic estimation into what has been described as a 'hypothetico-deductive' framework. From this point of view, the hypothesis that change is (predominantly) divergent, or that the majority of characters are mutually congruent, represents a single statement that simultaneously 'explains' all instances of this sort. In contrast, each instance of homoplasy or non-congruence demands an individual *ad hoc* explanation. It is worth recalling, at this point, that it is with instances of homoplasy that the Comparative Method is largely concerned (see Coddington, Chapter 3), and, in a sense, it seeks also to provide collective, rather than separate, explanations for instances of homoplasy.

Modern ideas about evolutionary divergence appear to take their origin from Darwin's 'Principle of Divergence' (Darwin, 1859; see discussion in Ospovat, 1981; Friday, 1987) and they surely involve an assumption, albeit a rather weak one, about evolutionary processes. Perhaps, however, the weak nature of this assumption, that divergent change is predominant, is the very guarantee of its credibility. Certainly, if we wish to reject any attempt to model morphological evolution in terms of assumptions already as weak as those underlying the simple stochastic processes described above, then any generalizations about processes of evolutionary change stronger than the Principle of Divergence are unlikely to be convincing.

The Principle of Divergence, however, simply represents a statement about a large-scale characteristic of evolutionary change, and although such a statement about the general tendency for divergence might well be thought defensible on the grounds of ecological theory, for example, it is not possible to retrieve from the Principle information about the relative probabilities of individual evolutionary events. It is instructive to consider the prospect of embarking upon the simulation of evolutionary change (on a given tree topology) using only the fact that there is a general tendency for divergence, and with no statement (not even a stochastic one) about where, and to what extent, non-divergent change occurs. This is the very sense in which the Principle of Divergence can be described as weak and it surely cannot on its own justify the use of parsimony as a criterion for optimisation in an evolutionary context.

These considerations may help to illuminate the nature of some assumptions underlying the procedures of cladistic analysis.

THE NATURE OF CLADOGRAMS

A cladogram is a diagram of set membership and not an evolutionary tree. Although there must be well-defined rules for the construction of cladograms, if a cladogram is to be permitted to remain only a diagram of set membership, then these rules need not make reference to any aspect of the process of evolution. When viewed in this way, however, neither can a cladogram justifiably claim to make any statement about evolutionary history: its justification may be sought in the degree to which it represents the embedded, hierarchic structure of the data and in terms of efficiency of storage of this information. From this 'pattern' cladistic point of view the data are being interpreted in terms of a set-theoretic model. It seems perfectly legitimate to use the Principle of Parsimony as a criterion for optimization in this context and it is, perhaps, even tempting to stop with the cladogram as some sort of systematic 'summary diagram' and to leave its relationship to evolutionary history as undefined.

However, when the ultimate aim of a cladistic analysis is held to be an estimate of an evolutionary tree, then, for example, the criteria used to generate hypotheses of the polarity of character state change only make sense in an evolutionary context. In these circumstances, the cladogram cannot maintain the same logical detachment and the phenomenon of nesting of sets represents a deduction from the Principle of Divergence: that is, a deduction from weak assumptions about the processes of evolutionary change. Finding an implied nesting is *en route* to an evolutionary tree, which is what is really required. This deduction of the property of hierarchic nesting is a useful one to be able to make, but it does not, of itself, identify which particular nesting is to be favoured: hence there is a need for an accessory Principle of Parsimony. Parsimony will certainly help to resolve the issue; but, without further (model) assumptions that provide information about the relative probabilities of different reconstructions of evolutionary events, it cannot be relied upon to have done so in a way that necessarily best reflects the course of evolution.

Perhaps conventional cladistic methodology might be characterized as being able to accommodate the whole class of models that give rise to predominantly divergent change. This is a potentially rather large class. The accessory use of parsimony would then ensure that we are somewhere near a reasonable estimate of the pattern of past events, but in the absence of extrinsic information it could not tell us how near. From this point of view, conventional cladistic methodology represents a relatively coarse approach to evolutionary reconstruction (and a correspondingly undemanding one in terms of commitment to theory). With real data it often appears by no means unconvincing that a rather less than most parsimonious solution might well be correct.

If for real morphological data, we cannot, in principle, know when we are right, then we might as well regard the Principle of Parsimony simply as an operational aid, apply it rigorously, and live happily in ignorance with the potential of error. The fact that we were wrong, in any particular case, might come to light on collection and analysis of further data, but without the guarantee of statistical consistency we cannot be at all sure of that. It might be argued that the rigorous use of parsimony at least ensures that we establish a minimum estimate of the amount of homoplasy implicit in some data, and that this estimate is a useful (and suitably conservative) one to have. This may be so, but the amount is not necessarily the feature of sole interest for the Comparative Method: it is also the distribution of individual instances among species, and that will not necessarily be well-estimated.

In the light of these difficulties, two further tactics are of value. The first is simply to accept that independent cladistic studies of the same group are highly desirable: they may substantially agree on at least a small set of possible solutions, in which case our confidence increases that we are on the right track (or that we are being massively misled). The second, and perhaps more important, is to examine the distribution of topologies, ranked by length, which can give an indication of the stability of a solution. Where the minimum length pattern (or set of patterns) is relatively isolated it is likely to remain stable on incorporation of further data. Such distributions are already quite commonly presented. It bears repeating that in the absence of a demonstration of statistical consistency for cladistic methodology, however, both these tactics must remain formally suspect.

RE-ENTER ADAPTATION?

Since it is difficult to see how to set up convincing stochastic models of evolutionary change for morphological characters (and especially to see how to set numerical values for probabilities of evolutionary events) and difficult, also, to have complete conviction in the use of parsimony as a method of estimation, what is to be done?

Even if the Principle of Divergence is too weak to justify convincing support for one solution (or a small number of solutions) out of many, then at least the different instances of homoplasy implied by competing reconstructions can be evaluated using criteria of adaptive credibility.

The Comparative Method puts emphasis on gaining evidence, from phylogenetic trees, of independent instances of evolutionary innovation, so that these instances may be used to evaluate theories of adaptation. It is, however, possible to reverse this logic, with some important consequences for the procedures of phylogenetic estimation. There has been a reluctance to use convictions about adaptation in phylogenetic studies. From the point of view of cladistic methodology, this reluctance might be best understood as an unwillingness to introduce any more elements of theory into the methodology than are strictly necessary. Simplicity of assumptions (simplicity of the model) is a laudable principle (and this, incidentally, appears to be another sense in which the term 'parsimony' is sometimes used) but, perhaps, it can be taken too far. In a desire to ensure that hypotheses of evolutionary relationship are as independent from theory as possible, and that they involve the minimum recourse to ideas about evolutionary process that are difficult, if not impossible, to make wholly objective, perhaps one possible tactic is not getting the formal emphasis it might.

Adaptation is a valid and necessary biological concept in its own right (Williams, 1966, 1992; Dunbar, 1982) and it is surely the case that hypotheses of adaptation can often be held with greater conviction than can the Principle of Parsimony, itself, for example. The latter becomes simply a working tool to enable the generation of a set of plausible evolutionary trees for further evaluation by the former. Ideas about adaptation, as well as emerging from cladistic analysis, might also be used to evaluate competing hypotheses of relationship between species.

Effectively, this amounts to a suggestion that some assessment of the relative probabilities of encountering convergent change in different characters might be defensible on adaptational grounds. This argument emerges as one in defence of character weighting, which is then seen in its correct role as an *a priori* statement about the probability of finding convergence in a given character, rather than as an *a priori* statement about the probability of finding individual instances of state change in that character. In practice, quietly, this seems to be the route that many take to resolve which tree pattern to prefer out of a number separated by only little in terms of parsimony. Perhaps it should be recognized as legitimate, if not inevitable. What is involved is reciprocal illumination rather than circularity of logic, because there is an obvious prohibition on extracting from the tree the same adaptational convictions that went into it. This suggestion rests, however, on the claim that some instances of adaptation are relatively compelling.

An example may help here, and the following one is based on molecular data. Incidentally, this example illustrates the importance of thinking in a functional way about molecular sequences, and is chosen also because it is relevant to recent discussions about how best to frame models of processes of molecular evolutionary change.

It has long been known that in DNA G:C base-pairings are thermodynamically more stable than are the pairings A:T, and thermodynamic measurements have been made to quantify this (Aboul-ela, Koh & Tinoco, 1985). It might, on these grounds alone, be suggested that organisms with high body temperatures or those that live in hotter environments could be expected to take advantage of this physical fact. Recent experimental evidence strongly supports this suggestion with the finding that frequencies of the bases G and C are positively correlated with environmental temperature in organisms that are apparently only distantly related phylogenetically (Dixon, Simpson-White & Dixon, 1992; Williams *et al.*, 1993). It appears virtually certain that, for some of these organisms, the possession of relatively high levels of G and C represents independent instances of adaptation to high temperatures. It has been claimed that such bias in base composition confounds the reconstruction of evolutionary relationships, using most current approaches, because organisms that are, on other evidence, unlikely to be closely related tend to be grouped together (Lockhart *et al.*, 1992). Some progress is now being made with this problem (Steel, Lockhart & Penny, 1993) which, it has been suggested, may be at least part of the reason for the controversial grouping together of birds and mammals on some evolutionary trees based on molecular sequence data (Bishop & Friday, 1988).

Information of this sort could be used to support the elimination of some out of a number of competing tree patterns; whereas other instances of adaptation are less obvious and more likely to come to light as a result of mapping additional data onto already existing evolutionary trees.

THE ROLE OF MOLECULAR DATA

How should molecular data be viewed in the light of the arguments presented here? Perhaps they will indeed prove, as so often suggested, to be the main source of independent phylogenies on the basis of which theories of adaptation can be assessed. They have three possible advantages when considered against morphological data. The first is that the boundaries of the characters are more clearly defined, although this does not, of course, necessarily extend to the clarity with which it is possible to establish comparability of characters between sequences.

The second possible advantage is that some molecular data might be regarded as having changed over time in ways that are neutral, or nearly neutral, in terms of natural selection; and that this, in itself, should lead more readily to formulation of convincing models of evolutionary change for these molecules. This is an argument that, understandably, is rarely encountered with reference to morphological characters and it is also becoming increasingly difficult to defend for molecular data as more and more is learnt about molecular function, and especially about interactions within genomes. It has sometimes been the case for molecular data that neutrality was assumed unless evidence could be presented to the contrary. It is now rather more questionable to accept neutrality as the default option, but it may be that third base positions of codons, pseudogenes and stretches of intron sequences will turn out, as so widely hoped, to be particularly amenable candidates for evolutionary modelling.

The third favourable characteristic of molecular data is their very quantity. Even if selection should prove to be a virtually ubiquitous feature of molecular life, then, by the argument already suggested, perhaps the evolutionary behaviour of the data can be modelled by quite simple and tractable stochastic models. This is an empirical issue and is waiting, for its investigation, on better statistical techniques for the evaluation of the sorts of models concerned. Currently, analyses are being widely conducted without real attention being given to the validity of the few, simple models in general use. Doubtless it will not be long before it is established whether or not such sorts of models are generally applicable. There will then be some hope of a critical demonstration of the role of molecular data in obtaining reliable estimates of evolutionary trees.

ACKNOWLEDGEMENTS

I thank Paul Eggleton and Dick Vane-Wright for their invitation to contribute to the enjoyable and productive meeting and for their inexhaustible patience. I thank Stephen Wood for discussions about the nature of cladograms and David Williams for positive critical comments.

REFERENCES

ABOUL-ELA, F., KOH, D. & TINOCO Jr., I., 1985. Base-base mismatches. Thermodynamics of double helix formation for $dCA_3XA_3G+dCT_3YT_3G$ (X,Y=A,C,G,T). *Nucleic Acids Research, 13*: 4811–4824.

BISHOP, M.J. & FRIDAY, A.E., 1985. Evolutionary trees from nucleic acid and protein sequences. *Proceedings of the Royal Society of London, Series B, 226*: 271–302.

BISHOP, M.J. & FRIDAY, A.E., 1988. Estimating the interrelationships of tetrapod groups on the basis of molecular sequence data. In M.J. Benton (Ed.), *The Phylogeny and Classification of the Tetrapods, Vol. 1: Amphibians, Reptiles, Birds*, 33–58. Oxford: Clarendon Press.

CAVALLI-SFORZA, L.L. & EDWARDS, A.W.F., 1967. Phylogenetic analysis: models and estimation procedures. *Evolution, 32*: 550–570.

CODDINGTON, J.A., 1994. The roles of homology and convergence in studies of adaptation. In P. Eggleston & R.I. Vane-Wright (Eds), *Phylogenetics and Ecology*, 53–78. London: Academic Press.

DARWIN, C., 1859. *On the Origin of Species*. London: John Murray.

DIXON, D.R., SIMPSON-WHITE, R. & DIXON, L.R.J., 1992. Evidence for thermal stability of ribosomal DNA sequences in hydrothermal-vent organisms. *Journal of the Marine Biological Association of the United Kingdom, 72*: 519–527.

DUNBAR, R.I.M., 1982. Adaptation, fitness and the evolutionary tautology. In King's College Sociobiology Group (Eds), *Current Problems in Sociobiology*, 9–28. Cambridge: Cambridge University Press.

EDWARDS, A.W.F. & CAVALLI-SFORZA, L.L., 1963. The reconstruction of evolution. *Heredity, 18*: 553.

EDWARDS, A.W.F. & CAVALLI-SFORZA, L.L., 1964. Reconstruction of evolutionary trees. In V.H. Heywood & J. McNeill (Eds), *Phenetic and Phylogenetic Classification*, 67–76. London: Systematics Association.

FARRIS, J.S., 1970. Methods for computing Wagner trees. *Systematic Zoology, 19*: 83–92.

FARRIS, J.S., 1973. A probability model for inferring evolutionary trees. *Systematic Zoology, 22*: 250–256.

FARRIS, J.S., 1983. The logical basis of phylogenetic analysis. In N.I. Platnick & V.A. Funk (Eds), *Advances in Cladistics, 2*, 7–36. New York: Columbia University Press.

FELSENSTEIN, J., 1973. Maximum-likelihood and minimum-steps methods for estimating evolutionary trees from data on discrete characters. *Systematic Zoology, 22*: 240–249.

FELSENSTEIN, J., 1978. Cases in which parsimony or compatibility methods will be positively misleading. *Systematic Zoology, 27*: 401–410.

FELSENSTEIN, J., 1981. A likelihood approach to weighting and what it tells us about parsimony and compatibility. *Biological Journal of the Linnean Society, 16*: 183–196.

FELSENSTEIN, J., 1985. Phylogenies from gene frequencies: a statistical problem. *Systematic Zoology, 34*: 300–311.

FRIDAY, A.E., 1987. Models of evolutionary change and the estimation of evolutionary trees. In P.H. Harvey & L. Partridge (Eds), *Oxford Surveys in Evolutionary Biology, 4*, 61–88. Oxford: Oxford University Press.

GOLDMAN, N., 1990. Maximum likelihood inference of phylogenetic trees, with special reference to a Poisson process model of DNA substitution and to parsimony analyses. *Systematic Zoology, 39*: 345–361.

HENNIG, W., 1966. *Phylogenetic Systematics*. Urbana: University of Illinois Press.

KLUGE, A.G. & FARRIS, J.S., 1969. Quantitative phyletics and the evolution of anurans. *Systematic Zoology, 18*: 1–32.

LOCKHART, P.J., HOWE, C.J., BRYANT, D.A., BEANLAND, T.J. & LARKUM, A.W.D., 1992. Substitutional bias confounds inference of cyanelle origins from sequence data. *Journal of Molecular Evolution, 34*: 153–162.

OSPOVAT, D., 1981. *The Development of Darwin's Theory*. Cambridge: Cambridge University Press.

STEEL, M.A., LOCKHART, P.J. & PENNY, D., 1993. Confidence in evolutionary trees from biological sequence data. *Nature, 364*: 440–442.

THOMPSON, E.A., 1975. *Human Evolutionary Trees*. Cambridge: Cambridge University Press.

WILLIAMS, G.C., 1966. *Adaptation and Natural Selection*. Princeton, NJ: Princeton University Press.

WILLIAMS, G.C., 1992. *Natural Selection: Domains, Levels, and Challenges*. Oxford: Oxford University Press.

WILLIAMS, N.A., DIXON, D.R., SOUTHWARD, E.C. & HOLLAND, P.W.H., 1993. Molecular evolution and diversification of the vestimentiferan tube worms. *Journal of the Marine Biological Association of the United Kingdom, 73*: 437–452.

Comparing real with expected patterns from molecular phylogenies

PAUL H. HARVEY & SEAN NEE

CONTENTS

Keywords: Phylogeny – ecology – evolution – birds – molecular phylogeny – community structure.

Abstract

Molecular phylogenies of high quality are becoming increasingly available. This paper asks what those phylogenies can tell us about evolutionary processes. Because molecular phylogenies contain no information about lineages that have gone extinct, we develop appropriate null models based on Markov processes to reveal how molecular phylogenies are likely to compare with the real phylogenies that generated them. Altering rates of lineage splitting and extinction influences tree structure in clearly-defined ways, and the effects of mass extinctions can be revealed. A comparison of evolutionary models with Sibley & Ahlquist's molecular phylogeny of the birds, suggests that rates of effective cladogenesis (lineage birth minus death rates) have decreased through time. The same avian phylogeny is used to reveal and examine an unexpected pattern in community structure: species from tribes that radiated at different times show different relationships between population density and body size in natural communities. These differences are likely to result from ecological processes relating to guild structure.

Phylogenetics and Ecology
ISBN 0-12-232990-2

INTRODUCTION

Each passing month witnesses a massive accumulation of nucleotide base and amino acid sequence data that are being used to resolve phylogenetic relationships. There is a vigorous and healthy debate concerning the accuracy of those molecular-based phylogenies. The debate is vigorous in part because of ignorance: we need sound statistical techniques for multiple sequence alignment; we need to know the relative rates of change of particular molecular characters, say transversions to transitions, of particular amino acids to others; we need to determine whether the merits of maximum likelihood sufficiently outweigh those of parsimony for us to have to wait until likelihood techniques are further refined or new breeds of supercomputers have become available before we should proceed with tree recon-struction; and we need to know the quirks of molecular clocks. The debate is healthy because it is forcing rigour into the exercise, and there can be no doubt that techniques are improving with time.

But what can evolutionary biologists do with these new molecular phylogenies? In this chapter, we assume that molecular phylogenies are correct and develop a series of statistical null hypotheses to test against the data. As should be evident from the preceding, we know that molecular phylogenies are inaccurate. However, as their accuracy improves, so will our need to use them. After all, none of us questions the value of G.G. Simpson's attempts to trace the tempo and mode of evolution from an imperfect fossil record.

We start by asking what molecular phylogenies look like and how, in principle, they differ from the real thing. The answer, of course, is that molecular phylogenies are drawn in the main from extant species. Except for freaks of fortune, chemically preserved fossils, molecular material does not exist for extinct lineages. We develop a null model for the expected shape of molecular phylogenies. Analytical and simulation studies show that information about extinction may well be preserved in certain molecular phylogenies, and we might even find evidence for whether taxa that left no fossil record underwent mass extinction events.

We go on to compare our models against Sibley & Ahlquist's (1990) admittedly imperfect avian phylogeny, and invoke ecological explanations for the patterns we find. Finally, we use Sibley & Ahlquist's phylogeny to reveal and investigate phylo-genetically-based patterns in natural communities of birds living today.

A NULL PHYLOGENY

The pioneering work of Gould et al. (1977, and subsequent papers) erected null phylogenies to compare with real phylogenies. Their basic null model, which remains undisputed, was one in which, at any point in time, each lineage had the same probability of dividing into two (b, for birth) or going extinct (d, for death) as does every other lineage. This is often called the Markov model. Birth and death rates are not necessarily equal and, indeed, unless the birth rate exceeds the death rate, the whole phylogeny would inevitably go extinct. Such null phylogenies allow us to trace the change in the number of lineages against time. That luxury is not available for molecular phylogenies which provide us with no information about lineages that have gone extinct.

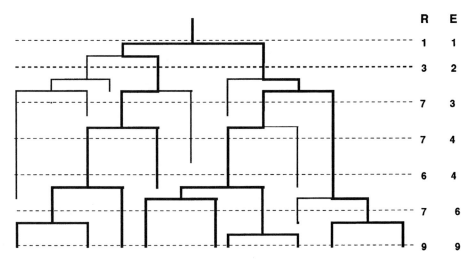

Figure 1 A hypothetical phylogeny. Those lineages that give rise to descendants in the present day are marked in bold. The column under the heading R (REAL) denotes the actual number of lineages at the points in time denoted by the dashed lines. The column under the heading E (ESTIMATED) gives the number of lineages which will give rise to present-day descendants at the same points in time. The latter column is the number of lineages that can be revealed from a molecular phylogeny.

Consider Fig. 1 which shows a phylogeny with lineages that gave rise to extant descendants at the present day picked out in bold. At each of seven time periods the number of lineages is recorded under the heading R (REAL). Several of those lineages went extinct so that, at the present day, nine still exist. Using molecular data, we can estimate when each pair of those nine species last shared a common ancestor, which would result in the bold-lined tree in the figure. Since we have no record of extinct lineages, the numbers under the heading E (ESTIMATED) provide a true estimate or an underestimate of the number of lineages at each specified time. Note that, from the origin of the phylogeny, E must increase through time, whereas R can both increase or decrease. In practice, molecular phylogenies gives us the estimated numbers with approximate times. Our aim here is to compare them with the real numbers, using a Markov model with specified lineage birth (b) and extinction rates (d). In the extreme, when d = 0, the real and estimated numbers of lineages are the same – all lines would be bold on the figure.

It turns out to be a fairly straightforward exercise to gain analytical expressions for the real and expected number of lineages at any time unit (Harvey, Nee & May, in press). Here we summarize the results of two computer simulations which bring out two major issues. In Figure 2a d = 0.5b, and in Fig. 2b, d = 0.75b. Note first, when d is small compared with b, the real and estimated plots are almost linear if the number of lineages axis is logarithmically scaled. Note second, as d is increased from 0.5b to 0.75b the expected plot becomes more curved. That degree of curvature can be used to estimate the relative values of birth *versus* extinction. This result means that under a Markov model of evolution, molecular data can be used to estimate the relative values of b and d.

The reason for the increased curvature with increased d is as follows. Early in time, ancestral lineages each tend to be represented by one or few descendant

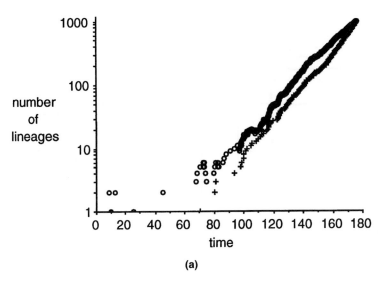

(a)

Figure 2a A simulated phylogeny produced according to the Markov process, with a per lineage death rate (d) being one half the birth rate (b). The open symbols (o) give the real number of lineages at each time period, while the crossed symbols (+) give the estimated number traced back from present-day species as in Fig. 1. Note that the estimated numbers lie along a near straight line with approximately the same slope as that describing the real numbers. The small tick on the time axis in this and subsequent simulation figures denotes the point where the estimated number of lineages is one.

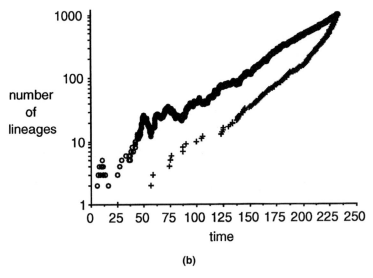

(b)

Figure 2b A simulated phylogeny produced according to the Markov process, with a per lineage death rate (d) being three-quarters the birth rate (b). The open symbols (o) give the real number of lineages at each time period, while the crossed symbols (+) give the estimated number traced back from present-day species as in Fig. 1. Note that in early times the estimated numbers lie along a near straight line with approximately the same slope as that describing the real numbers, but later the line curves upwards.

lineages. If those descendent lineages go extinct, the ancestral lineage leaves no representatives in the future. However, as time progresses, some ancestral lineages will become represented by so many descendant lineages, that it becomes exceedingly likely that at least one representative of the ancestral lineage will survive to some specified point in the future. Compare two lineages, one of which arose in the distant past and another of which arose in the recent past. If the lineage that arose in the distant past has survived to the recent past, it will be represented by many descendent lineages at the point in time when the lineage that arose in the recent past appeared. Now, the lineage that arose in the recent past is far more likely to go extinct between the recent past and the present, than are all the descendent lineages of the one that arose in the distant past. It follows that, as time proceeds, a higher proportion of 'secure' lineages will accumulate. In other words, b tends to dominate over d later in time, and the slope of the curve will get steeper; if b is already much larger than d, as in Fig. 2a, this effect is not so obvious. The slope of the real curve is b–d, which is also the slope of the estimated curve early in time, but later on the slope of the estimated curve gets nearer to b. The single most recent common ancestor of the extant lineages is expected to have occurred in the more recent past than the most recent common ancestor of all extant and extinct lineages.

MASS EXTINCTIONS

The previous section used a null model in which extinction rates are the same in all lineages, and in which extinction rates stay the same through time. The most notorious exceptions to those two assumptions occur during mass extinctions. At

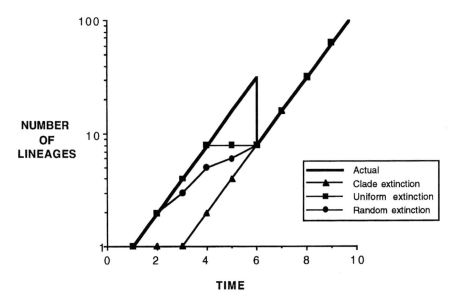

Figure 3 Each branch of a phylogeny splits into two at each time unit, but at time 6 three-quarters of the 32 extant lineages go extinct. The real number of lineages at any time is given by the bold line (labelled Actual). According to whether the extinct lineages form a complete clade (Clade extinction), are one of each pair of most closely related sister taxa (Uniform extinction), or are a random selection of lineages (Random extinction), the estimated phylogeny produced from relationships among extant species shows different patterns.

some times, large numbers of species suddenly go extinct. What would such mass extinctions look like on a reconstructed molecular phylogeny? The answer not only depends on the number of species going extinct but also on the phylogenetic relationships among those species.

Consider two extreme cases. First, 'clade extinction', in which all members of a clade go extinct. After such an event, no descendent is left from and therefore no record is left of the whole clade. The other extreme is 'uniform extinction' in which one representative species of every lowest level taxon goes extinct – say one species from each genus. In between those extremes we have 'random extinction' where the species that go extinct are a phylogenetically random selection of those available.

The effect of each type of mass extinction is evident from Fig. 3 (Harvey & Nee, 1993). Here we use a model phylogeny in which every lineage bifurcates at each time interval. At time 1 there is 1 species, at time 2 there are 2 species, so that by time 6 there are 32 species. At this time a mass extinction event occurs which gets rid of 16 species; subsequently the numbers double again each time unit. The real phylogeny is shown by the bold line in the figure, and the effects of clade, uniform and random extinction are shown by the separate reconstructed phylogenies. Clade extinction is virtually impossible to detect if all the members of just one clade go extinct, while uniform extinction leaves a clear record. As expected, random extinction lies between the two.

MODEL PHYLOGENIES AND THE BIRD DATA

Birds have left a very poor fossil record, and it is not known whether they suffered mass extinctions. Sibley & Ahlquist (1990) produced a molecular phylogeny of the birds based on DNA–DNA hybridization studies involving some 1700 species. The sample of species included representatives from almost all, if not all of the families of birds (defined here as monophyletic groups of birds with a common ancestor occurring about 45 million years ago). Would we expect to detect mass extinction events in birds? Of course, the answer will depend on how large the mass extinctions might have been, and what type of mass extinctions were involved (clade, random or uniform). As we shall see, if the birds did evolve according to a Markov model, the relative values of b and d would also influence the chances of finding evidence for mass extinctions.

Figure 4a reports the results of a simulation study starting with a single lineage about 135 million years ago. There was no background extinction (i.e. d = 0), but three mass extinctions did occur as recorded for other taxa in the Phanerozoic. Following Jablonski's (1991) estimates for other taxa, we assume that 53% of then-extant species were lost in the Late Cenomanian (90.4 m.y.a.), 76% in the End-Cretaceous or Maastrichtian (65.0 m.y.a) and 35% in the Late Eocene or Priabonian (35.4 m.y.a.). b was scaled so that about 9500–10 000 were extant at the present, and extinction was random with respect to phylogenetic relationships.

After considerable stochastic wiggling while the number of lineages is small, the pattern in the real phylogeny settles down, and we can see each of the three mass extinctions. The estimated phylogeny, reconstructed from the present-day species, is coincident with the real phylogeny back until the most recent mass extinction; it has to be because no lineages went extinct. Changes in slope of the plot for the

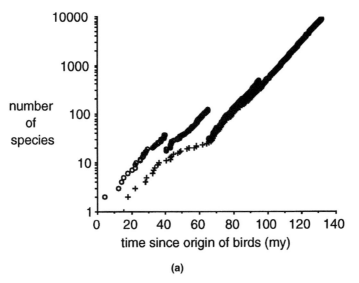

(a)

Figure 4a A phylogeny of the birds simulated according to a Markov model with no background extinction (d = 0) and three mass extinction events as described in the text. The real phylogeny is denoted with open symbols (o) and that reconstructed from relationships among extant species is denoted by crosses (+).

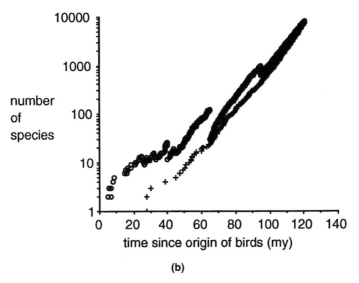

(b)

Figure 4b A phylogeny of the birds simulated according to a Markov model with background extinction (d = 0.5b) and three mass extinction events as described in the text. The real phylogeny is denoted with open symbols (o) and that reconstructed from relationships among extant species is denoted by crosses (+).

estimated phylogeny occur with each mass extinction, the most evident being for the large Maastrichtian event.

Now we repeat the same exercise with the same mass extinctions getting rid of the same percentage of extant species at the same dates, but with d = 0.5b. That is, the probability of a lineage dying out at each time interval is one-half that of it branching. We have scaled b up accordingly so that the same number of species as in the previous simulation are extant at the present. Now the evidence for mass extinction is much less marked (Fig. 4b), even though mass extinctions of the same magnitude occurred in the two simulations.

How do these simulated results compare with the estimated phylogeny from Sibley & Ahlquist's (1990) data? The answer is that they do not compare at all well. The top line labelled A in Fig. 5 (after Nee, Mooers & Harvey, 1992), which plots the data for the first 122 lineages on an arbitrary time scale (see below), is neither a straight line which we should have expected with no background extinction (d = 0), nor is it a line of increasing slope which would indicate a degree of background extinction. Instead, the rate of increase in lineage numbers decreases through time, so that the curve begins to flatten out. If the data are correct, they imply either that the rate of lineage division (b) was decreasing through time, or that the rate of lineage extinction (d) has been increasing through time, or both. If we define b–d as the rate of effective cladogenesis, the results may indicate density dependent cladogenesis. This ecological possibility is discussed by Harvey *et al.* (1991) and Nee, Mooers & Harvey (1992).

It is always possible that the data are not correct, and specifically that the molecular clock changes its rate of ticking with time. If that were so, it would be possible to apply a transformation to the time axis which differentially lengthened earlier time units (or shortened later ones) so that the curve became either a straight line or curved upwards. One way to tell whether the clock runs true is to compare the molecular data with dated fossils. That is not possible for the birds, although it can be done to some extent with carnivores and primates because the fossil record

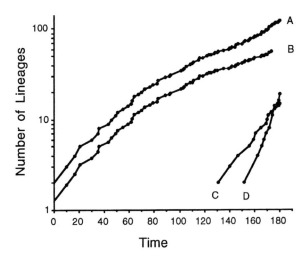

Figure 5 Estimated number of lineages plotted against time for Sibley & Ahlquist's (1990) molecular phylogeny of the birds – data are taken from the origin of the birds down to the family level because representatives of some lower taxa were not included in the analysis. Graph A plots the complete data set. If the tree and time axes are correct (time is measured here in units of genetic distance), then net cladogenesis has been decreasing through time. Curve B plots the complete data set, with the exception of the Passeri and Ciconiiformes (C and D), displaced down at the Y intercept for ease of comparison. Rates of cladogenesis in the Ciconiiformes and Passerines are high compared with other contemporary lineages.

is much richer for those groups than for birds (Wayne, van Valkenburgh & O'Brien, 1991). Our preliminary analyses of the available data (Nee, Mooers & Harvey, 1992) suggest no need to recalibrate the clock, and therefore transform the time axis in Fig. 5, in the suggested way. Another way of checking the clock is to compare known dated biogeographic events with supposed dates of splits between lineages that are associated with those events. For example, the split between South America and Africa would have separated populations of flightless ratites (rheas and ostriches) at a known time. This, indeed, was the way Sibley & Ahlquist (1990) calibrated their tree in the first place. Cotgreave (1992) suggests that several key biogeographic events occurred at about the times given by the relevant nodes on Sibley & Ahlquist's phylogeny.

The plot in Fig. 5 uses arbitrary genetic distance units. The cut-off point at the right hand side of Fig. 5 when there are 122 lineages, defined as the time of origin of bird families, is at about 45 million years ago (using the translation between genetic distance units and real time estimated by Sibley & Ahlquist). Our simulation study suggests that, with constant rates of b and d (where $d = 0.5b$), that number of estimated lineages should have been achieved about 60 million years ago. But as we have said, the top curve in the figure is apparently decreasing in slope through time. If we extrapolate this decreased slope from 122 lineages over 45 million years to the present day there is no way we can arrive at the 9600 present-day species, a number provided by a steeper slope starting with 122 lineages about 60 million years ago! What is the cause of this apparent discrepancy?

At least part of the answer lies in a slight steepening of the curve evident over the last 20 or so arbitrary time units. That steepening does not result from a general increase in effective cladogenesis $(b-d)$ during that time period, but to the existence of two massive radiations, the Passeri (the Oscines or songbirds; about 4560 extant species) and the Ciconiiformes (which includes the Sandgrouse, Plovers, Gulls, Herons, Flamingos, Ibises, New World Vultures and Storks; about 1027 extant species). When those radiations are plotted separately (as curves C and D on Fig. 5), it is evident that they show very high rates of effective cladogenesis when compared with the rest of the birds (curve B, which is displaced down so that it can be compared with A).

Why did we single out the Passeri and the Ciconiiformes for special treatment in Fig. 5? Are there other lineages which show unusual rates of cladogenesis? We have used a statistical test to identify radiations. Consider a time frame, such as that from 130 to 180 units in Fig. 5. Under a pure birth process $(d = 0)$, each lineage existing at the earlier time will be represented by one or more descendent lineages at the later time. If we assume that the per lineage rate of cladogenesis is constant and treat cladogenesis as a pure birth process in continuous time, then the distribution of the number of descendent lineages at the later time should fit a geometric distribution (Nee, Mooers & Harvey, 1992). The distribution of the number of descendent lineages at time 180 for the lineages existing at time 130 fits a geometric distribution very well indeed, but with two exceptions (Fig. 6). The probability of finding a parent lineage with more than 14 descendent lineages in the sample, under the specified null model with a best fit b, is less than 1 in 200. In fact the Passeri and Ciconiiformes produced 15 and 19 descent lineages each. They clearly had unusually high rates of cladogenesis. Harvey & Nee (1992) describe and review methods for detecting unequal rates of cladogenesis from cladograms (phylogenetic trees with unknown branch lengths).

Why did the Passeri and the Ciconiiformes radiate so rapidly? The answer must

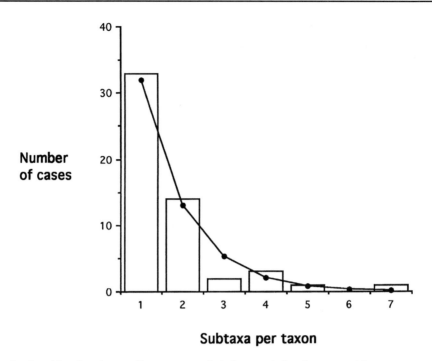

Figure 6 Consider that the per lineage rate of cladogenesis has been roughly constant over the time interval 130 to 180 in Fig. 5. Treating cladogenesis as a birth process in continuous time, the progeny distributions are expected to be geometric (Nee, Mooers & Harvey, 1992). This means that the Passeri and the Ciconiiformes, with 15 and 19 subtaxa respectively, are anomalous. When we remove these taxa as in this figure, the distribution (histogram) fits the geometric expectation (continuous line).

surely lie in the niches that became available at the time and the suitability of these particular taxa to occupy them (see Nee, Mooers & Harvey, 1992 for references to the literature on this topic). The two taxa are quite differently adapted: one to a territorial life in terrestrial habitats, and the other to an often-colonial social organization particularly along coastlines. We can only suggest that the radiations occurred at about the same time in response to new habitats brought about by continental drift and climatic change. Such a suggestion, however well researched the evidence for it, is open to the accusation that the sample size is two and any number of causal factors might be responsible for the radiations.

Fortunately, most evolutionary explanations can claim some generality and, since we have 122 lineages, is it possible to develop tests that use information from each lineage to determine whether particular characters or traits are associated with high or low rates of cladogenesis? High rates of cladogenesis will, on average, be associated with short branch lengths. How, then, might we go about finding statistical evidence for correlations between traits and branches? It would be possible to compare an estimated character state, say half-way along a branch, with the length of that branch. However, we should then enter into problems of phylogenetic association (Harvey & Pagel, 1991). For example, imagine we had data on Passerines and Ciconiiformes and that branch lengths for Passerines were shorter than those for Ciconiiformes. Since most Passerines are smaller than most Ciconiiformes, we should find an association between branch length and body size. Since more Passerines than Ciconiiformes live in terrestrial habitats, we should find an association between habitat use and branch length. Indeed, higher rates of cladogenesis would be found associated with any character that is more typical of Passerines than Ciconiiformes.

One way to resolve such problems is to examine differences between closely related lineages (e.g. Felsenstein, 1985). For the case we are dealing with, there are 122 lineages which are linked by 121 nodes. Each node has two independently evolving daughter lineages. For any metrically varying character of interest, say body size, we might ask whether the daughter taxon on the shorter branch tends to have had a larger or smaller character state than the other daughter taxon. If character states for ancestral taxa are estimated from daughter taxon values, the 121 comparisons can be made statistically independent of each other. Such comparisons reveal no statistically significant associations between differences in body size and differences in rates of cladogenesis (Harvey *et al.*, 1992; Nee, Mooers & Harvey, 1992; Mooers, Nee & Harvey, Chapter 11). However, they do reveal a correlation between changes in branch length and changes in age at first breeding, but that may result more from imperfections in Sibley & Ahlquist's (1990) reconstructed tree than a biological relationship between age of first breeding and rates of speciation or extinction. The topic is dealt with elsewhere in this volume (Mooers, Nee & Harvey, Chapter 11).

COMMUNITY STRUCTURE AND PHYLOGENETIC RELATIONSHIPS

Sibley & Ahlquist's (1990) avian phylogeny has been used to reveal and investigate one particularly intriguing pattern in the way bird communities are structured. Smaller bodied species can occur at much higher population densities than larger bodied species (Damuth, 1987), and within natural communities there are sometimes negative relationships across species between population density and body size. However, Nee *et al.* (1991) found that although there was an overall negative relationship across species between body size and the number of birds breeding in Britain, within tribes the same relationship was frequently positive. They suggested that a possible explanation for this pattern is that members of a tribe are frequently close competitors, and larger bodied species may win in direct competition for resources against their smaller bodied competitors.

If their explanation was correct, Nee *et al.* (1991) reasoned that those tribes which had been evolutionarily separated for longest would be the most likely to constitute complete guilds, and therefore would be the ones most likely to show positive relationships between population density and body weight. Accordingly, they predicted that those tribes which rooted most deeply into the phylogenetic tree of birds would be the ones most likely to show a positive relationship. For example, the Picid woodpeckers found in Britain have no close relatives, whereas the Acrocephaline warblers do. Nee *et al.*'s (1992) prediction turned out to be correct, particularly for non-passerines. Subsequently, however, Cotgreave & Harvey (1991) found that for 90 bird communities from around the world, the date at which a tribe first started to radiate (Y on Fig. 7) was a significantly better predictor of positive relationships between body size and population density than the date at which a tribe rooted into the phylogenetic tree of birds (X on Fig. 7). Those non-passerine tribes which started to radiate in the more distant past were more likely to show a

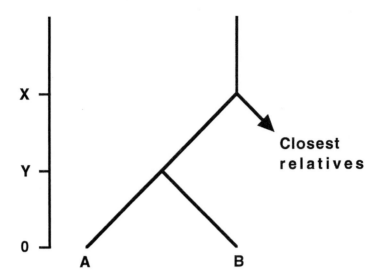

Figure 7 A tribe of birds consisting of two extant species (A and B) shared their most recent common ancestor with each other Y time units ago and their most recent common ancestor with other birds X time units ago. When natural communities are sampled, the measure Y is a better predictor than is X of the correlation between population density and body size within tribes.

positive relationship between population density and body weight. Cotgreave (1992) points out that these patterns, whatever their cause, are likely to be more pronounced in non-passerines than passerines because the latter have a much smaller range of dates of either radiation (Y) or rooting into the tree (X).

Cotgreave (1992) takes these analyses further. Austin (1962) provides descriptions of all the groups of birds in the world. Cotgreave points out that those tribes, (a) with more species and (b) with species having more diverse habits are more likely to require more space to describe. It is certainly true that there is a highly significant correlation between the number of species in a tribe and the number of pages Austin devotes to the tribe ($r^2 = 0.67$, $N = 29$). However, it is also true that those tribes which take more pages to describe than would be expected from the number of constituent species, are the ones in which tribes tend to show negative size-abundance correlations ($r = 0.46$, $N = 29$, $P = 0.002$). As Cotgreave (1992) writes "This is support, at least for the suggestion that tribes in which the species are in the same guild tend to show positive relationships, if not for the idea that the same is true of tribes which form complete guilds".

Currently, these patterns are being investigated further in the same and in other communities. Whatever the ultimate outcome, analyses based on a preliminary molecular phylogeny of the birds have revealed previously unknown differences in population density that will require explanations invoking both ecological and evolutionary process.

ACKNOWLEDGEMENTS

This work was supported by SERC grant number GR/H53655 and Wellcome Trust grant number 38468.

REFERENCES

AUSTIN, O.L., 1962. *Birds of the World.* New York: Golden Press.

COTGREAVE, P., 1992. 'Comparative Studies of Community Structure'. Unpublished D.Phil. thesis, University of Oxford.

COTGREAVE, P. & HARVEY, P.H., 1991. Bird community structure. *Nature, 353*: 123.

DAMUTH, J., 1987. The interspecific allometry of population density in mammals and other animals: the independence of body mass and population energy use. *Biological Journal of the Linnean Society, 31*: 193–246.

FELSENSTEIN. J., 1985. Phylogenies and the comparative method. *American Naturalist, 125*: 1–15.

GOULD. S.J., RAUP, D.M., SEPKOSKI, J.J., SCHOPF, T.J.M. & SIMBERLOFF, D.S., 1977. The shape of evolution: a comparison of random and real clades. *Paleobiology, 3*: 23–40.

HARVEY, P.H. & NEE, S., 1993. New uses for new phylogenies. *European Review, 1*: 11–19.

HARVEY, P.H. & PAGEL, M.D., 1991. *The Comparative Method in Evolutionary Biology.* Oxford: Oxford University Press.

HARVEY, P.H., NEE, S., MOOERS, A.Ø. & PARTRIDGE, L., 1991. These hierarchical views of life: phylogenies and metapopulations. In R.J. Berry and T.J. Crawford (Eds), *Genes in Ecology*, 123–137. Oxford: Blackwell.

HARVEY, P.H., NEE, S. & MAY, R.M., 1994. Phylogenies without fossils. *Evolution*, in press.

JABLONSKI, D., 1991. Extinctions: a paleontological perspective. *Science, 253*: 754–757.

MOOERS, A.Ø., NEE, S. & HARVEY, P.H., 1994. Biological and alogrithmic correlates of phenetic tree pattern. In P. Eggleton & R.I. Vane-Wright (Eds), *Phylogenetics and Ecology*, 233–251. London: Academic Press.

NEE, S., READ, A.F., GREENWOOD, J.J.D. & HARVEY, P.H., 1991. The relationship between abundance and body size in British birds. *Nature, 351*: 312–313.

NEE, S., MOOERS, A.Ø. & HARVEY, P.H., 1992. The tempo and mode of evolution revealed from molecular phylogenies. *Proceeding of the National Academy of Sciences, USA., 89*: 8322–8326.

SIBLEY, C.G. & AHLQUIST, J.E., 1990. *Phylogeny and Classification of Birds: a Study in Molecular Evolution.* New Haven, CT.: Yale University Press.

WAYNE, S.J., VAN VALKENBURGH, B. & O'BRIEN, S.J., 1991. Molecular distance and divergence time in carnivores and primates. *Molecular Biology and Evolution, 8*: 297–319.

11

Biological and algorithmic correlates of phenetic tree pattern

ARNE Ø. MOOERS, S. NEE & PAUL H. HARVEY

CONTENTS

Keywords: Bird phylogeny – UPGMA – cladogenesis – diversification – DNA hybridization.

Abstract

We show that there is a nonrandom pattern of branch production above the family level in Sibley and Ahlquist's UPGMA (Unweighted Pair Group Method using Arithmetic Averages) tree based on DNA hybridization data. The lengths of daughter-branches emanating from a node are positively correlated with the lengths of the parent branches leading to the node. This may be due to different clades having different rates of effective cladogenesis (lineage birth rates minus death rates). These different rates may be due to biological attributes of the clades exhibiting them, such as body size or generation time, or to hidden biases in the tree. Given the empirical observation of an apparent difference in the amount of change in DNA of bird species with differing ages to maturity, we investigate the possible effects this would have on both the branch lengths and tree topology of the UPGMA tree. We find that, under certain conditions, differing heritable rates of molecular divergence can affect both the branch lengths and the tree topologies of UPGMA trees systematically. We offer some evidence that this may be the case for the Sibley and Ahlquist tree and conclude with a discussion of other tests of biological correlates of diversity.

Phylogenetics and Ecology
ISBN 0-12-232990-2

INTRODUCTION

Sibley & Ahlquist's molecular phylogeny of the birds (1990) is by far the most extensive produced to date. It is based on cross-species DNA–DNA hybridization data using over 1700 of the 9600 species of birds. Sibley & Ahlquist attempted to spread the taxonomic representation of the species in their sample as much as possible, so that representatives from each family and almost all tribes were included. Their study is important for several reasons. First, it tells us a lot about birds. For example, it produced the unexpected but now uncontested conclusion that the Australian passerines are a separate radiation from other passerines. Second, it attempted to employ a modern molecular technique under standardized conditions so that the data set could be analysed as a whole. DNA–DNA hybridization has now become less important for molecular systematics than has direct comparison of nucleotide sequence data, but there are currently no phylogenies based on specific sequence data that approach the magnitude of the one presented by Sibley & Ahlquist. Accordingly, their phylogeny provides a potentially important model system for deciding how best to use molecular phylogenies to understand evolutionary processes.

Elsewhere, we have reported novel analyses (Harvey *et al.*, 1992; Nee, Mooers & Harvey, 1992), and Harvey & Nee (Chapter 10) develop others in the more general context of evolutionary models involving rates of lineage splitting and extinction. The purpose of the present chapter is to ask whether rates of effective cladogenesis (lineage birth rates minus death rates) are heritable as well as variable; that is, are parent- and daughter-branch lengths correlated? If they are, we can ask what characters or traits are associated with an increased tendency for cladogenesis. This type of analysis addresses the question of species selection: why do some species leave more descendent species than others? As we shall see, there are several potentially important patterns in the tree, but their biological intepretation is not as clear as we might have hoped.

HERITABILITY OF BRANCH LENGTH

We define the ancestral branch leading directly to a node as the parent branch, and the descendent branches from the same node as the daughter branches. If some clades are more prone to splitting (either as a consequence of higher lineage birth rates or lower lineage death rates) than others, we can investigate heritability of branch lengths (Harvey *et al.*, 1992; Nee, Mooers & Harvey, 1992). This investigation is based on the fact that branch length will be inversely proportional to the average rate of branch production. If the different clades that make up the tree differ in their average rate of branch production, then the parent and daughter branches within clades will have correlated lengths, being subject to the same average rate of branch splitting. If however the average rate of splitting is not a property of entire clades, but is a transient property of a given lineage, there should be no such correlation. The null hypothesis is that parent-branch length is not a significant predictor of average daughter-branch length. An obvious approach, then, is to regress mean daughter-branch length on parent-branch length (A. Burt, personal communication). However, there are some biases in the data which force us to consider only a specific subset of the available branch length data.

Parent- and daughter-branch lengths may not be independent of each other for reasons other than differential rates of splitting. The time from the origin of the phylogeny to the present provides a constraint. If that time is t, and parent-branch length is t_p, then daughter-branch length td has to be equal to or less than $t-t_p$. This means that long parent branches cannot give rise to long daughter branches, resulting in an artifactual negative heritability of branch lengths. This negative heritability would arise because not all daughter branches in a phylogeny will have split if branches from the present to parental nodes are included in the analysis. However, if we merely remove from consideration those daughter branches that have not split, we are still biasing the data because we are selectively removing potentially long branches (those that have not split since t_p) while leaving shorter branches (those that have split since t_p).

The problem of non-bifurcated daughter branches towards the tip of the tree means that daughter branches which occur near the root of the tree are more free to vary in length. With the Sibley & Ahlquist (1990) data set, there is another reason why branches near the tip of the tree can cause bias: many lower-level taxa were not sampled when the tree was constructed. For example, if only one member of a tribe was sampled, the daughter branch which that species caused to terminate might well have been shorter if other members of the tribe had been sampled (one or more of which was more closely related to the outgroup). This latter effect causes daughter branches near the base of the tree to be longer, while the previous effect will shorten them. We also know from our previous analyses of the same data set (Harvey *et al.*, 1992; Nee, Mooers & Harvey, 1992), that effective cladogenesis slows down later in the Sibley & Ahlquist tree, thereby causing daughter branch lengths to be longer.

Given the above problems, we have taken a pragmatic approach to changes in daughter-branch lengths near the tip of the tree. First, we have considered nodes only at the family level (as delineated by the taxonomy of Sibley & Monroe, 1990) and above for analysis. Because 171 of the 174 families considered by Wetmore (1960) were sampled, this means few if any higher nodes are missing. Assuming that each taxon delimited by Sibley & Monroe (1990) is monophyletic, we have further reduced the sample set by excluding all nodes from which one or more daughter taxa were not represented in Sibley & Ahlquist's (1990) data set. Using these procedures we are left with 61 nodes for which we have unbiased estimates of both parent- and daughter-branch lengths. Our second approach to possible biases resulting from a systematic slowdown of rates of cladogenesis through time is to include the height of the parent node as an additional independent variable in a multiple regression of daughter on parent-branch length.

All three measures: parent-branch length, mean daughter-branch length and height of node were logarithmically transformed prior to analysis. The results of a multiple regression analysis show that parent-branch length is a significant predictor of average daughter-branch length while height of node is not (parent-branch length t = 5.358, $P = 0.0001$; height of node t = 1.14, $P = 0.259$; $N = 61$). Figure 1 plots mean daughter-branch length on parent-branch length, with both lengths corrected for node height. Rates of cladogenesis are variable and branch length is a heritable trait among lineages, supporting the non-parametric results of Harvey *et al.* (1992). In the folowing section we consider why this might be so.

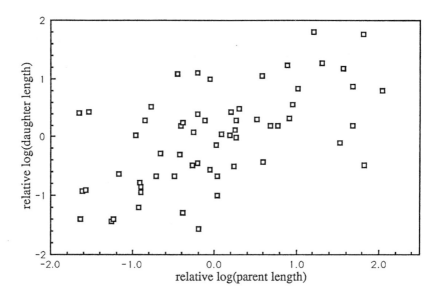

Figure 1 Plot of mean daughter-branch length on parent-branch length above the family level of the Sibley & Ahlquist (1990) UPGMA tree. Parent- and daughter-branch lengths (in ≡ T50H units) have been logarithmically transformed and corrected for their height on the tree.

CORRELATES OF PHYLOGENETIC PATTERN

We have documented elsewhere methods for studying correlations between phylogenetic tree structure and species' attributes (Nee, Mooers & Harvey, 1992). In a test akin to the independent contrasts method (Felsenstein, 1985; Harvey & Pagel, 1991), one can look for independent associations between branch length differences and character trait differences throughout the tree. After reconstructing character states at internal nodes under a given model of evolution, we test for association: is the difference in length of sister branches emanating from an ancestral node associated with the difference in character traits reconstructed at the ends of those branches? If so we have evidence of a relationship between the trait and rates of cladogenesis. Each bifurcating ancestral node yields one comparison, and so there are $N-1$ independent comparisons for a wholly bifurcating tree with N terminal taxa. In practice, there will be fewer comparisons given that, at some internal nodes, sister-branch lengths and/or sister-clade reconstructions will be equal. Discontinuous variables could be tested likewise; nodal reconstructions would be based on different assumptions of how traits evolve (e.g. parsimoniously) and then the association could be looked for only where sister clades differed in state (Mitter, Farrell & Wiegman, 1988; Zeh, Zeh & Smith, 1989).

This test of association showed no evidence of a relationship between one species' attribute, that of body size, and cladogenesis among birds (Nee, Mooers & Harvey, 1992). However any underlying pattern may have been obscured by an unmet assumption of the algorithm used by Sibley & Ahlquist to construct their tree: the algorithm (Unweighted Pair Group Method using Arithmetic Averages (UPGMA), Sokal & Michener, 1958) assumes that the rates of molecular change are equal throughout the tree while the evidence suggests otherwise. Sibley & Ahlquist

(1990: 18) state: "The correlation between branch length and delayed maturity is clear; we have found no species with delayed maturity that did not have shorter branch lengths when compared with related species that breed at one year of age". Age at first breeding (AFB) is a good general measure of generation time (Lewontin, 1965) and the two are highly correlated among birds (Sibley & Ahlquist 1990: 178). The work of Sibley & Ahlquist quoted above concerns trees of three species only and is a test of rate constancy (Swofford & Olsen, 1990). No one has yet examined how this observed branch shortening might affect the structure of the entire tree.

The relative branch length shortening reported by Sibley & Ahlquist (1990) is expected using the UPGMA algorithm for DNA–DNA hybridization data if bird species with long generation times had experienced less molecular evolution than sister species with shorter generation time. Adaptive scenarios could be envisioned for this difference, or it might be that most molecular change occurs at meiosis. DNA of species with older AFB would hybridise more fully with a given species than species with younger AFB because of less accumulated molecular change. This stronger hybridisation would mean that the DNA–DNA hybridisation distance between them would be lower and so the branch linking the two species would be shorter.

The above phenomenon could have two opposing effects translated up the tree, illustrated in Fig. 2. Figure 2a shows a hypothetical tree comprising four species (A, B, C and D). Species A and B each first breed when four years old, species C first breeds at two years, and species D breeds at one year of age. The branch lengths in Fig. 2a correspond to actual time: clades (AB) and (CD) each split at the same time (at 10 units). This assumes that generation time does not affect the rate of effective cladogenesis. Now let us assume that actual divergence is some inverse function of age at first breeding, of the form

$$\text{Divergence} = \frac{\text{time}}{\text{AFB}^x} \tag{1}$$

where 'time' is the time since two clades last shared a common ancestor and AFB is age at first breeding (both in arbitrary units). We ignore the case where $x = 0$ and AFB has no effect on divergence. Consider two other cases. In the first, let $x = 0.5$, and let the AFB of the ancestor be estimated as the mean of its daughters' values. This produces the approximate divergences in Fig. 2b. These measures can be translated into the distance matrix of Fig. 2b by simply summing the branch lengths between pairs of clades. The UPGMA algorithm, because it forces clades (AB) and (CD) to be the same distance from the root (since the rate of evolution is considered the same throughout the tree), will produce the phenogram in Fig. 2c. With $x = 0.5$, UPGMA has taken equal times of divergence ($i = j$ in Fig. 2a) and produced a phenogram with a longer branch leading to clade (AB) ($i' > j'$), even given less real divergence along this branch (2b). This lengthening is due to UPGMA dividing the total divergence between clades (the mean distance between the members of clades (AB) and (CD), represented by the mean of the four cells in the top right portion of the matrix of 2b) equally between the lineages.

Now consider the case when the effect of age at first breeding on divergence is stronger, e.g. let $x = 1$ in Equation (1) (the case when all genetic change occurs at meiosis). Such a relationship, coupled with the phylogeny of Fig. 2a, would produce the divergences represented in the tree and matrix of Fig. 2d. With this matrix,

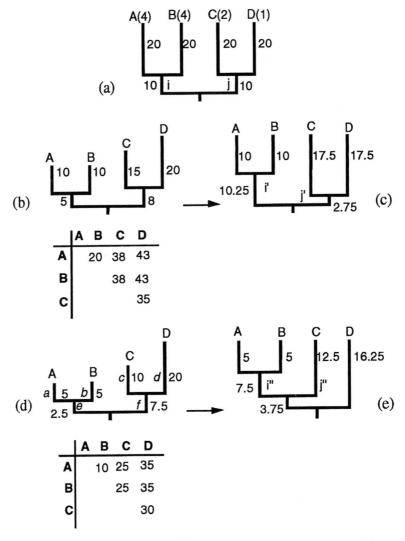

Figure 2 A hypothetical phylogeny with differing rates of divergence among lineages. (a) The 'true' phylogeny. Branch lengths are in arbitrary time units. Species A and B have AFB (age at first breeding) = 4, species C has AFB = 2, species D has AFB = 1 (arbitrary units). Ancestral AFB is mean of daughter values. (b) Same tree as in (a) but branch lengths denote actual divergence. Here divergence is proportional to (AFB) − 0.5. The divergences have been summed to produce the distance matrix. (c) The UPGMA phenogram based on the distance matrix of (b): branch i' > j'. (d) The same tree as in (b) but here divergence is proportional to $(AFB)^{-1}$. The divergences are summed to produce the corresponding distance matrix. (e) The UPGMA phenogram based on the distance matrix of (d). In this case i" < j"; the older AFB clade (AB) is on the shorter sister branch.

however, UPGMA would produce the phenogram in Fig. 2e. Here the topology is different from that of the true phylogeny of Fig. 2a, and now the branch leading to the clade with the higher AFB is shorter than than its sister branch (i" < j" in Fig. 2e). The ladder arrangement also means that the branch leading to the higher AFB clade (ABC) will also be shorter than its sister branch leading to species D. The necessary condition for a change in topology from the balanced four member tree of Fig. 2a to the ladder type tree of Fig. 2e (using the branch length designations of Fig. 2d), assuming $c < d$, is

$$\frac{a+b}{2} + e + f < d \tag{2}$$

Condition (2) groups clade C with clade (AB). The assumption $c < d$ is necessary to produce a bifurcating tree and may be considered trivial since exactly equal molecular divergence must be rare.

From the above example it can be seen that, if the rate of divergence is a property of lineages, the UPGMA algorithm may produce differing patterns of branch lengths. For those portions of the tree where condition (2) holds, clades that diverge more slowly may be found at the end of shorter branches. Where there are differences in divergence rates but condition (2) does not hold, slower diverging clades may be found at the end of longer branches. Unfortunately, to evaluate condition (2) one needs a true phylogeny and actual divergences along all branches and with all this information the exercise becomes moot. Also, the above predictions assume that there is no association between divergence and rates of effective cladogenesis: in the example phylogeny (Fig. 2a) the slower and faster diverging clades split at the same time (at 10 units). If there is an association between divergence and rates of effective cladogenesis then the expected branch length patterns may change. This possible interaction and the uncertainty surrounding the relative occurrence of condition (2) makes hypothesis testing difficult. In particular, a lack of association between branch length and age at first breeding is not evidence for the tree being an unbiased reflection of the phylogeny. If there are lineage-specific rates of divergence negatively correlated with generation time but condition (2) holds at only one-half of the nodes, no association is expected. Likewise, a significant association could be attributed to several factors, including the relative occurrence of condition (2) and lineage-specific rates of effective cladogenesis.

TESTING FOR BIAS

If condition (2) holds often enough, we should find a negative association between branch length and AFB: older AFB clades should be on shorter branches. Using ages at first breeding from Bennett (1986) and the sources cited in Sibley & Ahlquist (1990: table 16), 82 family-level measures of minimum age at first breeding were created by averaging up through taxonomic levels (for procedure see Harvey & Mace, 1982). Given that over 90% of the variation in age at first breeding is found at or above the family level (Bennett, 1986), the method of reconstruction of family level measures should be reasonably robust to particular assumptions about the way the character evolves.

Above the family level, nodal values of AFB were calculated following Felsenstein (1985). Branch lengths leading to 78 families denoted by an asterisk in Appendix 1 could not be measured directly because of possible biases due to missing lower taxa, and had to be estimated in the following manner. The height on the tree of the node defining each subfamilial categorical rank (subfamily, tribe, subtribe and genus), as defined by Sibley & Monroe (1990), was recorded for each bird family.

The median value of the heights defining a taxonomic rank was then taken as the best estimate of the expected height of that rank. Within each of the 78 families, the first subfamilial rank denoted by Sibley & Monroe (1990) was given this estimated height, and this height was used to estimate the branch length, unless the actual (measured) branch length was shorter. These estimated branch lengths were used only to calculate higher nodes; for the tests of association, comparisons using these estimated branch lengths were ignored.

The test of association between branch lengths at and above the family level and AFB showed no significant relationship (24 cases where the shorter sister branch ended in the younger AFB clade : 32 cases where the shorter branch ended in the older AFB clade; $P = 0.18$, one-tailed sign test).

The above test has as its null hypothesis that there is no relationship between the rates of cladogenesis (measured as branch length) and generation time (measured as AFB). However, given that smaller bodied taxa often contain more species than their larger bodied relatives (Van Valen, 1973; May, 1986; Kochmer & Wagner, 1988; Nee, Mooers & Harvey, 1992), smaller bodied taxa might have higher rates of effective cladogenesis and so would be found on shorter branches. Across bird families, age at first breeding is positively correlated with body size (Z-transformed Spearman Rank Correlation Coefficient = 6.515; $P = 0.0001$, $N = 82$). This suggests that, apart from any bias in the tree, clades with older ages at first breeding might be found on longer branches, and so our null hypothesis is no longer one of no association. We can remove this bias by using a measure of AFB that is independent of body size.

We produced a measure of AFB that is independent of body size by first calculating the evolutionary relationship between age at first breeding and body size above the family level and then using this relationship to calculate residual values of age at first breeding across families. The 'evolutionary relationship' between two variables can be thought of as the expected change in the measured value of one trait for a unit change in the value of another trait over time, independent of the lineage on which the change occurs. Felsenstein's independent contrasts method (Felsenstein, 1985; see also Harvey & Pagel, 1991) allows the sign and magnitude of changes in traits to be compared over a phylogenetic tree. Pagel (1993) offers a further discussion on the form of the evolutionary relationship. Data on body weight in grams for 3378 species of birds from 135 families were collated from the literature (Severenty, 1936; Haverschmidt, 1948; Sutter, 1955; Sutter & Cornaz, 1963; Winkel, 1968; Dawson & Fisher, 1969; Humphrey et al., 1970; Crowe, 1978; Karr, Willson & Moriarity, 1978; Ridley, 1982; Brough, 1983; Dunning, 1984; Bennett, 1986; Fry, Keith & Urban, 1988; Stiles & Skutch, 1989). Where possible, the mean of male and female weights was used and all weights were logarithmically transformed prior to analyses. Eighty-two family level measures were then calculated in the same way as the AFB scores (see Appendix 1).

Using the computer program C.A.I.C. (Comparative Analysis by Independent Contrasts 1.2), written by A. Purvis, we calculated 81 independent contrasts between body weight and age at first breeding, using the formulae presented in Pagel (1992). The independent contrast relationship is presented in Fig. 3. The slope of the least squares (Model 1) regression through the origin may be considered an estimate of the evolutionary relationship between the two variables. While the least squares approach does make the assumption that there is no error variance in the independ-

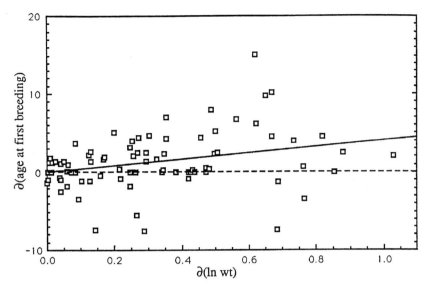

Figure 3 Contrasts in log(weight(g)) and Age at First Breeding (months) for 82 families of birds. All contrasts in the independent variable were constrained to be positive. The slope of the least squares regression through the origin is 4.18 (t = 4.461, N = 81, P = 0.0001).

ent variable (Sokal & Rohlf, 1981), which is clearly violated for this analysis, it alone has the necessary attribute of producing residuals in Y that have a zero correlation with X (Harvey & Pagel, 1991). The least squares slope was used to calculate residual AFB scores for each of the 82 families by first calculating the predicted age at first breeding for each family using the slope and then subtracting this predicted value from the actual age at first breeding. This procedure assumes the phylogeny is true as given, a weak assumption in this case. In order to test for unexpected bias in the contrast slope it was compared with the slope generated by using the family level data alone: the Model 1 regression slope across families of 4.00 (t = 7.36, df. = 80, P = 0.0001) is not different from the contrast slope (t-test for differences between slopes: t << 1.00, df. = 159, ns; Zar, 1984) and the set of residuals generated using the two regressions match almost exactly (Spearman Rank Correlation: Rho = 0.997, P = 0.0001).

Following the production of the res(AFB) scores, internal nodes were again estimated and the test of association performed. Estimated branch lengths were again not included in the sign test. The result shows a clear negative association between residual age at first breeding and branch length: there are nearly twice as many cases where the shorter branch ends in the larger residual age at first breeding than the smaller residual (38:21, P = 0.015; one-tailed sign test). Combined with the empirical observation that bird species with later ages to maturity have slower rates of molecular evolution, this result suggests that condition (2) is common.

DISCUSSION

It has long been recognized that when considering the distribution of subtaxa within taxa, one or a few taxa contain a disproportionate number of subtaxa (Willis & Yule,

1922). The distribution is common to many diverse taxonomic assemblages from forminiferans (Burlando, 1990) to angiosperms (Willis & Yule, 1922; Cronk, 1990) and vertebrates (Anderson, 1974; Bock & Farand, 1980). Recent work has moved from taxonomies to phylogenetic trees. The results presented here and elsewhere (Harvey *et al.*, 1992, Nee, Mooers & Harvey, 1992) refer to a phenetic tree constructed with a mathematical algorithm, such that "cognitive and perceptual biases" which might affect taxonomic classifications (Minelli, Fusco & Sartori, 1991: 95; see also Walters, 1986) can be discounted. If the UPGMA tree is assumed to reflect phylogenetic pattern, then our results suggest nonrandom effective cladogenesis.

It may be that intrinsic aspects of the biology of clades are associated with patterns of effective cladogenesis (Brooks & MacLennan, 1991). If these aspects can be identified and the pattern of association shown to hold in a number of independent cases, we might be tempted to suggest a causal link between them and the pattern. For example, we might consider small size as an attribute that promotes effective cladogenesis. If so, we would expect that across the Aves, smaller-bodied sister clades would split more often, a hypothesis not strongly supported by our studies of the UPGMA tree (Nee, Mooers & Harvey, 1992). This, we now argue, may be due to the effect of generation time on tree structure itself, independent of its effect on effective cladogenesis.

Marzluff & Dial (1991) present evidence that subtaxa with short generation times (early age at first reproduction and short lifespan) tend to be the most diverse across 33 different taxonomic assemblages, and they suggest that there is a causal link between generation time and diversity. This seems a plausible hypothesis, given that low rates of reproduction and long generation times are shared attributes of extinction-prone species (Wilson & Willis, 1975). Marzluff & Dial's study (1991) is an important contribution; there are, however, at least three problems with their analyses. (1) Age at first reproduction (their measure of generation time) for higher taxa was calculated wrongly: simple averages across species were taken, thus biasing higher taxa estimates towards speciose subtaxa (Harvey & Mace, 1982). This means that estimates of age at first breeding above the genus may not be representative. (2) Marzluff & Dial assume that, within an assemblage, subtaxa of equal rank are the same age (e.g. all the genera of Salamanders diverged at the same time). Other things being equal, we might expect older taxa to be more diverse; if so, considering all subtaxa of equal rank as having the same expected diversity is potentially hazardous (Cracraft, 1984). (3) In Marzluff & Dial's analysis which compares age at first breeding and diversity among sister taxa, the assumption of equal-aged taxa is avoided but here degrees of freedom are overstated. Their Appendix 2 contains 17 sets of sister groups, from which no more than 17 comparisons between relative diversity and relative first age of reproduction are possible. Eleven of the comparisons support a negative association, five support a positive association, and one supports no association. The probability of rejecting a hypothesis of no overall association is $P = 0.33$, not the $P = 0.004$ reported by Marzluff & Dial based on their 29 non-independent comparisons. Andy Purvis (personal communication) has pointed out another flaw: Even though Dial & Marzluff had already presented evidence (1988) that body size was strongly correlated with diversity, body size was not considered as a covariate in the later study.

The above problems make evaluation of Marzluff & Dial's results (1991) difficult.

However a binomial test on the sign of the Pearson product-moment correlations between diversity and age at first reproduction (data from their Table 1) is significant at the $P = 0.05$ level, even when reducing sample size so that each clade is used only once in the test (15 of the 19 correlations are negative, $P = 0.02$). If we assume that the subtaxa are of equal age, then this result supports Marzluff & Dial's contention. If there are no systematic differences among clades in effective cladogenesis then older clades will tend to be more diverse and this result suggests that older clades may have earlier ages at first breeding. In either case, age at first breeding may be implicated in phylogenetic pattern. The results presented above for the Aves, using branch length data as a measure of divergence rates, contradict Marzluff & Dial's hypothesis: clades with longer AFB seem to be associated with shorter branches. However, it may be that branch lengths of different portions of the Sibley & Ahlquist tree (1990) are not comparable, and the association is an artifact of the UPGMA algorithm.

Sibley & Ahlquist's UPGMA tree (Sibley & Ahlquist, 1990) and the taxonomy that they derive from it (Sibley, Ahlquist & Monroe, 1988; Sibley & Monroe, 1990) have come under criticism from several quarters (e.g. Lanyon, 1985; Houde, 1987; Cracraft, 1988; Mayr, 1989; Sarich, Schmid & Marks, 1989). The criticism has centered chiefly on data analysis, including both the choice of the measure of divergence (their $\equiv$ T50H) and the use of UPGMA as the clustering algorithm. There is now general acceptance that rates of DNA evolution differ among lineages (mammals: Brownell, 1983; Wu & Li, 1985; Britten, 1986; Catzeflis *et al.*, 1987; Seino, Bell & Li, 1992; birds: Houde, 1987; Sheldon, 1987; Sibley & Ahlquist, 1990; for a review see Gillespie 1991). It seems that clades with shorter generation times may have higher rates of DNA evolution (Kohne, 1970; Wu & Li, 1985; Catzeflis *et al.*, 1987; Sibley & Ahlquist, 1990). It is also clear that, if there are rate differences among lineages, distance measures made using DNA divergence scores will not be appropriate for clustering algorithms such as UPGMA which make the assumption of equal rates (Swofford & Olsen, 1990). Taken together, these suggest that the UPGMA tree may be flawed in some consistent way. Two possible manifestations of a general bias in the tree above the family level due to a generation time effect are represented by Fig. 2c,e.

Colless (1970) presented conditions necessary for a phenogram to correctly reflect cladistic structure. His analysis suggested that clustering algorithms such as UPGMA would often produce the correct cladistic topology even in the face of varying rates of evolution, although he had little to say about branch length biases. This result was based on the assumption that low rates of evolution on one lineage would be compensated to some degree by a higher rate in its sister lineage. In Fig. 2, if clade A were evolving slowly, clade B could be evolving at a fast enough rate such that equation (2) need not hold. Colless considered this a plausible assumption based on perceptions of varying rates of evolution following speciation events. However, if the rate of evolution is 'heritable' between parent and daughter lineages then sister lineages may both have relatively slow or relatively fast rates of evolution and so there may be little compensation and resulting divergences may be such that the topology could change (as in Figures 2d,e). Generation time is plausibly one such heritable trait.

We have documented a positive relationship between body size and age at first breeding (Fig. 3). If smaller bodied clades have higher rates of effective cladogenesis than larger bodied clades (and there is theory to suggest why they might; see Hutchinson & MacArthur, 1959; Dial & Marzluff, 1988; Fitzpatrick, 1988; Kochmer &

Wagner, 1988), larger bodied clades with older ages at first breeding than their sister clades should be on longer branches than their sister clades This is the same expectation as the branch length bias presented in Fig. 2b,c. Both the expectations of effective cladogenesis and that of a branch length bias produced by UPGMA are the same: large bodied clades with older ages at maturity should be on longer branches *vis a vis* their sister clades. There is no evidence for such a relationship and when the putative effect of body size is removed, there is a strong relationship suggesting that the topology (and not just the branch lengths) of the UPGMA phenogram is consistently biased. The test is of course based on a particular model of trait reconstruction which may not be correct. This is the case for any and all studies where ancestral character states are estimated.

There is another condition necessary for the branch swapping of Fig. 2: for condition (2) to hold, branches e and f must be short compared with branches a to d. In Fig. 2a the individual species were considered to have diverged individually for twice as long as their immediate ancestors (the branches leading to species are 20 time units long, the sister branches leading from the root only 10 units long). This condition would hold if cladogenesis occurred in spurts (e.g. producing clade (ABCD) and then producing the two clades (AB) and (CD)), followed by a period of divergence along branches without any effective cladogenesis. Over this pattern we must posit a faster rate (over the long term) of cladogenesis among smaller-bodied clades. If we are to explain the patterns in the Sibley & Ahlquist tree (1990) as being in part due to branch swapping as in Fig. 2, we must offer some plausible cause for such a cyclical pattern of effective cladogenesis.

The hypothesis that the UPGMA tree of Sibley & Ahlquist (1990) exhibits incorrect topology is difficult to evaluate. We have few *a priori* null models of what phylogenies look like (Harding, 1972; Simberloff *et al.*, 1981; Savage, 1983; Slowinski & Guyer 1989; Nee *et al.*, 1992). The branch swapping suggested here leads to the production of ladders in the tree structure (Fig. 2e). This ladder pattern is also consistent with the idea of radiations (nonrandom diversification), where one clade undergoes successive splitting. Ladders seem a common feature of published phylogenies (Colless 1982; Guyer & Slowinski, 1991). We can think of few models of evolution which would result in large ladders: after a lineage splits into two only one of the two resulting lineages undergoes effective cladogenesis. Perhaps models of cladogenesis are not necessary and ladders are a result of tree construction biases.

Nee *et al.* (1992) present a general method for testing whether the differences in the number of terminal taxa among clades can be explained by a stochastic branching process: a particular case for two sister clades was derived by Slowinski & Guyer (1989). Other methods which compare known tree topologies to expectations based on random splitting have recently been presented by Kirkpatrick & Slatkin (1993). One of these methods simply measures the mean number of nodes separating the root and each of the tips and compares this with expectations based on random branching patterns. These expectations (which they call $\bar{N}$) can be derived analytically. By this test, the Sibley & Ahlquist UPGMA tree is significantly more asymmetrical than expected by chance ($\bar{N} = 12.98$, upper 95% confidence interval for the expected $\bar{N}$ based on random branch splitting $= 10.75$; number of clades $= 137$), confirming the results presented above. A generation time effect on tree topology is also consistent with this pattern.

In summary, the UPGMA tree of the birds does show nonrandom pattern (see also

Harvey *et al.*, 1992; Nee *et al.*, 1992; Harvey & Nee, Chapter 10). Here we show that branch length acts like a heritable trait. This may be due to biological attributes of clades which change their effective rates of cladogenesis, such as body size or generation time. We must, however, be aware of and search for possible biases in the tree construction algorithms, as demonstrated above (see also Savage, 1983; Shao & Sokal, 1990). Only then can we begin to hazard what our patterns may result from.

ACKNOWLEDGEMENTS

We acknowledge support from SERC (GR/H53655), the Wellcome Trust (grant 38468) and NSERC (Canada).

REFERENCES

ANDERSON, S., 1974. Patterns of faunal evolution. *Quarterly Review of Biology, 49*: 311–332.

BENNETT, P.M., 1986. Comparative Studies of Morphology, Life History and Ecology Among Birds. Unpublished Ph.D thesis, University of Sussex.

BOCK, W.J. & FARRAND, J.F., 1980. The number of species and genera of recent birds: a contribution to comparative systematics. *American Museum Novitates*, (2703): 23 pp.

BRITTEN, R.J., 1986. Rates of DNA sequence evolution differ between taxonomic groups. *Science, 231*: 1393–1398.

BROOKS, D.R. & McLENNAN, D.A., 1991. *Phylogeny, Ecology and Behavior.* Chicago: University of Chicago Press.

BROUGH, T., 1983. *Average Weights of Birds.* Worplesdon, UK: Ministry of Agriculture, Fisheries and Food, Aviation Bird Unit.

BROWNELL, E., 1983. DNA/DNA hybridization studies of muroid rodents: symmetry and rates of molecular evolution. *Evolution, 37*: 1034–1051.

BURLANDO, B., 1990. The fractal dimension of taxonomic systems. *Journal of Theoretical Biology, 146*: 99–114.

CATZEFLIS, F.M., SHELDON, F.H., AHLQUIST, J.E. & SIBLEY, C.G., 1987. DNA–DNA hybridization evidence of the rapid rate of muroid rodent DNA evolution. *Molecular Biology and Evolution, 4*: 242–253.

COLLESS, D.H., 1970. The phenogram as an estimate of phylogeny. *Systematic Zoology, 19*: 352–362.

COLLESS, D.H., 1982. Phylogenetics: the theory and practise of phylogenetic systematics II (book review). *Systematic Zoology, 31*: 100–104.

CRACRAFT, J., 1984. Conceptual and methodological aspects of the study of evolution, with some comments on bradytely in birds. In N. Eldredge & S.M. Stanley (Eds), *Living Fossils*, 95–104. New York: Springer Verlag.

CRACRAFT, J., 1988. DNA hybridization and avian phylogenies. In M.K. Hecht, B. Wallace & G.T. Prance (Eds), *Evolutionary Biology, 21*, 47–96. New York: Plenum Press.

CRONK, Q.C.B., 1990. The name of the pea: a quantitative history of legume classification. *New Phytologist, 116*: 163–175.

CROWE, T.M., 1978. Limitation of population in helmeted guineafowl. *Suid Afrikiaanse Tydskrit vir Natuurnavorsing, 8*: 117–126.

DAWSON, W.R. & FISHER, C.D., 1969. Responses to temperature by the spotted nightjar (*Eurostopodus guttatus*). *Condor, 71*: 49–53.

DIAL, K.P. & MARZLUFF, J.M., 1988. Are the smallest organisms the most diverse? *Ecology, 69*: 1620–1624.

DUNNING, J.B., 1984. *Body weights of 686 species of North American birds.* Western Bird Banding Association Monograph 1, Cave Creek, Arizona.

FELSENSTEIN, J., 1985. Phylogenies and the comparative method. *American Naturalist, 125*: 1–15.

FITZPATRICK, J.W., 1988. Why so many passerine birds? A response to Raikow. *Systematic Zoology, 37*: 71–76.

FRY, C.H., KEITH, S. & URBAN, E.K., 1988. *The Birds of Africa.* London: Academic Press.

GILLESPIE, J.H., 1991. *The Causes of Molecular Evolution.* Oxford: Oxford University Press.

GUYER, C. & SLOWINSKI, J.B., 1991. Comparison of observed phylogenetic topologies with null expectations among three monophyletic lineages. *Evolution, 45*: 340–350.

HARDING, E.F., 1972. The probabilities of rooted tree-shapes generated by random bifurcation. *Advances in Applied Probability, 3*: 44–77.

HARVEY, P.H. & MACE, G.M., 1982. Comparisons between taxa and adaptive trends: problems of methodology. In Kings College Sociobiology Group (Eds), *Current Problems in Sociobiology*, 343–361. Cambridge: Cambridge University Press.

HARVEY, P.H. & NEE, S., 1994. Comparing real with expected patterns from molecular phylogenies.In P. Eggleton & R.I. Vane-Wright (Eds), *Phylogeneiics and Ecology*, 219–223. London: Academic Press.

HARVEY, P.H. & PAGEL, M.D., 1991. *The Comparative Method in Evolutionary Biology.* Oxford: Oxford University Press.

HARVEY, P.H., NEE, S., MOOERS, A.Ø. & PARTRIDGE, L., 1992. These hierarchical views of life: phylogenies and metapopulations. In R.J. Berry & T.J. Crawford (Eds), *Genes in Ecology*, 123–137. Oxford: Blackwell.

HAVERSCHMIDT, F., 1948. Birdweights from Surinam. *Wilson Bulletin, 60*: 230–239.

HOUDE, P., 1987. Critical evaluation of DNA hybridization studies in avian systematics. *The Auk, 104*: 17–32.

HUMPHREY, P.S., BRIDGE, D., REYNOLDS, P.W. & PETERSON, R.T., 1970. *Birds of Isla Grande (Tierra del Fuego).* Washington: Smithsonian Institute.

HUTCHINSON, G.E. & MACARTHUR, R.H., 1959. A theoretical model of size distributions among species of animals. *American Naturalist, 93*: 117–125.

KARR, J.R., WILLSON, M.F. & MORIARITY, D.J., 1978. Weights of some Central American birds. *Brenesia, 15*: 249–257.

KIRKPATRICK, M. & SLATKIN, M., 1993. Searching for evolutionary patterns in the shape of a phylogenetic tree. *Evolution, 47*: 1171–1181.

KOCHMER, J.P. & WAGNER, R.H., 1988. Why are there so many kinds of passerine birds? Because they are small. A reply to Raikow. *Systematic Zoology, 37*: 68–69.

KOHNE, D.E., 1970. Evolution of higher organism DNA. *Quarterly Review of Biophysics, 33*: 327–375.

LANYON, S.M., 1985. Molecular perspectives on higher-level relationships in the Tyrannoidea (Aves). *Systematic Zoology, 34*: 404–418.

LEWONTIN, R.C., 1965. Selection for colonization ability. In H. Baker (Ed.), *The Genetics of Colonizing Species*, 77–94. New York: Academic Press.

MARZLUFF, J.M. & DIAL, K.P., 1991. Life history correlates of taxonomic diversity. *Ecology, 72*: 428–439.

MAY, R.M., 1986. The search for patterns in the balance of nature: Advances and retreats. *Ecology, 67*: 1115–1126.

MAYR, E., 1989. A new classification of the living birds of the world. *The Auk, 106*: 508–512.

MINELLI, A., FUSCO, G. & SARTORI, S., 1991. Self-similarity in biological classifications. *Biosystems, 26*: 89–97.

MITTER, C., FARRELL, B. & WIEGMANN B., 1988. The phylogenetic study of adaptive zones: has phytophagy promoted insect diversification? *American Naturalist*, 132: 107–128.

NEE, S., MOOERS, A.Ø. & HARVEY, P.H., 1992. Tempo and mode of evolution revealed from molecular phylogenies. *Proceedings of the National Academy of Sciences, USA, 89*: 8322–8326.

PAGEL, M., 1992. A method for the analysis of comparative data. *Journal of Theoretical Biology, 156*: 431–442.

PAGEL, M., 1993. Seeking the evolutionary regression coefficient: an analysis of what comparative methods measure. *Journal of Theoretical Biology, 164*: 191–205.

RAIKOW, R.J., 1986. Why are there so many kinds of passerine birds? *Systematic Zoology, 35*: 255–259.

RIDLEY, M.W., 1982. A review of the ecology and behaviour of Button Quails. *World Pheasant Association Journal, 8*: 50–61.

SARICH, V.M., SCHMID, C.W. & MARKS, J., 1989. DNA hybridization as a guide to phylogenies: a critical analysis. *Cladistics, 5*: 3–32.

SAVAGE, H.M., 1983. The shape of evolution: systematic tree topology. *Biological Journal of the Linnean Society, 20*: 225–244.

SEINO, S., BELL, G.I. & LI, W.H., 1992. Sequences of primate insulin genes support the hypothesis of a slower rate of evolution in humans and apes than in monkeys. *Molecular Biology and Evolution, 9*: 193–203.

SEVERENTY, D.L., 1936. Feeding methods of Podargus. *Emu, 36*: 74–90.

SHAO, K. & SOKAL, R.R., 1990. Tree balance. *Systematic Zoology, 39*: 266–276.

SHELDON, F.H., 1987. Rates of single-copy DNA evolution in herons. *Molecular Biology and Evolution, 4*: 56–69.

SIBLEY, C.G. & AHLQUIST, J.E., 1990. *Phylogeny and Classification of Birds: a Study in Molecular Evolution*. New Haven, CT: Yale University Press.

SIBLEY, C.G., AHLQUIST, J.E. & MONROE, B.L., 1988. A classification of the living birds based on DNA-DNA hybridization studies. *The Auk, 105*: 409–423.

SIBLEY, C.G. & MONROE, B.L., 1990. *Distribution and Taxonomy of the Birds of the World*. New Haven, CT.: Yale University Press.

SIMBERLOFF, D.S., HECHT, K.L., McCOY, E.D. & CONNER, E.F., 1981. There have been no statistical tests of cladistic biogeographic hypotheses. In G. Nelson & D.E. Rosen (Eds), *Vicariance Biogeography: a Critique*, 40–63. New York: Columbia University Press.

SLOWINSKI, J.B. & GUYER, C., 1989. Testing the stochasticity of patterns of organismal diversity: an improved null model. *The American Naturalist*, 134: 907–921.

SOKAL, R.R. & MICHENER, C.D., 1958. A statistical method for evaluating systematic relationships. *University of Kansas Scientific Bulletin, 38*: 1409–1438.

SOKAL, R.R. & ROHLF, F.J., 1981. *Biometry*, 2nd edition. New York: W.H. Freeman.

STILES, F.G. & SKUTCH, A.F., 1989. *A Guide to the Birds of Costa Rica*. London: Christopher Helm.

SUTTER, E., 1955. *Turnix maculosa obiensis* subsp. nova. *Journal für Ornithologie, 96*: 220–221.

SUTTER, E. & CORNAZ, N., 1963. Uber die Korperentwiklung und das Schwingenwachstum junger Spitzschwung-Laufhuhnchen *Turnix sylvatica* Dussumier. *Der Ornithologischer Beobachter, 60*: 213– 223.

SWOFFORD, D.L. & OLSEN, G.J., 1990. Phylogeny reconstruction. In D.M. Hillis & C.J. Moritz (Eds), *Molecular Phylogenetics*, 411–501. Sunderland: Sinauer.

VAN VALEN, L., 1973. Body size and numbers of plants and animals. *Evolution, 27*: 27–35.

WALTERS, S.M., 1986. The name of the rose: a review of ideas on the European bias in Angiosperm classification. *New Phytologist, 104*: 527–546.

WETMORE, A., 1960. *A Classification for the Birds of the World*. Washington: The Smithsonian Institution.

WILLIS, J.C. & YULE, G.U., 1922. Some statistics of evolution and geographical distribution in plants and animals, and their significance. *Nature, 109*: 177–179.

WILSON, E.O. & WILLIS, E.O., 1975. Applied biogeography. In M.L. Cody & J.M. Diamond (Eds), *Ecology and Evolution of Communities*, 522–534. Cambridge, MS: Belknap.

WINKEL, W., 1968. Volierenbeobachtungen zur Biologie des Schwartzbrustlaufhuhnchen (*Turnix suscitator*). *Gefierdete Welt, 1968*: 141–144.

WU, C.I. & LI, W.H., 1985. Evidence for higher rates of nucleotide substitution rates in rodents than in man. *Proceedings of the National Academy of Sciences, USA, 82*: 1741–1745.

ZAR, J.H., 1984. *Biostatistical Analysis*. Englewood Cliffs, NJ: Prentice Hall.

ZEH, D.A., ZEH, J.A. & SMITH, R.L., 1989. Ovipositors, amnions and eggshell architecture in the diversification of terrestrial arthropods. *Quarterly Review of Biology, 64*: 147–168.

APPENDIX 1

Reconstructed log(weight) and Age at First Breeding for 135 Sibley & Ahlquist defined families. Branches leading to families denoted by an asterisk were estimated using the taxonomy of Sibley & Ahlquist (1990). Branches leading to families denoted by a double asterisk were estimated using medians as described in text.

Family name (grams)	Log(weight) (months)	Age at First Breeding
Struthionidae	11.5	33
Rheidae**	9.67	24
Casuaridae	10.49	36
Apterygidae	7.46	36
Tinamidae**	6.54	12
Cracidae**	7.15	24
Megapodidae**	7.16	24
Odontophoridae**	5.36	12
Numididae**	7.31	
Phasianidae	6.42	16.9
Anhimidae	7.56	30
Anseranatidae	7.64	36
Dendrocygnidae**	6.62	12
Anatidae**	7.16	25.7
Turnicidae**	3.82	5
Ramphastidae*	4.86	
Lybiidae**	3.67	
Megalaimidae**	4.37	
Indicatoridae**	2.99	
Picidae**	4.26	12
Bucconidae**	3.67	
Galbulidae**	3.55	
Upupidae**	4.19	12
Phoeniculidae**	3.76	
Rhinopomastidae**	3.44	
Bucorvidae	8.23	72
Bucerotidae**	6.48	
Trogonidae*	4.25	18
Coraciidae	4.9	12
Meropidae**	3.54	12
Momotidae**	4.63	
Todidae**	1.77	12
Alcedinidae*	2.78	12
Cerylidae**	4.55	12
Dacelonidae**	4.44	
Coliidae	3.93	12
Crotophagidae	4.77	
Neomorphidae**	4.75	
Opisthocomidae	6.7	24
Coccyzidae*	4.29	
Centropodidae**	5.46	
Cuculidae**	4.2	12
Psittacidae**	4.79	20.1
Apodidae**	3.34	24

Continued

Family name (grams)	Log(weight) (months)	Age at First Breeding
Hemiprocnidae**	4.23	
Trochilidae**	1.57	12
Musophagidae*	5.85	
Strigidae**	5.84	15.8
Tytonidae**	6.31	12
Aegothelidae**	4.15	
Podargidae**	5.69	
Batrachostomidae**	3.33	
Nyctibiidae**	5.67	
Steatornithidae	6.01	
Eurostopidae**	4.47	
Caprimulgidae*	3.95	12
Columbidae**	5.18	12
Rallidae**	5.3	12.9
Eurypygidae	5.4	
Otididae**	7.39	31.3
Cariamidae**	7.47	
Rhynochetidae	6.51	24
Psophidae**	6.99	
Heliornithidae*	5.87	
Gruidae*	8.08	48.2
Pteroclidae**	5.49	12
Jacanidae**	4.73	24
Rostratulidae**	4.6	
Thinocoridae*	5.09	
Pedionomidae	4.11	
Scolopacidae**	4.65	19.4
Charadridae*	4.91	26
Burhinidae**	6.37	36
Chionididae**	6.04	
Glareolidae*	4.66	12
Laridae*	5.8	39.7
Falconidae**	5.69	27.8
Sagittaridae	8.31	
Accipitridae*	6.96	43.1
Podicipedidae**	6.14	24
Phaethontidae**	6.25	60
Sulidae**	7.46	37.3
Anhingidae**	7.15	24
Phalacrocoracidae**	7.35	28.7
Ardeidae**	6.15	18.9
Scopidae	6.05	
Phoenicopteridae	7.7	34.7
Threskiornithidae**	7.02	38
Ciconiidae*	8.23	49.9
Pelecanidae*	8.8	37.9
Fregatidae**	7.06	48
Spheniscidae**	8.36	36.7
Gaviidae**	8.06	27.7
Procellariidae*	5.95	50.3

Continued

Family name (grams)	Log(weight) (months)	Age at First Breeding
Acanthsittidae**	1.9	
Eurylaimidae**	3.87	
Pittidae**	4.39	
Furnariidae*	3.39	
Formicariidae**	4.09	
Conopophagidae**	3.02	
Rhinocryptidae**	3.73	
Thamnophilidae**	3.02	
Tyrannidae*	3.13	14.9
Climacteridae*	3.44	12
Menuridae	5.32	33
Ptilonorhynchidae**	4.94	52.5
Maluridae*	2.58	24
Meliphagidae**	3.11	
Pardalotidae*	2.43	
Eopsaltridae**	2.75	
Irenidae*	3.79	
Pomatosomidae**	5.69	
Laniidae*	3.7	23.7
Vireonidae**	2.67	12
Corvidae*	3.82	23.4
Bombycillidae*	3.57	
Cinclidae	4.18	12
Muscicapidae*	3.41	12
Sturnidae*	4.33	12.9
Sittidae*	2.81	12
Certhidae*	2.44	12
Paridae*	2.53	12
Aegithalidae*	1.87	12
Hirundidae**	2.88	12
Regulidae**	1.79	12
Pycnonotidae**	3.54	
Cisticolidae**	2.47	
Zosteropidae**	2.31	12
Sylviidae*	3.07	12
Alaudidae**	3.36	12
Nectariniidae*	2.45	
Melanocharitidae*	2.4	
Paramythiidae	3.95	
Passeridae*	3.02	13.2
Fringillidae*	3.09	12

Sexual size dimorphism and comparative methods

SÖREN NYLIN & NINA WEDELL

CONTENTS

Keywords: Size dimorphism – comparative methods – phylogeny – sexual selection – sperm competition – mating system – insects – birds – mammals.

Abstract

The 'traditional' methods for statistical analysis of comparative data, for example correlations and ANOVAs, have been criticized on several grounds. The response has been to use phylogenies to find independent events in evolution, and a number of methods now exist. Four major categories of comparative methods can be distinguished: the four combinations of non-directional (equilibrium) and directional (transformational) methods applied to discrete and continuous types of data. Presently there seem to be two kinds of phylogenetic ecology, being performed in two different research traditions. Many ecologists belong to the 'actualistic' or 'equilibrium' tradition, whose aim it is to explain the states of recent organisms. This tradition favours research on continuous traits and has no strong objection to non-directional methods. Many researchers at the interface between taxonomy and evolutionary ecology belong to a second, 'historical' tradition. This tradition favours research on discrete traits, using directional methods that can address evolutionary sequences and the origin of states. A review of comparative studies on sexual size dimorphism in animals illustrates what different methods are able to 'see'. For example, a lekking, sexually monomorphic bird species which evolved from an ancestral state of non-lekking and female-biased size dimorphism would be seen as evidence against sexual selection hypotheses by non-directional methods, but as supporting evidence by directional methods. The question of whether sperm competition has been important for the evolution of large body size in males is addressed. In general, it is suggested that non-significant results using phylogenetic methods should not be seen as falsifying a hypothesis, and that significant results using non-phylogenetic methods (and inflated degrees of freedom) should not be discarded outright. The statistical significance and explanatory power of comparative methods depend on many factors, including the ecological range studied and the taxonomic level chosen for analysis. For this reason, the generality and consistency of comparative trends across taxa should be seen as a more important criterion than statistical significance alone.

INTRODUCTION: A CHOICE OF COMPARATIVE METHODS

Evolutionary ecology could be described as the study of patterns of adaptation, and of the evolutionary processes creating and maintaining them. One of the strongest methods available to this branch of evolutionary biology, applied fruitfully ever since Darwin, is the 'comparative method', where species inhabiting different environments or displaying different traits are compared. Comparative methods look for either associations between environmental variation and variation in traits of organisms, revealing potential adaptations to the environment, or associations between variation in two or more traits of the study organisms. Together, these two types of associations may give clues as to what evolutionary processes originated these traits, and what maintains them today.

As evolutionary biology has matured, methods for statistical analysis of comparative data have been applied to an increasing degree. These methods include qualitative tests, for example when comparing frequencies of species displaying a given discrete trait, but also quantitative techniques such as correlation analysis of continuous data. Further, Pagel & Harvey (1988) and Harvey & Pagel (1991) made the distinction between 'non–directional' and 'directional' comparative methods. Earlier, Huey (1987) called them 'equilibrium' and 'transformational' methods. Non-directional methods look horizontally over the endpoints of phylogenies, finding the patterns of association seen in recent species: they compare only descendent taxa. Directional methods look vertically along the branches of phylogenies, in an attempt to discern how the present states came to be: they compare ancestors with their descendants. Thus they focus on changes in one trait which may have been associated with a change in another, that is on evolutionary sequences and on correlations between inferred evolutionary changes in traits, rather than on correlations between present states of traits.

All traditional comparative methods, including observation without statistics (the technique so successfully employed by Darwin) belong to the non-directional category. In general, non-directional comparative methods are not very good tests of hypotheses about evolutionary processes. This is because they are really aimed at explaining present 'states' of organisms (or their maintenance through natural selection) rather than at explaining evolutionary 'events' (or evolutionary origins; cf. O'Hara, 1988). As a necessary complement, experimental techniques and field experiments have therefore been used to test hypotheses of process. These techniques have similar limitations, however, in that they can only address current function and maintenance. Directional comparative methods which make use of phylogenies could, at least potentially, test theories of evolutionary process more directly by inferring historical events.

In recent years, the traditional comparative methods have been criticized on several grounds. In particular, species are not statistically independent units and therefore unsuitable as data points (e.g. Ridley, 1983; Clutton-Brock & Harvey, 1984; Felsenstein, 1985). This confounding effect of phylogeny, that is that related species living in similar environments are likely to be similar due to common ancestry as well as to common adaptation, has long been realized (see Harvey & Pagel, 1991, for review of reasons). However, it was not always taken into account during the period (late 60s to late 80s) which Brooks & McLennan (1991) called the 'eclipse of history' in ecology.

During the last decade, many methods for dealing with phylogeny in ecology have been developed. They can be divided into four groups: the four combinations of non-directional or directional methods applied to discrete or continuous data (Fig. 1). Two early approaches addressing the statistical problem of similarity by descent were applied to discrete and continuous traits, respectively. Both had the same general aim: to avoid inflating the degrees of freedom (and significance values) that results from counting each species as an independent observation. Ridley (1983) suggested the use of phylogenies to determine when a change in a trait association occurred, counting all species inheriting the new trait association as a single observation. The statistical technique suggested by Ridley is, in principle, an addition to the category of non-directional qualitative techniques, because associations seen in extant species are counted regardless of what state preceded it and thus regardless of the sequence of evolutionary events that created the associations.

To deal with quantitative variation, a non-phylogenetic method was advocated by Stearns (1983). It was suggested that one should look for the taxonomic level where most of the variation occurs, and perform an analysis of variance at this level. Later, Bell (1989) suggested that a nested analysis of covariance at several taxonomic levels is more appropriate and informative. These methods cure or reduce some of the problems with inflated degrees of freedom, but suffer from the objection that taxonomic levels are arbitrarily defined (taxa of equal rank are not necessarily comparable) and therefore problematic as categorical levels in statistical analysis (Wanntorp et al., 1990).

An important (also non-directional) addition to these quantitative methods was Felsenstein's (1985) suggestion to use phylogenetically independent contrasts, that is comparing pairs of relatively closely related species which, for example inhabit

TYPE OF DATA

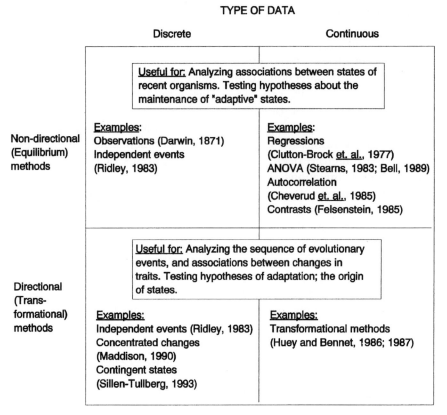

	Discrete	Continuous
Non-directional (Equilibrium) methods	<u>Useful for:</u> Analyzing associations between states of recent organisms. Testing hypotheses about the maintenance of "adaptive" states. <u>Examples:</u> Observations (Darwin, 1871) Independent events (Ridley, 1983)	<u>Examples:</u> Regressions (Clutton-Brock <u>et. al.</u>, 1977) ANOVA (Stearns, 1983; Bell, 1989) Autocorrelation (Cheverud <u>et. al.</u>, 1985) Contrasts (Felsenstein, 1985)
Directional (Trans-formational) methods	<u>Useful for:</u> Analyzing the sequence of evolutionary events, and associations between changes in traits. Testing hypotheses of adaptation; the origin of states. <u>Examples:</u> Independent events (Ridley, 1983) Concentrated changes (Maddison, 1990) Contingent states (Sillen-Tullberg, 1993)	<u>Examples:</u> Transformational methods (Huey and Bennet, 1986; 1987)

Figure 1 A classification of comparative methods into four major categories according to: (1) type of data and (2) whether the method analyses only data on recent organisms (non-directional methods) or makes of use of character optimization techniques to reconstruct ancestral states, and study evolutionary sequences and changes along branches of phylogenies (directional methods). The terminology is from Pagel and Harvey (1988) and Harvey and Pagel (1991). Terminology in parentheses from Huey (1987). Papers cited are sources for methods or (for traditional methods) examples of their use in the study of sexual size dimorphism. The method of Ridley (1983) has both non-directional and directional aspects, depending on the statistical method applied.

different environments. A number of such comparisons can be collected and statistically analysed. The simplest analysis is to find the direction of each comparison and perform a sign test (e.g. Burt, 1989). It is also possible to make more use of the quantitative information, that is the amount of difference between contrasts, although this may create problems (see below).

Most techniques in use today are variations on one of these general themes. With co-authors, we pointed out the dichotomy between qualitative and quantitative techniques that could be seen in the methods and research we reviewed (Wanntorp *et al.*, 1990). To a large extent, this dichotomy remains today, and is manifest in two general 'schools' or categories of work. The first category belongs to a tradition emphasizing history and evolutionary origin, typically consisting of taxonomists and ecologists interested in phylogenetic reconstruction of adaptive processes. They mainly deal with qualitative traits that can be mapped onto a phylogeny using the technique of character optimization. This tradition was manifested in a book by Brooks & McLennan (1991; see also Nylin, 1991).

Sillén-Tullberg's (1988) study of the association between warning coloration and gregariousness in butterfly larvae used Ridley's (1983) methodology for inferring changes in traits. Her analysis tested not if the association exists (a non-directional hypothesis, suitable for Ridley's statistical test), but what the evolutionary sequence of events that created the association had been. Her study demonstrates how directional studies can test process theories. If, as had been suggested by others, kin selection among gregarious butterfly larvae had facilitated the evolution of warning coloration, changes to gregariousness would be expected to have preceded shifts to aposematism (this is a directional hypothesis). This was not found in any case, whereas the opposite sequence was found repeatedly. There was no statistical method available at the time that dealt explicitly with directional, qualitative hypotheses (except for simply counting the numbers of each sequence), but it is also typical of the historical tradition that statistics did not figure very prominently in Sillen-Tullberg's study.

Other researchers have focused on finding present-day patterns in nature, and on how they are maintained by natural selection. Non-directional comparative methods like correlations and analyses of variance have been used extensively in this field, often ignoring the problem of statistical dependence between data points. The use of phylogenies to avoid this problem, when applied to continuous data, was manifest in a book by Harvey & Pagel (1991). The lack of overlap with Brooks & McLennan (1991) illustrates the existence of two largely separate fields of 'phylo-genetic ecology'. This was graphically illustrated by reviews of the Harvey & Pagel book, written by and for ecologists, by the taxonomists Carpenter (1992) and Coddington (1992).

In this paper we consider one trait (sexual size dimorphism) and processes which may be associated with it, concentrating on one which is often neglected in this context: sperm competition (Parker, 1970). We investigate how the associations between sexual size dimorphism and potentially correlated traits or environments have been treated by various qualitative and quantitative, non-directional and directional comparative methods. Since different methods have been applied to the problem, it is instructive to see how they deal with the same trait, which patterns they are able to 'see', and which questions they can answer. We address whether evidence favours the idea that sperm competition has been important in the evolution of male body size, and which other processes may have been important.

SEXUAL SIZE DIMORPHISM IN ANIMALS

The fact that sexes differ in size and weight in many animals has stimulated a great deal of research (e.g. Trivers, 1972; Ralls, 1976; Clutton-Brock, Harvey & Rudder, 1977; Shine, 1978, 1979, 1988, 1989, 1991; Payne, 1984; Arak, 1988; Wiklund & Karlsson, 1988; Cheverud, Dow & Leutenegger, 1989; Hedrick & Temeles, 1989; Höglund, 1989; Fairbairn, 1990; Kappeler, 1990; Björklund, 1991; Elgar, 1991; Wiklund & Forsberg, 1991; Wiklund, Nylin & Forsberg, 1991). In a few taxa (e.g. mammals and birds) males are generally the larger sex (Darwin, 1871; Greenwood & Wheeler, 1985). This is often attributed to intrasexual selection in the form of direct competition for access to females (Darwin, 1871; West-Eberhard, 1983). In most animals, however, females are heavier. This includes some mammals (Ralls, 1976) and birds (e.g. Wheeler & Greenwood, 1983), other vertebrates including amphibians (Shine, 1979), and many insects and other invertebrates (Darwin, 1871). Such patterns could be due to a number of possible causes of stronger selection for large size in females, for example because large females are better parents (Ralls, 1976), because females are territorial and/or carry males in mating (Fairbairn, 1990) or because fecundity is more dependent on size in females than in males (a potentially more general explanation: e.g. Darwin, 1871; Williams, 1966; Trivers, 1972; Arak, 1988; Wiklund & Karlsson, 1988; but see also Shine, 1988).

Sexual size dimorphism is a trait of both sexes taken together. When females are larger than males, this can be because a larger size has been favoured in females, because a smaller size has been favoured in males, or both. Sexual size dimorphism is essentially the sum of many selection pressures, including both sexual and natural selection, and other factors such as allometry, affecting size in one or both sexes (Ralls, 1976; Cheverud, Dow & Leutenegger, 1985; Arak, 1988; Bjorklund, 1991; Webster, 1992). In addition, size and traits affecting size may be phylogenetically constrained (e.g. Cheverud, Dow & Leutenegger, 1985). This makes size dimorphism an unusually complex trait, and it is often hard to pick out a single factor of particular importance (e.g. Cheverud, Dow & Leutenegger, 1985; Greenwood & Adams, 1987; Kappeler, 1990; Trail, 1990). It can be argued that this is where phylogenetic methods can be of great importance, because it may be possible to disentangle the effects of many factors by viewing them in a historical context. In our brief review of the subject, we concentrate on comparative methods which have been used to sort out associations with sexual size dimorphism.

A REVIEW

Primates and other mammals: is there a correlation between mating system and sexual size dimorphism?

Darwin (1871) used the primates, including Man, as one example where larger male size has probably been favoured by sexual selection among males for access to females. He arrived at this conclusion from simple observations of male size bias and male aggressive competition in primates and other taxa. Today, this is still the most generally accepted explanation for male-biased size dimorphism in mammals. In this tradition, Clutton-Brock & Harvey (1977) and Alexander et al. (1979) reported a positive correlation between relative male size and the 'socionomic sex ratio', that is the number of males per female in primate troops (Fig. 2). The same pattern was found in ungulates and pinnipeds (Alexander et al., 1979). The studies differed in that Alexander et al. (1979) included only species where there is generally a single male per group, reasoning that not all males may breed in multi-male groups. The clearest result from the two studies was that the sexes differed more in polygynous than in monogamous species. This was a success for the sexual selection category of explanations for the evolution of sexual size dimorphism, because one competing explanation (ecological niche separation; Selander 1966, 1972; Shine, 1989) would probably predict the opposite pattern (Harvey & Pagel, 1991).

There were clear problems with the primate results, however. In Alexander et al. (1979) species were used as independent observations, meaning that the reported

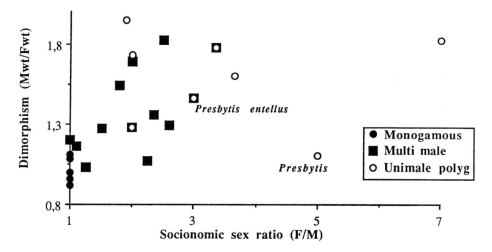

Figure 2 Sexual dimorphism (male weight/female weight) in primates and 'socionomic sex ratio', that is the number of females per male in a troop. Data from Clutton-Brock and Harvey (1977). Different symbols represent mating system according to legend. Data on mating system from Smuts *et al.* (1987) and Sillén-Tullberg & Möller (1993).

significance value of $P < 0.001$ ($N = 22$ species) has little meaning. Statistical significance was dependent on a cluster of monomorphic, monogamous species (seven species of *Hylobates* (!) and three prosimians). This perhaps demonstrates the difference between monogamous and polygynous mating systems, but the reported trend towards relatively larger males in more polygynous species is not present in the remaining data (i.e. among degrees of polygyny). Such trends can be better seen in the data from ungulates and pinnipeds, but again the number of independent observations is not clear.

Clutton-Brock & Harvey (1977) were explicitly aware of the problem with statistical non-independence of data points. Since there was often little variation in mating system within genera, they based correlations on generic means rather than on species means, except when a species differed from the rest of the genus. This is a non-directional technique somewhat similar to the method of performing an analysis of variance at the taxonomic level where most of the variation is present (Stearns, 1983). The method partly solves the problem of inflated degrees of freedom, but significance values should not be relied upon, since genera are not statistically independent observations, either. For example, the prosimian primates are all relatively monomorphic (Kappeler, 1990, discusses the reasons for this pattern), the Cercopithecinae (e.g. macaques and baboons) are all relatively dimorphic (Clutton-Brock & Harvey, 1977; Harvey & Clutton-Brock, 1985), whereas the Callitrichidae (tamarins and marmosets) are probably all monogamous (Smuts *et al.*, 1987).

The reported significance for the correlation between socionomic sex ratio and sexual size dimorphism is therefore questionable. Also, as in Alexander *et al.* (1979), it is dependent mostly on a cluster of monogamous, relatively monomorphic genera (e.g. *Aotus, Saguinus, Callithrix, Hylobates*), whereas the trend in the remaining range of sex ratios is far from clear (Fig. 2).

Primates: autocorrelation methods

In another study dealing with sexual size dimorphism in primates, Cheverud, Dow & Leutenegger (1985) used an autocorrelation model to separate 'phylogenetic' from 'specific' effects. Again, this is conceptually similar to an analysis of variance at different levels in the taxonomical hierarchy, but this technique can use actual phylogenies to remove the 'phylogenetic' portion of variance. In the analysis, Cheverud, Dow & Leutenegger, included data on sexual weight dimorphism, weight itself, mating system (monogamous/polygynous), habitat and diet. They concluded that 50% of the variation in weight dimorphism among primates is 'due to phylogeny', 36% to weight itself (scaling, or allometry) and only minor amounts of variation is due to habitat (4%), diet (2%) or mating system (2%). This is an interesting, although not uncomplicated result.

The high proportion of variance that could be attributed to 'phylogeny' (in fact, the taxonomic hierarchy, which is not likely to be a very good reflection of actual phylogeny in this or any similar case) demonstrates that size dimorphism has been a conservative trait in primate evolution. This result clearly shows the non-independence of species data points and the importance of viewing sexual size dimorphism in a phylogenetic context. The scaling effect of size itself (larger primate species are relatively more dimorphic) was reported also by Clutton-Brock & Harvey (1977) and Leutenegger (1978). Cheverud, Dow & Leutenegger (1985) view the proportions of total variance in sexual size dimorphism that are due to 'phylogeny' and 'size' as potentially non-adaptive, that is as constraints on adaptation.

It can be argued that, if related species share a character association, this may be due not to 'phylogenetic inertia' but to a shared selective regime under which the association is evolutionarily stable (Grafen, 1989). This does not remove the statistical problem; data points from related species would still be non-independent, because of the shared selective regimes (Ridley, 1989). Similarly, an allometric relationship is not necessarily non-adaptive. The scaling effect of size on phylogeny has been interpreted in a framework of sexual selection theory. In small species, overall size may be less important than, for example speed and agility, or size of canines in male-male conflicts (Clutton-Brock, 1985; Kappeler, 1990; see Webster, 1992, for a review of several possible adaptive and non-adaptive explanations for the allometry). The scaling effect is by no means universal, even among primates. It is not seen among monogamous species (Leutenegger, 1978, 1982), or among small primates (Kappeler, 1990), supporting the idea of a positive feedback between size and polygyny (Leutenegger, 1978, 1982).

Although autocorrelation methods are of interest for comparing inertia and/or evolutionary plasticity, applied to a particular trait association such as sexual size dimorphism and mating system, they do not really resolve the question of whether a significant, adaptive, association exists (see also Harvey & Pagel, 1991).

Birds: dimorphism as a discrete character

As in mammals, the prevalent pattern in birds is male-biased size dimorphism. Darwin (1871) attributed this to sexual selection, noting also that males of polygamous species tend to be more brightly coloured and larger than the females, whereas monogamous species are more sexually monomorphic. Numerous exceptions, however, have provoked discussion.

Höglund (1989) pioneered the application of explicitly phylogenetic methods to explain sexual dimorphism in birds. He tested the correlation between mating system (lekking or non-lekking) and sexual dimorphism in size and plumage. Lekking can be viewed as an extreme form of polygamy, in that males in lekking species compete in some way (sometimes aggressively) for access to females. Using species as independent observations, both correlations were significant. Höglund then proceeded by applying Ridley's (1983) method of counting only phylogenetically independent origins of trait associations, a method which is designed for discrete characters. Consequently, Höglund had to recode dimorphism as present or absent, using an arbitrary cut-off point. Using this method, the sample size dropped from 113 to 35 and neither correlation was significant. Höglund concluded from this that the assumption that males are larger and more brightly coloured than females in lekking species "is not supported by a phylogenetic analysis".

However, an analysis such as Höglund's (1989) can only look for present associations between lek mating systems and male-biased sexual size dimorphism. These traits must have evolved independently many more times than the number of times when one trait has evolved without the other, or when one of them has been lost but not the other. Also, by recoding continuous characters as discrete, information is lost and only cases of strong dimorphism will count. Finally, counting only what is inferred to be phylogenetically independent events is a conservative method of finding associations (since even related species may have evolved independently to an unknown degree), although necessary in order to avoid inflating the degrees of freedom. Despite these considerations, the trend to be tested partly survived the analysis (size dimorphism: $P > 0.18$; plumage dimorphism: $P > 0.23$). Significance limits are after all arbitrary, and we would prefer to view the pattern found by Höglund as a trend in the predicted direction, a tendency which only weakly supports the association, rather than as no support. Höglund stressed that there seemed to be cases where lekking had promoted dimorphism, although the correlation did not seem to be strong across all taxa studied.

One strength of Ridley's (1983) and similar methods is that they focus on evolutionary changes in characters, which is desirable from the point of view of the 'historical' tradition. They are also at least potentially directional methods, that is the inferences about when changes occurred can be used to study the evolutionary sequence of events, and the number of times that each sequence is found can be counted and analysed statistically. Höglund (1989) found no evidence for the directional hypothesis that evolution of lek mating systems has preceded the evolution of sexual size or plumage dimorphism. In the case of size dimorphism, he did find four instances when lekking seemed to have preceded dimorphism, but seven instances of either the reverse sequence or the simultaneous loss of both traits (the inclusion of the latter type of events in this category, supposedly falsifying the hypothesis, is however not self evident).

Birds: methods for continuous traits

For continuous traits, the most common practice has been to make use of one of the family of 'independent contrasts' methods (e.g Felsenstein, 1985; Grafen, 1989; Burt, 1989; Pagel & Harvey, 1989; see also Harvey & Pagel, 1991). Some need fully resolved phylogenies, knowledge of branch lengths, or are affected by critical assumptions such as dependence on an evolutionary model (generally Brownian motion). The latter problem becomes most critical when it comes to correlating two continuous traits with each other. Some of these 'independent contrasts' methods make use of comparisons nested within each other, producing several contrasts from a single clade, but such contrasts are not likely to be independent under all evolutionary models (Carpenter, 1992). Even if no nested clades are used, contrasts for two continuous traits are not likely to be independent. Clades should often be characterized by typical ranges of values of both traits, perhaps also including typical evolutionary responses of one trait to a change in the other, so that contrasts drawn from within these clades are not truly independent of each other. This would tend to exaggerate correlations between the two traits (M. Björklund and B. Tullberg, personal communication).

A simpler situation is to correlate a discrete character (e.g. mating system) with a continuous one (e.g. sexual size dimorphism). If only non-nested contrasts between pairs of clades that differ in mating system are used, these contrasts may be independent, although the possibility remains that there may be clade-specific responses to changes in the discrete trait. Oakes (1992) used a contrast method to reanalyse Höglund's (1989) size data, and found statistically significant differences ($P = 0.026$ in a paired t-test) between mating systems, supporting the hypothesis that lekking favours male-biased size dimorphism. Although few contrasts could be made, Oakes also compared lekking species with closely related non-lekking polygynous species ($P = 0.146$ in a paired t-test). In Oakes' words, this "trend does seem to suggest that lekking birds may be more sexually size dimorphic than their polygynous non-lekking relatives". Note, however, that the significance level for this 'trend' is similar to that for the result which Höglund considered 'no support'.

The contrasts tests reported above were non-directional, testing association but not causality, but Oakes (1992) also used the directional method of Maddison (1990). This method takes the distribution of states in the phylogeny into account, and in this case calculates whether sexual size dimorphism has been more likely to evolve on branches of the phylogeny that have been reconstructed to have a lekking state. Since the method is designed for qualitative data, Oakes could apply it only to Höglund's discrete form of the data. The results of this analysis supported the process theory (large male size is favoured in species with lek mating systems) by suggesting that sexual size dimorphism has been more likely to evolve in branches of the phylogeny where birds lek ($P = 0.064$–0.003, depending on how equivocal branches are assigned). This result can be questioned, since Oakes could use Maddison's method only by joining Höglund's separate phylogenies into a single bird phylogeny. Höglund's phylogenies only represented lekking taxa and enough related non-lekking taxa to resolve which evolutionary changes most probably had occurred in each clade. This is not a problem for methods such as Ridley's (1983) or for contrasts methods, which focus only on portions of phylogenies where an evolutionary change in a character of interest is reconstructed to have occurred. Maddison's directional method, which makes use of the number of branches representing each character state, should however not be uncritically applied to a biased sample of bird phylogeny (over-representing lekking taxa and their close relatives), as if this sample represented the birds as a whole. There is a strong possibility that this affects both character reconstruction and the relative number of branches with each character state (Sillén-Tullberg, 1993).

Another possibility would be to apply a method that is both quantitative and directional, that is focus on correlations both in amount and direction of change in characters (cf. Huey & Bennett, 1986, 1987; Björklund, 1991). When one discrete trait is supposed to influence another continuous trait, we can 'simply' use a phylogeny to determine the amount and direction of change in the continuous trait associated with switches in the discrete trait. Björklund (1991) did this for the association between sexual dimorphism and mating system in grackles (*Quiscalus*: Icterinae). Two of the species in this genus are monogamous, three are 'regularly polygynous' (care for offspring, defend females) and two are 'highly polygynous' (do not care for offspring, defend resources for females). The change to a higher degree of polygyny in the last two species was accompanied by a significant change to more pronounced male-biased dimorphism in both size and tail length (although Björklund, 1991, cautioned that the former change was small enough to have been caused by drift alone). A monogamous species which has as its probable sister-species one of the highly polygynous grackles was found not to have evolved a higher degree of dimorphism, but rather the opposite.

The main problem with explicitly directional methods such as the one used by Björklund (1991) is that they make use of comparisons between ancestors and descendants. Character states at interior nodes in the phylogeny (ancestral states) have to be reconstructed from the states in recent taxa, using the principle of parsimony. This means that, for statistical purposes, the comparisons between ancestors and descendants are not made between independently derived sets of data (Harvey and Purvis, 1991).

Recently, Webster (1992) analysed an extended data set on grackles using non-directional methods and a continuous measure of male polygyny (harem size). He showed a strong positive relationship between harem size and relative male size that held true in both species regressions, generic regressions and nested analyses of covariance. There were also positive relationships between body size and dimorphism (as in the primates), and between body size and harem size. The relationship between body size and dimorphism disappeared when controlling for harem size, suggesting that the allometry resulted purely from the association between body size and harem size. Webster (1992) argued that these patterns favour adaptive explanations for the allometry, that is larger species tend to be more polygynous and therefore more dimorphic.

Insects and other invertebrates: general patterns and case-studies

The general pattern in insects, and indeed in most invertebrates and poikilotherm vertebrates, is that the female is the larger sex (Darwin, 1871; Greenwood and Wheeler, 1985). There are exceptions, and some of these involve species where there is strong male–male competition or where males carry females during mating, as Darwin noted. His opinion about the general pattern of female-biased size dimorphism in the "lower classes of the animal kingdom" was that "the greater size of the females seems generally to depend on their developing an enormous number of ova", that is a fecundity-advantage hypothesis (fecundity is often proportional to body size in invertebrates).

Concerning insects, he found more probable an explanation put forward by Wallace (cited in Darwin, 1871), that small male size is advantageous because it promotes the earlier emergence of males (protandry; see also Singer, 1982). Protandry is the likely evolutionarily stable strategy in most insect populations (e.g. Wiklund & Fagerström, 1977; Parker & Courtney, 1983). Alternatively, protandry could be an incidental effect of sexual size dimorphism, the "developmental-constraints hypothesis" (Thornhill & Alcock, 1983; but see Nylin *et al.*, 1993). This set of (not mutually exclusive) hypotheses highlights how sexual dimorphism is a trait of both sexes, not one of them in isolation, and how it is also interrelated with other parts of the life-history, such as age at sexual maturation.

In modern terms, a fecundity-advantage hypothesis could be stated thus: 'females will be the larger sex when a small increment in female body size results in a relatively higher increment in reproductive success, taking costs of longer development time or higher growth rates into account, than does the same increment in male body size' (cf. Trivers, 1972, 1985; Arak, 1988; see also Shine, 1988). However, it may be dangerous to jump from generalities to conclusions about particular species. For example, Greenwood & Adams (1987) discussed the case of *Gammarus* amphipods. In many amphipods, males are larger. This was interpreted as an effect of male–male competition (Ward, 1983), but Greenwood & Adams (1987) claimed that a more likely explanation was that males carry the females for several days, in pre-copula. Of course, nothing prevents both explanations from being true simultaneously, and comparative methods are needed to sort out their relative importance.

Elgar's (1991) study of the relationship between size dimorphism and sexual cannibalism in orb-weaving spiders (Araneidae) is an example of a pattern which should be rather taxon-specific. It has long been hypothesized that sexual cannibalism in spiders selects for small male size, because small males may have a better chance to avoid being eaten before copulation. Elgar (1991) found comparative evidence for the influence of sexual cannibalism on size dimorphism: spiders that mate at the central hub of the orb-web (which means that the male has to traverse the web to be mated) are strongly sexually dimorphic. In other species, a male constructs a mating thread to which he attracts the female. In such species, large males may in fact be less likely to be cannibalized and more likely to achieve matings. These species were less sexually dimorphic.

The analysis was done by calculating sexual dimorphism as the residuals from the regression of log male body length on log female body length, in order to obtain a measure which was independent of size itself, that is of any allometric pattern present. These measures were subjected to an analysis of covariance where it was shown that the location of the courtship had a significant effect on dimorphism among species. Elgar (1991) also noted a taxonomic association: subfamilies differed in courtship location and also differed significantly in dimorphism in an analysis of covariance. In an attempt to control for confounding phylogenetic effects he analysed *Argiope*, which is polymorphic with respect to courtship location, and showed that the pattern held true within this genus. This does not solve the problem of statistical dependence, but shows that there is at least some degree of independence between evolutionary events. Elgar's study explains some of the variation in sexual size dimorphism within the orb-weaving spiders, but not the typical female-bias itself. It could be added that Darwin (1871) noted that male spiders seldom seem to fight over females, "nor, judging from analogy, is this probable; for the males are generally much smaller than the females, sometimes to an extraordinary degree".

Fairbairn (1990) studied sexual size dimorphism in temperate waterstriders

(Hemiptera: Gerridae), in which females tend to be larger than males. As in the primates, males were relatively larger in larger species. There was a significant allometry among 12 investigated species, dimorphism decreasing with increasing size as males became larger and more similar to females. Fairbairn favoured a non-adaptive explanation for this allometry. A covariance analysis showed no significant effects of phylogeny (taxon) among the 12 species, but Fairbairn (1990) allowed for the possibility that the basic female bias in dimorphism may reflect descent from a common ancestor rather than adaptation within the taxon.

Concerning adaptive hypotheses, Fairbairn rejected the "developmental constraints' hypothesis", that is dimorphism was not associated with protandry. She found support for three selective processes: dispersal by flight may reduce dimorphism, reduced loading during prolonged pairing may increase dimorphism (because females carry males in copula), whereas sexual selection may act in both directions. In species where males defend oviposition sites, males are relatively large and there is little, or even reversed dimorphism. In female-territorial species there is strong female bias. However, these patterns were confounded by the basic allometry, and Fairbairn (1990) concludes, along with Cheverud, Dow & Leutenegger (1985), that "sexual size dimorphism is most strongly affected by potentially non-adaptive factors".

In a similar study in tephritid fruit flies, Sivinski & Dodson (1992) investigated evidence for various adaptive and non-adaptive hypotheses. Again, males are smaller than females, and again this was not associated with protandry. The same allometry as in the water-striders was found, and Sivinski & Dodson argued that this was evidence against the 'fecundity-advantage' hypothesis. All other things being equal, selection for increased fecundity in females should tend to give rise to the opposite pattern. Very fecund females of large absolute size should be associated with relatively small males, because males should be more uniform across species. However, it seems possible that, in this case and in the waterstriders, the reason for the basic female-bias in size could be a fecundity advantage in the ancestors of the clades. The fecundity advantage may also have helped maintain female-bias within the clades, even though its effects are hidden by the general allometry among fruit flies and waterstriders. Nothing prevents large species from having both relatively large males compared to the female (due to one adaptive or non-adaptive process responsible for the allometry) and at the same time females which are larger compared to other species (not to the male) and also more fecund. If so, there is no support for the fecundity advantage hypothesis, but not really evidence against it either. In any case, Sivinski & Dodson (1992) concluded that sexual size dimorphism in tephritids has probably been influenced by non-adaptive allometries, but also by selection for distance flight and sexual selection (some male-territorial species have less or reversed dimorphism).

In favour of the 'fecundity-advantage' hypothesis, Wiklund & Karlsson (1988) found that maximum lifetime fecundity was positively correlated with sexual dimorphism among 14 species of satyrine butterflies with female-biased size dimorphism. They ruled out effects of either loading constraints (since females are the carrying sex in all investigated species), protandry (phenology was not correlated with dimorphism) or sexual selection (no variation in mating system among the 14 species). Analyses were done as straight-forward correlations with species as data points, which puts significance values into doubt but probably not the general pattern, since nine genera were represented in the data set.

Finally, another study on butterflies is of considerable interest because of its treatment of phylogeny. Wickman (1992) was primarily interested in how male mate-finding tactics (perching/territorial *versus* searching/patrolling) has affected the evolution of body 'design' for flight (e.g. mass ratios and wing shape), but sexual dimorphism in body and wing size was also studied as part of the investigation. Wickman used an independent contrasts method which should be free of the non-independence problems, because one of the traits (male tactics) was discrete and no nested clades were used. Interestingly, he showed that, using species as data points, there were no significant differences in sexual dimorphism between species with different mating systems. However, when independent contrasts of closely related species (or clades) which differed in male tactics were compared, there were significant differences in sexual dimorphism for wing length and wing area, but not for body mass. Thus, male tactics have not affected the evolution of overall sexual size dimorphism, but has affected sexual dimorphism in traits related to flight design. This latter fact was not seen when species were used as data points, suggesting that in this case such a procedure did not result in inflated statistical significance, but rather the opposite.

Evidently, the evolutionary effects of male tactics on body design were hidden in the over-all comparisons by strong phylogenetic patterns in body design, such that related species had rather similar design regardless of male tactics. Only when related taxa with different tactics were compared was it possible to pick up slight departures from phylogenetic patterns.

Wickman (1992) also showed significant differences in body design (within males) between perching and patrolling species in the predicted directions for wing loading, wing area, thorax ratio and thorax mass, and this was true whether the comparison was made among species or contrasts. However, differences in abdomen ratio and aspect ratio (a wing shape variable) were significant only among contrasts, again suggesting that effects of common descent can hide patterns as well as artificially strengthen them. Wickman's study is of interest in the context of sexual dimorphism also because of another consideration. There were significant differences between species with different male tactics also in female body design, suggesting genetic correlations between the sexes (one of the general constraints on sexual dimorphism; Lande, 1987), and putting the specific evolutionary process into doubt. Wickman's approach was to use data from one sex as a covariate in the analysis of data from the other, in an analysis of covariance. He could show that the differences within males remained after removing covariance with females, whereas the opposite was not true. This strengthened the conclusion that butterfly male tactics have influenced the evolution of body design for flight and sexual dimorphism in flight-related traits.

Insects: evidence for the importance of sperm competition

Sperm competition has been invoked as a process which could influence sexual size dimorphism in animals. For some insect taxa it has been shown that spermatophore size is strongly correlated with male body size among taxa. This is true for butterflies (Rutowski, Newton & Schaefer, 1983; Svärd & Wiklund 1986, 1989) and bushcrickets (Wedell, 1993). The comparative correlations have been done with species or generic means as data points, but butterfly patterns are strong even when species of several families are included, and bushcricket patterns hold regardless of taxonomic level used. Moreover, they are also found within species. This suggests a consistent pattern of strong relationship between male body size and spermatophore size.

Sperm competition may take place whenever sperm from two or more different males is present within the same female (Parker, 1970; Birkhead & Hunter, 1990). This could happen under most mating systems, except for systems of monogamy or male polygamy without any copulations with other males, but would be increasingly likely with increasing degree of polyandry (female promiscuity). A male's chances of fertilizing eggs, and therefore his reproductive success or fitness, should in such cases be dependent on the number of sperm that he can transfer to the female, which in turn should be strongly dependent on spermatophore or ejaculate weight. Given that ejaculate weight is dependent on male body weight, we could consequently expect selection for larger production of sperm relative to body mass, that is an increase in the general relationship between ejaculate weight and body weight, and/or larger and heavier males when sperm competition is important.

In butterflies there is evidence for a relationship between the intensity of sperm competition and large relative male size. Wiklund & Forsberg (1991) found that sexual size dimorphism (male/female forewing length) was positively correlated with female polygamy (as assessed by spermatophore counts on wild-caught females) across species in both pierid and satyrine butterflies (Fig. 3). There was no attempt to control for phylogeny, which means that the significance values could be questioned. This is true especially for the correlation among satyrines, which rested mostly on a single polyandrous species, the only satyrine in the data set which showed male-biased size dimorphism. On the other hand, the trend was seen in both of the investigated butterfly taxa. It could not be explained by direct male–

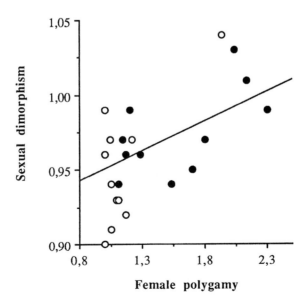

Figure 3   Male/female dimorphism in wing length and female polygamy (polyandry; number of spermatophores found in females) in pierid (closed circles) and satyrine (open circles) butterflies. Data from Wiklund and Forsberg (1991).

male competition (territoriality), by loading constraints (carrying sex) or by selection for protandry, as correlations with these variables were absent.

In Fig. 4 we have attempted to show part of the pierid data in a phylogenetic context. It can be seen that, when viewed in this way, the data does lend some support for a relationship between polyandry and size dimorphism. Most taxonomists agree that *Pieris rapae* and *P. napi* are not each other's closest relatives, the former being more closely related to the *P. brassicae* species group. Thus, the high degree of polyandry in *P. rapae* and *P. napi*, compared to relatives, has either evolved twice or has been lost once in *P. brassicae*. In either case, this section of the phylogeny shows two data points in favour of the investigated hypothesis. This is because *P. rapae* and *P. napi*, besides being polyandrous, also both have relatively large males compared to both *P. brassicae* and the outgroup. For a complete analysis, all data points should be examined together using a quantitative method such as that of Huey & Bennett (1986, 1987), correlating the amount and direction of change in both traits.

In Wedell's (1993) comparative analysis of bushcricket spermatophores, the degree of polyandry was not known in her data set of Australian and European species. However, larger spermatophores increased the time that passed before a mated female remated, resulting in higher fertilization success of males that provided such large spermatophores. Her results also supported two other steps in the hypothesized relationship between sperm competition and size dimorphism (relative

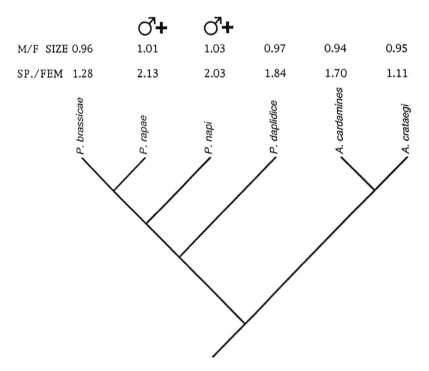

Figure 4 Male/female dimorphism in wing length and female polygamy (number of spermatophores found in females) superimposed on a hypothesis of phylogeny based on genetic distance trees by Geiger (1981). *Pieris napi* and *P. rapae* may have evolved to relatively larger male size (+) as a result of a high degree of female polygamy and sperm competition.

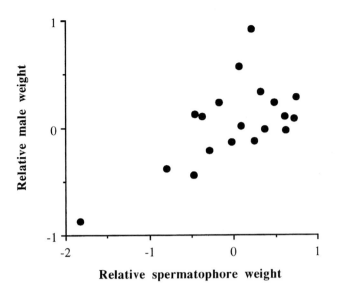

Figure 5   Relative male weight (sexual dimorphism; residuals from regressions of male weight on female weight) and relative spermatophore weight (residuals from regressions of spermatophore weight on male weight) in genera of bushcrickets. Data from Wedell (1993).

male size). First, as mentioned above, spermatophore weight was strongly positively correlated with male body weight across genera, species and within some species. Second, relative spermatophore weight (residuals from a regression on male weight) was positively correlated with relative male weight (Fig. 5; residuals from regressions on female weight, a measure independent of weight itself). There is experimental and comparative evidence that sperm competition is an important process in bushcrickets (Wedell, 1991), which suggests that large relative male weight could be selected for because this would result in large spermatophore weight. The statistical significance of these analyses (e.g. Fig. 5) resulted partly from treating genera as data points in regressions, and the appropriate phylogenetic analysis has not yet been performed.

Birds revisited: what about sperm competition?

As in butterflies and bushcrickets, the necessary prerequisites for an influence of sperm competition on sexual size dimorphism seem to be present in birds. First, there is strong evidence that sperm competition is important in birds (Möller, 1991; Birkhead & Möller, 1992). Second, testes size is strongly correlated with body size among bird species (Möller, 1991). However, results concerning the influence of sperm competition in birds presented so far has focused not on sexual size/weight dimorphism but on relative testes weight (residuals from the regression against body weight), that is body weight has been treated as a confounding variable that should be factored out of comparative analyses.

Using this method, Möller (1991) showed that relative testes weight is high in species with mating systems that should be associated with a high degree of sperm competition, e.g. polyandrous systems and communal monogamous systems. In the latter category of mating systems, extra-pair copulations (increasing the effective degree of polyandry and sperm competition) are likely to be more frequent than in solitary monogamous systems. As predicted, solitary monogamous species showed lower relative testes weights. The smallest testes were found in lekking species. In such species, many males compete for access to females, but it is assumed that once a pair has been formed, the female of that pair will not mate with any other male and sperm competition should therefore be uncommon (Birkhead & Möller, 1992).

That lekking species seem to be more sexually dimorphic than their non-lekking relatives, perhaps even when these relatives are polygynous (Höglund, 1989; Oakes, 1992), whereas they have relatively small testes (Möller, 1991), can be seen as evidence in favour of the view of sexual selection in birds which seems to be generally favoured today. This suggests that direct male–male competition (strong in lekking species) has influenced sexual size dimorphism, whereas sperm competition (weak in lekking species) has influenced only relative testes size. However, this view may stem partly from studies on the influence of direct male competition which have focused on measures of body size dimorphism, whereas researchers interested in sperm competition have focused on relative testes size, rather than on absolute measures of testes and body size.

Critical investigations designed to determine the relative influence of direct *versus* sperm competition on size dimorphism should score both types of variables and (since extra-pair copulations are so common in birds) preferably use real data on female polyandry rather than infer them from mating systems. The importance of using real data on female copulations is underlined by the fact that some lekking species, for example the ruff (*Philomachus pugnax*) and peafowl (*Pavo cristatus*), in fact show a considerable degree of polyandry (Birkhead & Möller, 1992). The former species is the most sexually dimorphic scolopacid in the data set of Oakes (1992), in a family of birds where the female often is the larger sex even in lekking species such as *Gallinago media*, and was mentioned by Höglund (1989) as one of the cases where he thought that lekking had really promoted dimorphism. The highly sexually size-dimorphic peafowl belongs to the pheasant subfamily of the Tetraonidae. In this family, male-biased size dimorphism is the rule and lek mating systems are common, suggesting the importance of direct male competition. However, the incidence of polyandry in peafowl suggests that sperm competition should be investigated as a factor that may have contributed to size dimorphism in this group of birds, and in other associations between lek mating systems and size dimorphism.

The results of the recent study by Webster (1992) are also suggestive in this context. Webster found a strong positive relationship between degree of male polygyny (harem size) and relative male weight in his data set of grackle species. However, all of the species with large harem sizes nest in colonies or aggregations, meaning that there is opportunity for polyandry and sperm competition. Webster actually found that nesting dispersion correlated with dimorphism independently of its association with mating system and/or body size, but could not explain this pattern.

Primates and other mammals revisited: what about sperm competition?

Sperm competition is also likely to be an important process in mammals (Short, 1979; Harcourt et al., 1981; Ginsberg & Huck, 1989; Möller & Birkhead, 1989; Birkhead & Hunter, 1990). In the primates, Harcourt et al. (1981) showed that absolute testes weight is positively correlated with male body weight across genera, but males of genera with multi-male breeding systems (where sperm competition is more likely) had larger testes for their body size than males of monogamous or single-male genera. The species level was not used because "closely related species within various genera have very similar values and cannot be treated as independent points" (Harcourt et al., 1981). Since genera are not independent observations, either, this is only a partial control for phylogeny and the statistical significance of the results could be questioned, even though the pattern is convincing.

As is the case for the birds, the connection between selection for large testes size and selection for large relative male size does not seem to have been made in comparative analyses, despite the fact that testes weight seems to be positively correlated with male weight within species (e.g. Bercovitch, 1989) and among taxa (Harcourt et al., 1981). Rather, male size has been factored out and testes size has been treated independently in studies looking for the influence of sperm competition (Harcourt et al., 1981), whereas size dimorphism has been treated in studies investigating the influence of direct male–male competition (e.g. Clutton-Brock & Harvey, 1977; Alexander et al., 1979). In the latter studies, the influence of sperm competition on body size itself was not investigated. We suggest that this should be done, as sperm competition could potentially explain some of the deviations from the 'expected' relationships between harem size or socionomic sex ratio on one hand, and sexual size dimorphism on the other. For example, it is obvious from Fig. 2 (data from Clutton-Brock & Harvey, 1977; many more data points could be added) that, for a socionomic sex ratio of one male per female, genera with multi-male groups (e.g. Pan, Lemur) are more dimorphic than monogamous genera. This may be easier to explain by selection for large male body size through sperm competition than through direct male–male competition. As another example, Presbytis entellus, which is sometimes found in multi-male groups, is more dimorphic than other Presbytis (which are uni–male polygamous), although the socionomic sex ratio is lower in P. entellus (Fig. 2).

Both processes could be important simultaneously and contribute to large male size in many taxa, since it is probable that copulations with other males occur also in 'uni-male' mating systems and that some aggressive competition also occurs in multi-male systems. In general, however, it is likely that (as Oakes, 1992, did for the birds) some distinction could be made between uni-male polygamous mating systems, where males compete for and often defend females, and multi-male mating systems, where frequency of copulations and number of mates may be favoured over aggressive competition. An example of the latter type among primates could be savanna baboons, Papio cynocephalus (Bercovitch, 1989). To assess the relative importance of both processes, these two types of mating systems should be rigorously compared, preferably comparing sister taxa. With present knowledge of phylogeny and mating systems, very few such contrasts can be made. It seems probable, however, that uni-male polygynous systems would come out on top for scores of size dimorphism among primates, as such species seem to be more dimorphic overall (Fig. 2).

DISCUSSION

Darwin (1871) proposed intra-sexual selection (direct competition) among males for access to females as an explanation of the general pattern of male-biased size dimorphism in birds and mammals. He inferred this process from observational evidence alone. It could to some extent be tested for mammals using predictions from the more elaborate sexual selection theory developed during the 70s, and although one might fail to be impressed by the quantitative fit of regressions and by statistical significances obtained by using species or genera as data points (Clutton-Brock & Harvey, 1977; Alexander *et al.*, 1979), the general pattern of relatively larger males in species where one male monopolizes many females is convincing.

Concerning the birds, most studies have focused on the pattern perceived by Darwin that lekking species are more dimorphic, again suggesting the process explanation that males of such species have to compete for access to females and have evolved towards stronger competitive ability. Höglund (1989) found only non-significant support for this pattern when phylogeny was controlled for, using Ridley's (1983) non-directional technique for discrete traits, and no support for the directional hypothesis that lekking has preceded dimorphism in evolution. However, he was of the opinion that lekking has favoured dimorphism in several bird taxa, only questioning the generality of this association. Oakes (1992) made use of the quantitative information and reanalysed Höglund's data using an independent contrasts technique. He found statistical support for the association between lek mating systems and male-biased size dimorphism. Applying Maddison's (1990) directional method for qualitative data, he also found support for the predicted evolutionary sequence, although his use of the method can be questioned (see above). In a directional, quantitative study of sexual size dimorphism in birds, Björklund (1991) found evidence that changes in mating system towards higher or lower degrees of polygyny in grackles (i.e. degree of monopolization of females) have been accompanied by the predicted changes in size dimorphism. Webster (1992) found even stronger support from grackles using non-directional methods with continuous measures of both dimorphism and polygyny.

As the above review shows, there is considerably less agreement concerning what processes have been important in shaping the pattern of female-biased size dimorphism seen in most other animals, although for these taxa most of Darwin's (1871) observations seem to hold. It seems probable that one important cause for the general female size bias is the correlation between large female size and high fecundity seen in most invertebrates and poikilotherm vertebrates (although we are not aware of any explicitly phylogenetic tests of this hypothesis). One of the most direct tests (Wiklund & Karlsson, 1988) obtained a positive result in that more female-biased butterflies were more fecund, but this analysis was performed as a straight forward species regression. Shine (1988) invoked comparative evidence from lizards in an attempt to refute the fecundity-advantage hypothesis, but the observation that taxa where fecundity may be related to size are no more female-biased in size than taxa with fixed clutch-sizes seems a rather weak refutation of the general importance of this process. This pattern could be masked by other processes hiding it, as in the case of the allometric pattern found in tephritid fruit flies, which Sivinski & Dodson (1992) argued had the wrong slope if the fecundity-advantage process has been important in this taxon. It is the nature of non-directional comparative methods that they make weak tests of process theories, because they do not really address the processes themselves but only the predicted associations, via a set of assumptions linking process to pattern. It is possible to disprove a process theory only by showing that predicted patterns are absent over a whole range of taxa, thus demonstrating the lack of general importance of a particular process.

In insects, as in mammals and birds, there is comparative evidence that male–male competition has influenced dimorphism. Other important processes include: sexually-biased selection depending on which sex carries during mating, and sexually-differential selection for flight requirements. Female-biased size due to selection for protandry, a process believed to be important by Darwin, has not been supported by comparative or experimental data (see references above and in Nylin *et al.*, 1993). Another important process, one of the few that Darwin did not think of, may have been sperm competition. The presence of sperm competition may have selected for large ejaculate size and therefore large male size. Species regressions show an association between degree of polyandry (and thus opportunity for sperm competition) and sexual size dimorphism in butterflies (Wiklund & Forsberg, 1991). In bushcrickets, species and genera with relatively large male size also deliver large spermatophores for a given male size, suggesting that sperm competition may be relatively more important in these taxa (Wedell, 1993). A preliminary investigation of these patterns suggests that these associations could hold in a quantitative phylogenetic test (Nylin and Wedell, in preparation).

It is not known to what degree sperm competition may have influenced patterns of size dimorphism in mammals and birds. Results so far (see above) strongly suggest that sperm competition is, and has been, an important process in both groups. It has influenced the evolution of relative testes size, but has probably been less important than direct male–male competition in favouring large relative male size in these taxa. This does not rule out the possibility that sperm competition also may have had a major influence and may have caused some of the deviations from patterns predicted by direct male–male competition. More data on polyandry in the field (important for sperm competition), not only degree of polygyny (important for male–male competition), as well as phylogenetic analyses that explicitly address this question, are needed.

The above review and discussion demonstrates how comparative methods have been applied to sexual size dimorphism in animals since Darwin. Our aim has not been to show the shaky foundations of comparative evidence on sexual size dimorphism, but rather how different methods have had success in demonstrating patterns and testing process theories. Patterns found using non-directional and even non-phylogenetic techniques often hold when they are tested more rigorously using phylogenies and improved methodologies. For this reason, we would advise against simply discarding results from older analyses (e.g. species counts or species regressions) just because the statistical significance of these results can be questioned. Instead, the results should be taken for what they are, observations of patterns for which we can seek improved understanding.

The same reasoning applies to phylogenetic tests, especially using qualitative methods. Such tests are inherently conservative, as similarities between related species are always attributed to common ancestry rather than common adaptive change. This reduces the degrees of freedom for statistical analysis in a way which may or may not be appropriate. Also, the classes of qualitative traits that are entered into the analyses are always arbitrary.

From an 'adaptationist' or 'equilibrium' point of view there is less to be objected against non-directional methods, than from the point of view of the 'historical' school of research. This is because such methods are good at revealing what associations between states, or between states and the environment, seem to be common and thus evolutionarily stable and adaptive (if we assume that non-adaptive associations would tend to disappear). Directional methods are not designed to achieve this particular goal, but on the other hand they can reveal how the associations were reached. Other advantages of the non-directional phylogenetic methods (especially pair-wise contrasts) over the directional are that they are generally not as dependent on knowledge of the exact topology of phylogenies, and suffer less from the fact that interior nodes in the trees have to be inferred from states in recent taxa (Harvey & Purvis, 1991). On the other hand, they do not make much use of knowledge of the phylogeny, even when it is present (Fig. 6).

The two traditions, then, to some extent have separate goals. Both goals are legitimate (see Harvey & Purvis, 1991, for a discussion), and one method may not suit both. It is of interest to examine how different methods handle a special case: a species may have evolved to a point of no difference between the sexes at all, from an ancestral state of female-biased size (Fig. 6). This must frequently happen in invertebrates and poikilotherm vertebrates, but also in the normally female-biased mammal and bird groups. If this happened in a species with intense male–male competition compared to relatives, it would be a data point in favour of the sexual selection hypothesis. It would, however, not be 'seen' by a qualitative, non-directional method such as Ridley's (1983), which would lump all instances of monomorphism, regardless of how they evolved.

Using an independent contrasts method, as Oakes (1992) did, we can pick up some such instances by using quantitative information and compare lekking species with related non-lekking taxa, seeking the answer to the question; 'are males relatively larger in lekking species?'. This is a different question than the one asked by Höglund (1989), which accounts for the difference in results. A monomorphic lekking species can now be distinguished from a related female-biased non-lekking species and become a data point in favour of the tested hypothesis. We would still not know, however, whether the lekking species evolved to monomorphism or the non-lekking species evolved to female bias. This is less important to the actualistic tradition (which is interested in what associations are stable) but essential to the historical tradition. To find the answer to a question such as 'have males evolved to relatively larger size as a result of male–male competition in lek mating systems?', we need directional methods (Björklund, 1991). In the example just mentioned, adding outgroups to each comparison would polarize them and one could infer the most probable direction of evolution. However, if we find support for predictions or not can depend on our choice of directional method. Qualitative directional methods can be used for analysing the sequence of events, but only when associated traits do not change 'simultaneously', that is between two nodes in a

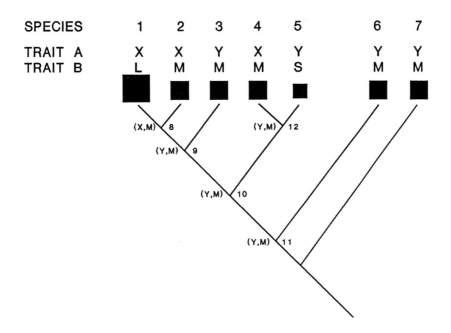

Figure 6 An illustration of what various comparative methods would 'see' for a given distribution of states and a given phylogeny. There are five species in the study group and two in the outgroup, each of which has state X or Y for trait A (X and Y could also be two discrete types of environments) and state L (large), M (medium) or S (small) for trait B. Trait B could represent any quantitative trait, but we will refer to it as size. The states L, M and S are not really discrete, but for simplicity some very similar values of B are all represented as M. S is only slightly smaller, whereas L is much larger. The prediction under study is that large size is associated with X. This is what various methods would see. (1) Traditional methods such as species regressions or simply comparing the means of B associated with X or Y would see weak support for the prediction among species 1–5, in that the mean of B is slightly larger in species with state X. The degrees of freedom are inflated, but on the other hand species that have not changed trait B at all (species 2–4 are similar to the outgroup) work against statistical significance for the predicted pattern by not adding anything to it. (2) A contrasts method sees stronger support for the prediction, because both contrasts that can be made (species 1–2 against 3, and species 4 against 5) support the prediction of relatively larger size associated with X. (3) To apply a qualitative test we need to interpret the size classes as discrete. If we interpret S as belonging to the same class as M, information is lost compared to contrasts methods. There are now three cases where the association between A and B has changed. In two cases Y changed to X (making use of character optimization and the outgroup it can be inferred that this happened between node 9 and 8, and between node 12 and species 4). In the first of these cases trait B later changed to larger size, which supports the prediction. In the second case B did not change, which is against the prediction. The lack of change in B when A changed between node 9 and 8 is also against the prediction. Qualitative methods would see less support than a contrasts methods would. (4) A qualitative, more explicitly directional technique would find a higher degree of support. There has been only one change to larger size, and this did happen on one of the X-branches, as predicted. Adding a third size-class, the only change to a smaller size happened on a Y-branch. The evolutionary sequences are as predicted. (5) In contrast, a quantitative, directional technique would see no support for the prediction. There is no correlation between changes in A and B, because in no case do they change together along the same branch of the phylogeny.

phylogeny. Such cases are considered ambiguous and create problems for methods in this category. Quantitative directional methods, on the other hand, make use of precisely such instances as strong evidence in favour of the predicted associations. They show correlations between coupled traits, but not necessarily sequences or potential causality. In this they are similar to non-directional methods, and, in fact, the distinction between directional and non-directional methods almost ceases to be useful when it comes to correlating two quantitative traits. These points, regarding what the different categories of methods would 'see', do not apply only to sexual size dimorphism, but are general for all inherently quantitative traits (Fig. 6).

In conclusion, when interpreting the results from a phylogenetic analysis of comparative ecological data, as in any statistical analysis, it is necessary to distinguish explicitly between statistical significance and explanatory power. If quantitative species data is entered into a regression analysis, there could be a high level of statistical significance (because of the inflated degrees of freedom) but very little explanatory power (poor fit to the regression). Such results have limited interest except as a single data point, showing the general trend in the data set, which could be added to other taxa in a more inclusive comparative analysis. Alternatively, both significance and explanatory power could be high. The significance would have to be treated only as a demonstration that there is indeed a pattern to the data. Otherwise, the high explanatory power is more interesting, since this shows that the regression has predictive power. Knowing the 'independent' factor (e.g. latitude or degree of protandry), we can predict something of the traits of a species before they have been measured.

The results of Wickman (1992) exemplify an important third case. There may be non-significant associations between variables when species are treated as independent data points, because associations may be hidden rather than amplified by phylogenetic patterns. As in Wickman's study, it may still be possible to find consistent patterns when contrasts of related taxa are compared. Hence, the significance and predictive power of comparative results also depend on the taxonomic level of comparisons. As a another example, the fact that sexual size dimorphism in primates has been strongly phylogenetically 'constrained' (Cheverud, Dow & Leutenegger, 1985) implies that many interesting patterns can be seen only by comparing closely related taxa. Such methods (the 'contrasts' family of methods), when used sensibly, can be powerful tools for finding consistent differences between ecological pairs of closely related taxa. However, such patterns (even if significant, and of considerable interest for the understanding of fine modifications of the studied traits) do not necessarily result in predictive or explanatory power at higher taxonomical levels, where the general level of the trait may be more or less constant over large clades. Because of the problems involved in interpreting significances of trends within taxa, which applies to all methods of phylogenetic analysis, we suggest that generality and consistency of trends across taxa should be seen as a more important criterion than statistical significance alone.

ACKNOWLEDGEMENTS

We thank Birgitta Sillén-Tullberg, Mats Björklund, Olof Leimar and Per-Olof Wickman for fruitful discussions during the early stages of the writing of this chapter. The comments of R.I. Vane-Wright, Jonathan Coddington and an anonymous reviewer improved the manuscript. This research was supported by grants from the Swedish Natural Science Research Council to S. Nylin and from the Royal Swedish Academy of Sciences to N. Wedell.

REFERENCES

ALEXANDER, R.D., HOOGLAND, J.L. HOWARD, R., NOONAN, K.M. & SHERMAN, P.W., 1979. Sexual dimorphism and breeding systems in pinnipeds, ungulates, primates and humans. In N.A. Chagnon & W.D. Irons (Eds), *Evolutionary Biology and Human Social Behaviour*, 402–435. North Scituate, MA: Duxbury Press.

ARAK, A., 1988. Sexual dimorphism in body size: a model and a test. *Evolution, 42*: 820–825.

BELL, G., 1989. A comparative method. *American Naturalist, 133*: 553–571.

BERCOVITCH, F.B., 1989. Body size, sperm competition, and determinants of reproductive success in male savanna baboons. *Evolution, 43*: 1507–1521.

BIRKHEAD, T.R. & HUNTER, F.M., 1990. Mechanisms of sperm competition. *Trends in Ecology and Evolution, 5*: 48–52.

BIRKHEAD, T.R. & MÖLLER, A.P., 1992. *Sperm Competition in Birds. Evolutionary Causes and Consequences.* London: Academic Press.

BJÖRKLUND, M., 1991. Evolution, phylogeny, sexual dimorphism and mating system in the grackles (*Quiscalus* spp.: Icterinae). *Evolution, 45*: 608–621.

BROOKS, D.R. & McLENNAN, D.H., 1991. *Phylogeny, Ecology, and Behavior. A Research Program in Comparative Biology.* Chicago: University of Chicago Press.

BURT, A., 1989. Comparative methods using phylogenetically independent contrasts. In: P.H. Harvey & L. Partridge (Eds), *Oxford Surveys in Evolutionary Biology, 6*: 33–53.

CARPENTER, J.M., 1992. Comparing methods. [Review of P.H. Harvey & M.D. Pagel, 1991, *The Comparative Method in Evolutionary Biology*, Oxford University Press] *Cladistics, 8*: 191–196.

CHEVERUD, J.M., DOW, M.M. & LEUTENEGGER, W., 1985. The quantitative assessment of phylogenetic constraints in comparative analyses: sexual dimorphism in body weight among primates. *Evolution, 39*: 1335–1351.

CLUTTON-BROCK, T.H., 1985. Size, sexual dimorphism, and phylogeny in primates. In W.L. Jungers (Ed.), *Size and Scaling in Primate Biology*, 51–60. New York: Plenum Press.

CLUTTON-BROCK, T.H. & HARVEY, P.H., 1977. Primate ecology and social organization. *Journal of Zoology, 183*: 1–33.

CLUTTON-BROCK, T.H. & HARVEY, P.H. 1984. Comparative approaches to investigating adaptation. In J.R. Krebs & N.B. Davies (Eds), *Behavioural Ecology: an Evolutionary Approach*, 2nd edition, 7–29. Oxford: Blackwell.

CLUTTON-BROCK, T.H., HARVEY, P.H. & RUDDER, B., 1977. Sexual dimorphism, socio-nomic sex ratio, and body weight in primates. *Nature, 269*: 797–800.

CODDINGTON, J.A., 1992. Avoiding phylogenetic bias. [Review of P.H. Harvey & M.D. Pagel, 1991, *The Comparative Method in Evolutionary Biology*, Oxford University Press] *Trends in Ecology and Evolution, 7*: 68–69.

DARWIN, C., 1871. *The Descent of Man, and Selection in Relation to Sex.* London: J. Murray.

ELGAR, M.A., 1991. Sexual cannibalism, size dimorphism, and courtship behavior in orb-weaving spiders (Araneidae). *Evolution, 45*: 444–448.

FAIRBAIRN, D.J., 1990. Factors influencing sexual size dimorphism in temperate water-striders. *American Naturalist, 136*: 61–86.

FELSENSTEIN, J., 1985. Phylogenies and the comparative method. *American Naturalist, 125*: 1–15.

GEIGER, H.J., 1981. Enzyme electrophoretic studies on the genetic relationships of the pierid butterflies (Lepidoptera: Pieridae) I. European taxa. *Journal of Research on the Lepidoptera, 19*: 181–195.

GINSBERG, J.R. & HUCK, U.W., 1989. Sperm competition in mammals. *Trends in Ecology and Evolution, 4*: 74–79.

GRAFEN, A., 1989. The phylogenetic regression. *Philosophical Transactions of the Royal Society of London, Series B, 326*: 119–156.

GREENWOOD, P.J. & ADAMS, J., 1987. Sexual selection, size dimorphism and a fallacy. *Oikos, 48*: 106–108.

GREENWOOD, P.J. & WHEELER, P., 1985. The evolution of sexual size dimorphism in birds and mammals: a hot-blooded hypothesis. In P.J. Greenwood, P.H. Harvey & M. Slatkin (Eds), *Evolution: Essays in Honour of John Maynard Smith*, 287–299. Cambridge: Cambridge University Press.

HARCOURT, A.H., HARVEY, P.H., LARSON, S.G & SHORT, R.V., 1981. Testis weight, body weight and breeding system in primates. *Nature, 293*: 55–57.

HARVEY, P.H. & CLUTTON-BROCK, T.H., 1985. Life history variation in primates. *Evolution, 39*: 559–581.

HARVEY, P.H. & PAGEL, M.D., 1991. *The Comparative Method in Evolutionary Biology.* Oxford: Oxford University Press.

HARVEY, P.H. & PURVIS, A.J., 1991. Comparative methods for explaining adaptations. *Nature, 351*: 619–624.

HEDRICK, A.V. & TEMELES, E.J., 1989. The evolution of sexual dimorphism in animals: hypothesis and tests. *Trends in Ecology and Evolution, 4*: 136–138.

HÖGLUND, J., 1989. Size and plumage dimorphism in lek-breeding birds: a comparative analysis. *American Naturalist, 134*: 72–87.

HUEY, R.B., 1987. Phylogeny, history and the comparative method. In M.E. Feder, A.F. Bennett, W. Burggren & R.B. Huey (Eds), *New Directions in Ecological Physiology*, 76–198. Cambridge: Cambridge University Press.

HUEY, R.B. & BENNETT, A.F., 1986. A comparative approach to field and laboratory studies in evolutionary biology. In M.E. Feder & G.V. Lauder (Eds), *Predator-prey Relationships: Perspectives and Approaches for the Study of Lower Vertebrates*, 82–96. Chicago: University of Chicago Press.

HUEY, R.B. & BENNETT, A.F., 1987. Phylogenetic studies of co-adaptation: preferred temperatures versus optimal performance temperatures of lizards. *Evolution, 41*: 1098–1115.

KAPPELER, P.M., 1990. The evolution of sexual size dimorphism in prosimian primates. *American Journal of Primatology, 21*: 201– 214.

LANDE, R., 1987. Genetic correlations between the sexes in the evolution of sexual dimorphism and mating preferences. In J.W. Bradbury & M.B. Aanderson (Eds), *Sexual Selection: Testing the Alternatives*, 83–94. Chichester: John Wiley.

LEUTENEGGER, W., 1978. Scaling of sexual dimorphism in body size and breeding system in primates. *Nature, 272*: 610–611.

LEUTENEGGER, W., 1982. Scaling of sexual dimorphism in body weight and canine size in primates. *Folia Primatologica, 37*: 163–176.

MADDISON, W.P., 1990. A method for testing the correlated evolution of two binary characters: are gains or losses concentrated on certain branches of a phylogenetic tree? *Evolution, 44*: 539–557.

MÖLLER, A.P., 1991. Sperm competition, sperm depletion, paternal care, and relative testis size in birds. *American Naturalist, 137*: 882–906.

MÖLLER, A.P. & BIRKHEAD, T.R., 1989. Copulation behaviour in mammals: evidence that sperm competition is widespread. *Biological Journal of the Linnean Society, 38*: 119–131.

NYLIN, S., 1991. The phylogenetic approach to ecology. [Review of D.R. Brooks & D.H. McLennan, 1991, *Phylogeny, Ecology and Behavior,* University of Chicago Press] *Evolution, 45*: 1731–1733.

NYLIN, S., WIKLUND, C., WICKMAN, P.-O. & GARCIA-BARROS, E., 1993. Absence of trade-offs between sexual size dimorphism and early male emergence in a butterfly. *Ecology 74*: 1414–1427.

OAKES, E.J., 1992. Lekking and the evolution of sexual dimorphism in birds: comparative approaches. *American Naturalist, 140*: 665–684.

O'HARA, R.J., 1988. Homage to Clio, or toward an historical philosophy for evolutionary biology. *Systematic Zoology, 37*: 142–155.

PAGEL, M.D. & HARVEY, P.H., 1988. Recent development in the analysis of comparative data. *Quarterly Review of Biology, 63*: 413–440.

PAGEL, M.D. & HARVEY, P.H., 1989. Comparative methods for examining adaptations depend on evolutionary models. *Folia Primatologica, 53*: 203–220.

PARKER, G.A., 1970. Sperm competition and its evolutionary consequences in the insects. *Biological Review, 45*: 525–567.

PARKER, G.A. & COURTNEY, S.P., 1983. Seasonal incidence: adaptive variation in the timing of life history stages. *Journal of Theoretical Biology, 105*: 147–155.

PAYNE, R.B., 1984. Sexual selection, lek and arena behaviour, and sexual size dimorphism in birds. *Ornithological Monographs, 33*: 1–52.

RALLS, K., 1976. Mammals in which females are larger than males. *Quarterly Review of Biology, 51*: 245–276.

RIDLEY, M., 1983. *The Explanation of Organic Diversity: the Comparative Method and Adaptations for Mating.* Oxford: Oxford University Press.

RIDLEY, M., 1989. Why not to use species in comparative tests. *Journal of Theoretical Biology, 136*: 361–364.

RUTOWSKI, L.R., NEWTON, M. & SCHAEFER, J., 1983. Interspecific variation in the size of the nutrient investment by male butterflies during copulation. *Evolution, 37*: 708–713.

SELANDER, R.K., 1966. Sexual dimorphism and differential niche utilization in birds. *Condor, 68*: 113–151.

SELANDER, R.K., 1972. Sexual selection and dimorphism in birds. In B. Campbell (Ed.), *Sexual Selection and the Descent of Man, 1871–1971*, 180–230. Chicago: Aldine.

SHINE, R., 1978. Sexual size dimorphism and male combat in snakes. *Oecologia, 33*: 269–277.

SHINE, R., 1979. Sexual selection and sexual dimorphism in the amphibia. *Copeia, 2*: 297–306.

SHINE, R., 1988. The evolution of large body size in females: a critique of Darwin's 'fecundity advantage' model. *American Naturalist, 131*: 124–131.

SHINE, R., 1989. Ecological causes for the evolution of sexual dimorphism: a review of the evidence. *Quarterly Review of Biology, 64*: 419–461.

SHINE, R., 1991. Intersexual dietary divergence and the evolution of sexual dimorphism in snakes. *American Naturalist, 138*: 103– 122.

SHORT, R.V., 1979. Sexual selection and its component parts, somatic and genital selection, as illustrated by man and the great apes. *Advances in the Study of Behaviour, 9*: 131–158.

SILLÉN-TULLBERG, B., 1988. Evolution of gregariousness in aposematic butterfly larvae: a phylogenetic analysis. *Evolution, 42*: 293–305.

SILLÉN-TULLBERG, B., 1993. The effect of biased inclusion of taxa on the correlation between discrete characters in phylogenetic trees. *Evolution, 47*: 1182–1191.

SILLÉN-TULLBERG, B. & MÖLLER, A.P., 1993. The relationship between concealed ovulation and mating systems in anthropoid primates: a phylogenetic analysis. *The American Naturalist, 141*: 1–25.

SINGER, M.C., 1982. Sexual selection for small size in male butterflies. *American Naturalist, 119*: 440–443.

SIVINSKI, J.M. & DODSON, G., 1992. Sexual dimorphism in *Anastrepha suspensa* (Loew) and other tephritid fruit flies (Diptera: Tephritidae): possible roles of development rate, fecundity, and dispersal. *Journal of Insect Behavior, 5*: 491– 506.

SMUTS, B.B., CHENEY, D.L., SEYFARTH, R.M., WRANGHAM, R.W. & STRUHSAKER, T.T. (Eds), 1987. *Primate Societies.* Chicago: University of Chicago Press.

STEARNS, S.C., 1983. The influence of size and phylogeny on patterns of covariation among life-history traits in mammals. *Oikos, 41*: 173–187.

SVÄRD, L. & WIKLUND, C., 1986. Different ejaculate delivery strategies in first versus subsequent matings in the swallowtail butterfly *Papilio machaon L. Behavioral Ecology and Sociobiology, 18*: 325–330.

SVÄRD, L. & WIKLUND, C., 1989. Mass and production rate of ejaculates in relation to monandry/polyandry in butterflies. *Behavioral Ecology and Sociobiology, 24*: 395–402.

THORNHILL, R. & ALCOCK, J., 1983. *The Evolution of Insect Mating Systems.* Cambridge, MS: Harvard University Press.

TRAIL, P.W., 1990. Why should lek-breeders be monomorphic? *Evolution, 44*: 1837–1852.

TRIVERS, R.L., 1972. Parental investment and sexual selection. In B. Campbell (Ed.), *Sexual Selection and the Descent of Man, 1871–1971*, 136–179. Chicago: Aldine.

TRIVERS, R.L., 1985. *Social Evolution.* Menlo Park, CA: Benjamin-Cummings.

WANNTORP, H.-E., BROOKS, D.R., NILSSON, T., NYLIN, S., RONQUIST, F., STEARNS, S.C. & WEDELL, N., 1990. Phylogenetic approaches in ecology. *Oikos, 57*: 119–132.

WARD, P.I., 1983. A comparative field study of the breeding behaviour of a stream and a pond population of *Gammarus pulex* (Amphipoda). *Oikos, 46*: 29–36.

WEBSTER, M.S., 1992. Sexual dimorphism, mating system and body size in New World blackbirds (Icterinae). *Evolution, 46*: 1621– 1641.

WEDELL, N., 1991. Sperm competition selects for nuptial feeding in a bushcricket. *Evolution, 45*: 1975–1978.

WEDELL, N., 1993. Spermatophore size in bushcrickets: comparative evidence for nuptial gifts as sperm competition device. *Evolution, 47*: 1203–1212.

WEST-EBERHARD, M.J., 1983. Sexual selection, social competition, and speciation. *Quarterly Review of Biology, 58*: 155–183.

WHEELER, P. & GREENWOOD, P.J., 1983. The evolution of reversed sexual dimorphism in birds of prey. *Oikos, 40*: 145–149.

WICKMAN, P.-O., 1992. Sexual selection and butterfly design – a comparative study. *Evolution, 46*: 1525–1536.

WIKLUND, C. & FAGERSTRÖM, T., 1977. Why do males emerge before females? A hypothesis to explain the incidence of protandry in butterflies. *Oecologia, 31*: 153–158.

WIKLUND, C. & FORSBERG, J., 1991. Sexual size dimorphism in relation to female polygamy and protandry in butterflies: a comparative study of Swedish Pieridae and Satyridae. *Oikos, 60*: 373–381.

WIKLUND, C. & KARLSSON, B., 1988. Sexual size dimorphism in relation to fecundity in some Swedish satyrid butterflies. *American Naturalist, 131*: 132–138.

WIKLUND, C., NYLIN, S. & FORSBERG, J., 1991. Sex-related variation in growth rate as a result of selection for large size and protandry in a bivoltine butterfly (*Pieris napi* L.). *Oikos, 60*: 241–250.

WILLIAMS, G.C., 1966. *Adaptation and Natural Selection.* Princeton, NJ: Princeton, University Press.

CHAPTER

13

Evolution of bird-pollination in some Australian legumes (Fabaceae)

MICHAEL D. CRISP

CONTENTS _____

Keywords: Bird-pollination – Australia – convergence – phylogeny – functional structures – permutation tests – corroboration – Fabaceae.

Abstract

Systematists are interested in adaptation as synapomorphy. Evolutionary sequences of adaptive features may be elucidated from cladograms, and from this elaboration of a feature and its adaptive functions may be inferred. Convergence in complex adaptive structures is problematic for phylogenetic reconstruction and classification, especially if correlated convergence due to underlying developmental constraint is suspected. Although permutation tests evaluate the degree of corroboration of a phylogenetic hypothesis within the limits of a given data set, independent evidence is the only kind of corroboration that may deal with correlated convergence.

These issues are explored in the present study of the evolution of morphological features related to bird-pollination in Australian legumes, particularly the tribe Mirbelieae. Bird-pollination has evolved several times independently from bee-pollinated ancestors in Australian pea-flowered legumes. In the closely related genera *Brachysema*, *Jansonia* and *Nemcia* evidence for the number of independent origins (one or two) is evaluated using cladistic analysis. Morphological characters related to the bird-pollination are mapped onto the cladograms, and their evolutionary sequences are inferred. Previous misclassification of these genera is identified: *Nemcia* appears to be paraphyletic with respect to *Jansonia*. Permutation tail probability tests are used to explore the degree of corroboration of these phylogenetic hypotheses.

INTRODUCTION

Traditionally adaptation, that is the evolution of structures and their related functions, has been studied by looking for non-historical correlations between them (Pagel & Harvey, 1988; Donoghue, 1989). Many species were sampled with the assumption that each provided independent evidence of the hypothesized relationship. However, this naive approach is now recognized to be invalid because closely related species are likely to have inherited an adaptive trait from a common ancestor, and therefore they do not provide independent samples. The historical factor in adaptation has been termed 'phylogenetic constraint' (Harvey & Purvis, 1991) or 'historical constraint' (Wanntorp *et al.*, 1990), but to a systematist it is simply synapomorphy (Coddington, 1988, Chapter 3; Donoghue, 1989; Nelson, in press). There are two contrasting approaches to dealing with the historical component of adaptation: (1) factoring out phylogeny, in effect treating adaptation as homoplasy (Pagel & Harvey, 1988; Faith, 1989; Harvey & Purvis, 1991; Coddington, Chapter 3; Faith & Belbin, Chapter 7; Pagel, Chapter 2), and (2) treating all adaptations as unique historical events (apomorphies), and investigating them as such (Coddington, 1988, Chapter 3; Donoghue, 1989; Linder, 1991; Wenzel & Carpenter, Chapter 4).

The first approach assumes that similar solutions to similar selective problems, arising convergently, are interesting, because they may reflect some underlying ecological law. It seeks repeated examples as tests of some postulated relationship among quantitative variables, such as trade-offs, for example adult weight and birth weight or size and fecundity, or allometry, for example optimal life histories (Harvey & Purvis, 1991). The second approach is more concerned with origins and sequences: using the known directionality of evolution that a phylogeny provides to ask whether a trait and its associated function were acquired simultaneously (true adaptation), or in some other sequence (Coddington, 1988; Donoghue, 1989). Moreover, a cladogram is a powerful tool for investigating extended evolutionary sequences in the transformation of characters (Donoghue, 1989; Linder, 1991). Time may tell whether one method is superior. Alternatively, both approaches may be valid, depending upon whether phylogenetic history or ecological law is of prime interest to the researcher (see also Coddington, Chapter 3).

The present study takes the homology approach. I analyse the history of adaptations to bird-pollination in related taxa, within a single plant family: the legumes (Fabaceae). Taking the view that independent originations of bird-pollination in a primitively insect-pollinated group are convergent, I assume that each should show unique structural features that may be identified by cladistic analysis, despite the superficial similarity resulting from convergence. As a systematist, I explore the problems that convergence, especially congruent convergence among several characters comprising a structural–functional complex, pose for phylogenetic analysis. Also, I emphasize that a phylogenetic study of adaptive features is limited by the reliability of the phylogeny used as the basis. Some authors (Harvey & Pagel, 1991; Wanntorp et al., 1990) have promoted the use of statistics in testing adaptive hypotheses, while appearing to accept without question uncorroborated cladograms or even naively transcribed non-phylogenetic classifications. While the importance of using a reliable estimate of phylogeny may now be acknowledged (Harvey, 1991), the means of obtaining one is not necessarily understood (Coddington, 1992). In the present study, I show how a hypothesis about origination of adaptive features depends crucially both upon a choice among alternative phylogenies, and upon the means of estimating the degree of support for a phylogeny.

POLLINATION SYNDROMES: PROBLEMS FOR SYSTEMATISTS

A major concern in my revisionary studies of genera of the Fabaceae has been the potential for convergence in characters related to bird-pollination. In related organisms, convergence is sometimes observed to be congruent, and may be explained by: (1) chance congruence or strong selection acting upon genetically similar organisms ('concerted evolution'; Trueman, 1993); (2) by complex structures being 'switched' on or off by one or few genes; or (3) by developmental constraints limiting possibilities for evolutionary transformation of phenotypes ('inside parallelism'; Brundin, 1976; 'latent homology', 'parallelism', 'orthogenesis', 'underlying synapomorphy'; reviewed by Cranston & Humphries, 1988; Nelson, in press); or (4) by a combination of these. Heterochrony, or small changes in the length and timing of developmental pathways, can have far-reaching effects on the adult phenotype (Alberch et al., 1979), and explicit phylogenetic analysis shows that heterochrony can give rise to large and repeated (convergent) morphological changes in evolution (Alberch & Gale, 1985; Boughton, Collette & McCune, 1991).

Schrire (1989) argues that constraints in developmental pathways, leading to canalization, limit the possibilities for modifying an existing floral structure for a new function ('functional complex'), and cites evidence for strong early constraints in the development of the papilionaceous flower. In other words, there are only limited ways in which a pea flower, which is designed for bee-pollination, can be modified to allow pollination by birds. Guerrant (1982) has shown empirically that flowers of *Delphinium* have been modified drastically to cause a switch from insect-pollination to bird-pollination, simply by changes in the timing of development. Such changes may be controlled by only one or two regulatory genes (Ambrios, 1988). Guerrant identified two species of *Delphinium*, which are not close relatives, in which this process appears to have occurred in parallel. Recent studies on *Antirrhinum* and *Arabidopsis* have identified some of the genes controlling floral development (Coen, 1991; Coen & Meyerowitz, 1991). Small mutations blocking the function of transposons in the genes have dramatic effects on the presence and arrangement of the four basic whorls in the flower (sepals, petals, stamens and carpels). Moreover, morphological features that are fundamental to the structure and function of the flower as a pollination unit, such as pigmentation, sexuality and zygomorphy, are under control of these homeotic genes. For example, in *Antirrhinum*, a mutation in the **cyc** function, controlled by two linked genes, alters the flower from a complex zygomorphic structure to a simple actinomorphic shape (Coen, 1991).

Thus a change in one gene may simultaneously transform what appears to be not one but several independent characters, but this appearance is misleading (Gottlieb, 1984). Moreover, such a mutation, having eliminated a long and complex developmental pathway, removes pre-existing constraints on floral morphology, allowing radical evolutionary reshaping of a flower, such as for a different mode of pollination. Such may be the basis of origination of a bird-pollinated species in a predominantly insect-pollinated group. Some systematists have shown examples of correlated convergence among functional characters which has biased phylogenetic reconstruction or classification. Goldblatt (1990) sank *Anapalina* into *Tritoniopsis* (Iridaceae) on the grounds that it had been defined purely on floral morphology adapted for bird-pollination, which had arisen several times in parallel in related taxa, whereas a number of 'synapomorphies' of the bracts, seeds and leaves (not apparently related to pollination syndrome) were held in common by *Anapalina* and *Tritoniopsis*. However, he did not carry out a cladistic analysis based upon parsimony to support this argument; nor did he cite developmental evidence that morphology related to pollination might be controlled by small genetic differences. Doyle & Donoghue (1992) suggested that functional correlation among reproductive characters in seed plants (epigeal seed germination, lack of a sarcotesta and siphonogamy) is convergent in conifers and anthophytes, and has biased cladistic analysis of seed plants using only extant taxa (Loconte & Stevenson, 1990). When Doyle and Donoghue included fossil taxa lacking this structural–functional complex in their data, the fossils usually took a phylogenetic position intermediate between the conifers and anthophytes, suggesting convergence in the latter taxa. These examples sound a warning to systematists. Scoring several such correlated characters as if they were independent might force a false grouping in a phylogenetic analysis through inadvertent weighting of what would be better treated as a single binary character (state of the controlling gene).

Convergent floral morphologies in distantly related families are not likely to be a problem, because they will differ in detail, for example in Armstrong (1979: fig. 4), the insect-pollinated species *Pimelea sylvestris* R. Br. (Thymeleaceae) and *Darwinia vestita* (Endl.) Benth. (Myrtaceae) are very similar, by comparison with the bird-pollinated species *P. physodes* Hook. and *D. macrostegia* (Turcz.) Benth., which are similar to each other but very different from the first two species. However, these similarities are superficial and have not caused taxonomic difficulty. Within a group of closely related taxa, however, taxonomic problems are more likely to arise. Red and green tubular flowers with a 'ligule', adapted for bird-pollination, have apparently evolved several times in the Rutaceae tribe Boronieae, and cladistic analysis using a broad range of characters (not just floral morphology) suggests that this parallelism had led taxonomists to misclassify species in genera related to *Phebalium* and *Eriostemon* (Armstrong, 1979, 1987). In higher taxa of plants which are pollinated by birds or specialized insects, taxonomists have used a far greater proportion (*c.* 40%) of floral characters than they have in classifying groups pollinated by unspecialized insects (15%) or abiotic agents (4%) (Grant, 1949: fig. 1). Perhaps, as Grant suggests, this indicates that structural adaptations for specialized animal pollinators are a potent force for isolation and speciation, according to the 'biological species concept.' However, it also reveals the high potential for taxonomic error, as well as circularity in using the same features both for classifying plants and then for making hypotheses about their evolution and adaptation (but see also Brooks & McLennan, Chapter 1).

IDENTIFYING POLLINATION SYNDROMES

Pollination syndromes can be identified either directly by observation of flowers and their pollinators, or by inference from characteristic floral features known to be associated with a particular class of pollinators and pollination mechanisms (Faegri & van der Pijl, 1978). Very few direct observations of bird-pollination in Australia have been published (for reviews, see Ford, Paton & Forde, 1979; Ford & Paton, 1986), and for Fabaceae, literature is almost entirely lacking (but see Porsch, 1927). Keighery (1980, 1982a) provides reviews for Western Australia, including observations of his own, but many of these appear not to have been published in detail. One reference (Keighery, 1982b) proved very difficult to locate and ultimately uninformative. I suspect that a great deal of useful information is buried in the notebooks of field naturalists, where it is relatively inaccessible to researchers (e.g., see caption to appendix in Ford *et al.*, 1979). For the legume tribe Mirbelieae, I have found only the following published records of observations of avian pollinators visiting flowers: *Ephthianura tricolor* Gould (Crimson Chat) visiting *Leptosema chambersii* F. Muell. (Frith, 1976: 510); *Certhionyx variegatus* (Lesson) (Pied Honeyeater) visiting *L. daviesioides* (Turcz.) Crisp (Porsch, 1927: 513; see also Fig. 2); *Phylidonyris novae-hollandiae* (Latham) (New Holland Honeyeater) and *Lichenostomus virescens* (Vieillot) (Singing Honeyeater) visiting *Jansonia formosa* (Keighery, 1984).

It has long been recognized that the general mode of pollination of a flower may be inferred from characteristic features of its morphology. All flowers may be classified into relatively few functional structures (Faegri & van der Pijl, 1979: p. 93), but such a generalized classification has low predictive value. However, when this approach is restricted to a particular class of pollinators, its predictive capacity increases, for example hummingbird flowers are more readily definable than generalized bird-pollinated flowers. Its predictive value also increases when there are constraints (whether phylogenetic, developmental or selective) on the number of possible solutions to a particular functional 'problem'. Thus one might predict that within the legume family, phylogenetic and developmental constraints ('burden': Riedl, 1978) would canalize the range of morphotypes that might evolve for bird-pollination from an insect-pollinated ancestor. Within Australia, where mainly Meliphagidae are available as pollinators (Ford & Paton, 1986), the range of morphotypes would be further restricted, according to constraints within the pollinator group.

Given that direct observations of bird-pollination are unavailable for most Australian legume species, it is not possible to analyse separately the morphology and function associated with this adaptation, as suggested by Coddington (1988). In the present study, pollination syndrome is inferred indirectly from floral morphology, as was done by Donoghue (1989) in his study of fleshy-fruitedness and the evolution of dioecy.

HISTORY OF BIRD-POLLINATION IN AUSTRALIA

The ancestral angiosperm flower was probably a generalized type pollinated by insects such as beetles, flies, micropterigid moths, sawflies or sphecid wasps (Crepet & Friis, 1987; Lloyd & Wells, 1992). In Australia, bird-pollinated flowers evolved from ancestors that were pollinated by either insects or mammals, possibly before the beginning of the Tertiary (Ford *et al.*, 1979; Paton, 1986b). Typically Australian scleromorphic genera within the Myrtaceae and Proteaceae had evolved by the Miocene (Lange, 1981; McLoughlin, 1993); in fact, proteaceous pollen genera are recognizable even in late Cretaceous pollen floras (Dettman & Jarzen, 1991). Thus by the mid-Tertiary, a similar level of differentiation may have occurred in families such as Epacridaceae, Loranthaceae, Rutaceae and Fabaceae, which today include many bird-pollinated taxa (Ford *et al.*, 1979; Paton, 1986b). On a global comparison, bird-pollination is unusually frequent in Australia, although the reasons for this are unclear (Ford *et al.*, 1979). Given the taxonomic diversity of extant bird-pollinated taxa, as well as their frequency, there has probably been a history of strong selection for bird-pollination, perhaps driven by a diverse and abundant group of pollinators.

Honeyeaters (Meliphagidae) comprise the major nectar-feeding group of birds in Australia; in fact they are the most species-diverse family of passerines there, including more than 70 species (Schodde & Tidemann, 1986; Ford *et al.*, 1979). Sibley & Ahlquist (1990) place the honeyeaters within the parvorder Corvida of the suborder Passeri (Oscines or songbirds). In their classification the Corvida is an endemic Australian taxon, and sister taxon to the parvorder Passerida, a predomi-nantly Eurasian group with Australian representatives; together, the Passerida and Corvida comprise the songbirds. Sibley & Ahlquist (1985, 1990) suggest that the Corvida arose from an Australian ancestor, while Christidis & Schodde (1991) extend that claim to the more inclusive taxon Passeri (for a contrary view, see Olson, 1988). However, various authors have criticized the DNA hybridization methods, assump-tions, data analysis and resulting phylogeny and classification of birds (Sarich, Schmid & Marks, 1989; Mindell, 1992, and references therein). Attempts to corrobo-rate Sibley and Ahlquist's phylogeny of the passerines using enzyme electrophoresis (Christidis & Schodde, 1991) and sequences of mitochondrial DNA (Edwards, Arctander & Wilson, 1990) have supported the monophyly of the songbirds but failed to reproduce their basal dichotomy into Corvida and Passerida. These disputes notwithstanding, it seems clear that the honeyeaters at least are monophyletic (perhaps with the inclusion of the chats) and have an autocthonous Australian origin (Sibley & Ahlquist, 1985, 1990; Christidis & Schodde, 1991). Honeyeaters may have appeared during the Miocene (Ford, Paton & Forde, 1979), but the pre-Miocene fossil record of Australian birds is very meagre (Olson, 1988), and the basis for this date is unclear, so it may be grossly inaccurate. According to the $\triangle$T50H values from DNA–DNA hybridization (Sibley & Ahlquist, 1985), the honeyeaters diverged from their sister-group {warblers + pardalotes} at about 50 Ma, during the Eocene when Australia was separating and moving north from Antarctica (BMR Palaeogeo-graphic Group, 1990). However, this date depends upon the assumption of a clock-like rate of molecular evolution, as well as the assumption that DNA–DNA hybridization distances are a sound basis for the reconstruction of phylogeny. Both these assumptions have been severely criticized (Sarich, Schmid & Marks, 1989). Nevertheless, as a first approximation, their $\triangle$T50H values are probably a better guide to relationships and ages of lineages than those based upon one or two fossils.

All the honeyeaters have brush-tongues which are well adapted to licking nectar (Ford, Paton & Forde, 1979; Paton, 1986b; Schodde & Tidemann, 1986). Within the Corvida, only the Meliphagidae are such specialized nectar-feeders and pollinators, although other, smaller groups of Australian songbirds have nectar-adapted tongues, e.g. chats, wood-swallows, white-eyes, babblers and sunbirds (Sibley & Ahlquist, 1990). Chats are brush-tongued casual nectar-feeders, and have been placed in a separate family (Epthianuridae: Schodde & Tidemann, 1986), but they are either monophyletic (Sibley & Ahlquist, 1990: fig. 374) or paraphyletic (Christidis & Schodde, 1991: fig. 2) with the honeyeaters, with which they probably shared a

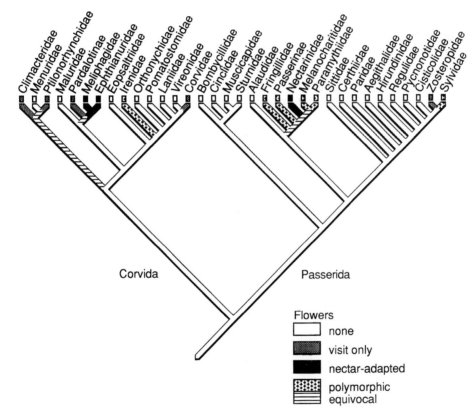

Figure 1 Phylogeny of the songbird suborder Passeri (Oscines), based upon DNA–DNA hybridization distances (Sibley & Ahlquist, 1990), showing the distribution among families of tongues adapted for nectar-feeding, and Australian taxa with members seen visting flowers. Only the Meliphagid–Epthianurid clade and the Nectariniidae consist entirely of specialist nectar-feeders. The phylogeny of Christidis & Schodde (1991), using protein electrophoretic data, differs in several details, but this does not greatly alter the inferred pattern of nectar-feeding. The most important difference is that their scheme shows the Meliphagidae–Epthianuridae as paraphyletic with repect to *Pardalotus* and *Acanthiza*, suggesting that specialized nectar–feeding either evolved independently within the Meliphagidae and Epthianuridae, or that it was secondarily lost within the latter genera.

specialized nectar-feeding ancestor. All the remaining nectar-adapted taxa are more distantly related to the honeyeaters (Fig. 1). Therefore nectar-feeding probably originated independently in those lineages, and their morphological and behavioural adaptations are probably convergent; for example, the tongue of the sunbirds (Nectariniidae) is curled in the reverse direction to that of honeyeaters, and lacks a brush tip (Frith, 1976: 590). Some of the non-honeyeater nectar-feeders (e.g. white-eyes, babblers, sunbirds) belong to the parvorder Passerida, the sister-taxon of the Corvida (Fig. 1; Sibley & Ahlquist, 1990). This group appears to be Eurasian in origin; its Australian representatives are relatively few and probably arrived only following the Miocene contact between Australia and south-east Asia (Sibley & Ahlquist, 1985; Christidis & Schodde, 1991). With only a short evolutionary history in Australia, these taxa are unlikely to have played a significant part in the early evolution of bird-pollination in this continent. The same applies to other songbirds that have been seen visiting flowers, but which are not specialized nectar-feeders and do not show morphological features related to nectar-feeding (pardalotes, trillers, orioles, strike-thrushes, treecreepers, thornbills, sitellas, bowerbirds and butcherbirds; Ford *et al.*, 1979).

Besides the honeyeaters, only the distantly related lorikeets (Loriidae) seem likely to have driven a co-evolutionary pollinator–flower radiation in Australia. Eucalypts have been suggested as the first bird-pollinated flowers in Australia, with lorikeets as their pollinators (Keast, 1976; Ford *et al.*, 1979), but these groups are respectively unspecialized in their pollination systems and feeding behaviour (lorikeets feed on pollen as well as nectar). Despite the inference made by Ford *et al.* (1979), there is no evidence for a historical connection between the evolution of the lorikeets and the origin of the Meliphagidae or between generalized bird-pollination in the Myrtaceae and the much more specialized bird-pollination systems in other families.

While casual nectar-feeding may have arisen early in the history of the Corvidae, specialization probably originated at the base of the Meliphagidae-Epthianuridae (Fig. 1; cf. Paton, 1986b: 40). Presumably it coincided with the origin of obligate bird-pollination in one or more plant taxa. Irrespective of the uncertain origins of bird-pollination in Australia, it seems most likely that the honeyeaters drove a radiation in which nectar-feeding birds co-evolved with host plants, and that this process was well underway before the mid-Tertiary. Paton (1986b) postulated a series of steps through which insect-pollinated flowers might have evolved into specialized bird-pollinated flowers. This process would be initiated when nectar-feeding birds casually visit insect-pollinated flowers, and would be followed by increased nectar production, then morphological specialization of the flowers, such as tubular and gullet shapes, plus other features as described in the following section. These specializations target the transfer of pollen to and from the bird to ensure efficiency of pollination, as well as excluding insects which are not effective pollinators. Although many of the honeyeaters are highly dependent upon nectar as a source of food, pollinator-flower relationships are seldom species-specific (Ford *et al.*, 1979; Hopper & Burbidge, 1986; Paton, 1986a). Regular nectar-feeders like *Phylidonyris* visit a wide range of flowers (in different families), and these are visited by many species of birds. Combining their demand for nectar as a food-source with their ability to switch hosts, the abundant honeyeaters must have exerted a potent selective force for the repeated evolution of bird-pollinated flowers within prim-itively insect-pollinated groups in Australia (Paton, 1986b). Where this has occurred in parallel among closely related plant taxa, such as in the same family or genus, independently evolved similar floral structures serving a similar function (bird pollination) have the potential to deceive plant systematists into defining non-monophyletic taxa, as discussed above.

POLLINATION SYNDROMES IN AUSTRALIAN LEGUMES

Legumes are not known before the upper Cretaceous, and unequivocal papilionoid legumes first appear in the lower Eocene fossil record (Herendeen, Crepet & Dilcher, 1992), by which time both angiosperms and their insect pollinators were well diversified and, in some lineages, highly specialized (Crepet & Friis, 1987). Unfortunately, the Australian fossil record is virtually uninformative about the history of papilionoid legumes in the Tertiary (Herendeen, Crepet & Dilcher, 1992), which has led to the controversial suggestion that legumes first entered Australia from the north following contact between the Australian plate and south-east Asia about 15 Myr ago (Raven & Polhill, 1981).

Legumes are mostly pollinated by bees, and in return they provide a major food resource, suggesting a long history of co-evolution (Leppik, 1966; Arroyo, 1981). Bee-pollination is most likely the primitive condition in the pea-flowered subfamily, Faboideae. Pea flowers show a series of features designed for bee-pollination (Arroyo, 1981): the standard-petal is prominent, erect, has nectar guides at the base, and attracts the pollinator; the remaining petals (wings and keel) are short and project horizontally at right angles to the standard, providing a landing platform; the keel encloses and protects the sexual parts; wings and keel interlock and, together with the rigid (usually joined) staminal column and two-lipped calyx, function as a tripping mechanism to transfer pollen actively to the abdomen of the bee.

In the endemic Australian pea-flowered tribe Mirbelieae, floral morphology is typical of the subfamily, and remarkably uniform. In addition, flowers of the Mirbelieae have unique and characteristic red and yellow markings, especially on the standard. These so-called 'egg-and-bacon' floral markings are phylogenetically primitive within the Mirbelieae, having been inherited from an ancestor held in common with their sister-taxon, the tribe Bossiaeeae (Crisp & Weston, 1987). Pollination of 'egg-and-bacon' flowers in genera of the Mirbelieae by Australian native bees has been demonstated by Gross (1992). The central spot on the standard, which is yellow in visible wavelengths, is strongly absorptive of UV light, to which bees are particularly sensitive (Gross, 1992).

Detailed descriptions of variations on the pea flower and its modes of pollination may be found in Faegri & van der Pijl (1978), Arroyo (1981) and Schrire (1989). Notwithstanding these variations, it is clear that "few characteristics of the papilionoid flower are not directly associated with tripping, and it goes without saying that this feature has dominated the architecture" (Arroyo, 1981: 735). Here is the suggestion of strong constraint on the architecture of the flower involving most of its parts, which are integrated into a complex pollination mechanism. Modifications to this structure–function complex are likely to be minimal, unless under strong selection for a completely different pollination mechanism, in which case floral morphology may be radically altered, within the constraints set early in development of the flower (Schrire, 1989; Tucker, 1989). Radical alterations have occurred when pea flowers have been adapted for bird-pollination, but common features provide evidence of constraint.

As suggested above, bird-pollinated flowers may be recognized by a common set of features, which are described in detail by Faegri & van der Pijl (1978), Ford *et al.* (1979) and Paton (1986b), and are only summarized here. Some features appear to be distinctively Australian (Ford *et al.*, 1979; Keighery, 1982a), presumably because of selection by regionally specific pollinator morphology and behaviour. Bird-pollinated flowers are brightly coloured, often red, although in Australia they may be green, black, white or yellow. Their structural blossom types may be tubular, flag-shaped or gullet-shaped. The corolla is large and strong, and provides no landing stage (the latter being provided by stems or low stature, allowing the bird to stand on the ground). Odour is usually absent, although in Australia, many *Banksia* species have a musty smell and also attract small mammals. Nectar is abundant and rich in energy. Anthers and stigma are distant from the nectary, consistent with large-sized pollinators.

Indirect evidence suggests parallelism in bird-pollinated pea flowers, and this probably reflects constraint as well as selection. The keel of papilionaceous flowers is commonly enlarged to serve as a drinking vessel for non-hovering birds (Arroyo, 1981), and in Western Australia, Keighery (1982a) suggests that red flower colouring and low plant stature have evolved in parallel in several families, including the 'pea-flowered' genera *Brachysema* and *Leptosema* (Fig. 2). The Sturt Desert Pea of the central Australian deserts, with its striking elongated red and black flowers, was placed in the genus *Clianthus*, together with one other (New Zealand) species. Yet,

Figure 2   Pied Honeyeater (*Certhionyx variegatus*) visiting *Leptosema daviesioides*. From Porsch (1927).

in characters of the inflorescence, calyx, fruit, indumentum, stipules and growth form, this species appears closely allied to a small group within *Swainsona*, and Thompson (1990) transferred it to that genus, as *S. formosa* (G. Don) J. Thompson. Apparently, the similarity between the flowers of *S. formosa* and those of *C. puniceus* Solander ex Lindley is a convergent adaptation to bird-pollination in an otherwise insect-pollinated group.

BIRD-POLLINATION IN THE AUSTRALIAN LEGUME TRIBE MIRBELIEAE

Crisp & Weston (1987, 1994) have reconstructed the phylogeny of the Fabaceae tribe Mirbelieae, which comprises a monophyletic group of about 30 genera endemic in Australia. When the occurrence of bird-pollination is mapped onto the cladogram of genera in the Mirbelieae (Fig. 3), a parsimonious interpretation suggests that the trait

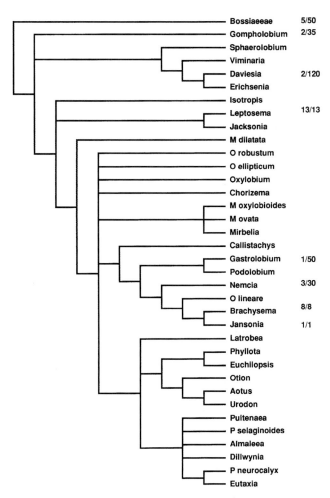

Figure 3 Cladogram of the Mirbelieae from Crisp & Weston (1994), showing distribution of bird-pollinated taxa. Numbers at right of taxon labels give number of bird-pollinated species out of total species. Bossiaeeae, the sister-group of the Mirbelieae (Crisp & Weston, 1987), is the outgroup.

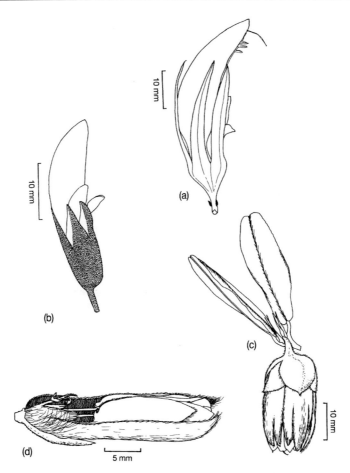

Figure 4 Floral morphology of bird-pollinated members of Mirbelieae. Note convergent similarities, e.g. resupination, elongated keel, reduced standard. (a) *Leptosema aphyllum* (Hook.) Crisp. (b) *Brachysema celsianum.* (c) *Jansonia formosa*, inflorescence. (d) *J. formosa* flower. (a) drawn by B. Osborne; (b)–(d) drawn by A. (Prowse) O'Dowd.

has originated independently at least five times within the group: in *Gompholobium, Daviesia, Leptosema, Gastrolobium,* and possibly in the common ancestor of *Nemcia, Brachysema* and *Jansonia.* Some of the morphological features which have evolved in parallel to serve the bird-pollination function in Rutaceae (Armstrong, 1987) are similar to those found in bird-pollinated taxa of the Mirbelieae, for example within the genera *Brachysema, Leptosema, Nemcia* and *Jansonia*: inflorescence reduced to one or a few flowers; flowers pendant or resupinate; corolla tubular; nectary conspicuous and producing abundant nectar; and red or green colouring in the flowers (Fig. 4). Whether these similarities are homologous, or convergent as a result of selection combined with constraint, can be evaluated using the phylogeny. For example, the floral morphology of *Leptosema* (Fig. 4a) is very similar to that of *Brachysema* (Fig. 4b). Bentham (1864) took morphological similarities in the flowers of these taxa to mean that they were closely related, and placed them in a single genus. However, cladistic analysis using additional characters shows that these taxa are not closely related (Fig. 3; see also Crisp, 1982;

Crisp & Weston, 1987), and it seems clear that their floral structures are convergent.

Brachysema, *Jansonia* and *Nemcia*

Brachysema, *Jansonia* and *Nemcia* form a clade (together with the single species, *Oxylobium lineare*), and all three genera include bird-pollinated species (Fig. 3). Therefore, it is tempting to suggest that bird-pollination arose once, in the single common ancestor of this group. However, *O. lineare* and most species of *Nemcia* are insect-pollinated, so homoplasy must have occurred: either bird-pollination originated once in the common ancestor and reverted to insect-pollination within *Nemcia* and *O. lineare*; or bird-pollination originated twice, once in the ancestor of *Brachysema* and *Jansonia*, and independently within *Nemcia*. Alternatively, the phylogenetic reconstruction may be incorrect. Monophyly of all genera had been assumed *a priori* in the analysis shown in Fig. 3 and has not been tested previously by analysis using the species as terminals. There is little point in using a doubtful phylogeny to make hypotheses about the origin and evolution of adaptive traits. Therefore, I re-analysed the relationships of these taxa at the species level, and tested the reliability of the cladograms by manipulating the data and using permutation tail probability tests.

Material and methods

A primary data set was compiled comprising 27 morphological characters (Appendix 1) sampled from herbarium specimens and 13 terminal taxa (Appendix 2), the latter including all eight species of *Brachysema*, the sole species of *Jansonia*, *Oxylobium lineare*, and two groups within *Nemcia*: insect-pollinated species and bird-pollinated species. *Callistachys*, comprising a single insect-pollinated species, was used as the outgroup, this being at the base of a small well-corroborated clade including the ingroup (Fig. 3). No synapomorphy was found for the insect-pollinated species of *Nemcia*, so this group may be paraphyletic. Therefore, a secondary data set was made by splitting *Nemcia* into smaller groups, although no new morphological characters could be found that might discriminate among these groups. These new terminals, any of which may be paraphyletic, are marked with an asterisk in Table 1. Multistate characters 6 (number of flowers per unit inflorescence), 17 (standard lamina shape) and 23 (petal colours) were coded as additive, and character 24 (ovule number) was coded as non-additive. The justification of the latter coding is that the analyses of all genera in the tribe showed this character to be very homoplasious, with transformations not following a simple monotonic sequence (Crisp & Weston, 1987, 1994). All the remaining characters were coded as binary. The data matrix comprising both primary and secondary data sets is shown in Table 1 (the primary data set includes only those rows [taxa] unmarked by an asterisk; the secondary data set comprises the whole of Table 1 except '*Nemcia* bee-pollinated'). Specimens are lodged in the Australian National Herbarium (CANB) and the Western Australian Herbarium (PERTH).

Parsimony analyses were carried out using PAUP ver. 3.0 (Swofford, 1991) and Hennig86 ver. 1.5 (Farris, 1989a; Fitzhugh, 1989).

The ultimate test of a phylogeny is the congruence or otherwise between

Table 1   Data matrix used in the cladistic analyses. Explanation of symbols: a = (01); b = ?; c = (12); d = (03); e = (123); * = informal species groups into which 'Nemcia insect-pollinated' group was subdivided for further analysis. Polymorphism is indicated by multiple values in parentheses. For further details on terminal groups, see Appendix 2 and text.

Ingroup

Brachysema bracteolosum	10010	10110	00010	03100	00311	00
B. celsianum	1a110	00101	01000	02111	11101	01
B. latifolium	10110	00100	00000	01101	10101	01
B. melanopetalum	1a110	10110	00011	03100	00211	1b
B. minor	10110	10100	00111	02100	00111	10
B. praemorsum	11100	00101	00000	02110	11e01	01
B. sericeum	10110	10110	00011	03100	00311	10
B. subcordatum	01110	00100	01011	02100	00221	10
Jansonia	11001	01000	10100	1d110	00c20	00
Nemcia bird-pollinated	01101	01000	11100	10000	00120	00
Nemcia bee-pollinated	aaaa1	aaa00	01100	a0000	00020	00
Nemcia 2-ovules	011a1	a0100	01100	00000	00020	00
N. alternifolia	100b1	10000	01100	00000	00020	b0
Nemcia 2-tone	01001	01000	01100	10000	00020	00
Nemcia 2-tone/2-ovules	01101	01000	01100	10000	00020	00
N. capitata	aaa11	0a000	01100	00000	00020	00
Oxylobium lineare	0a100	00000	01000	00000	00000	00

Outgroup

Callistachys	0a000	00000	01000	00000	00000	0a

cladograms constructed from independent data sets, or the robustness of a cladogram to the addition of new data (Ruse, 1979). This is the empirical basis of phylogenetic systematics (Nelson & Platnick, 1981: 165). Quite apart from this, tests have been proposed for evaluating the support for a cladogram within a data set. Such tests are applied either to trees or data sets, either wholly or in part. They include measures of consistency index and related indices, data decisiveness and skewness of the frequency distribution of treelengths, as well as the bootstrap and permutation tail probability (PTP) tests. Consistency index and retention index are measures of best fit based upon the principle of parsimony, and are useful only for comparing trees derived from the same data set (Farris, 1989b). The other measures are reviewed by Trueman (1993) and are not discussed in detail here. Most of these appear to address irrelevant questions (Trueman, 1993; Källersjö *et al.*, 1992). Only permutation tests appear to provide a truly statistical and relevant test, based upon the accepted parsimony criterion, of questions such as 'could a cladogram this short have arisen by chance alone?' (PTP tests; Archie, 1989; Faith & Cranston, 1990), or 'could a tree this shape have arisen by chance alone?' (T–PTP or topology-dependent PTP tests; Faith, 1991). On the premise that it is pointless using an estimate of phylogeny to evaluate hypotheses about evolution, unless the estimate of phylogeny is reliable, I applied permutation tests to the cladograms derived in this study. PTP tests were used to test for significant cladistic structure in the whole data sets, and T–PTP tests were used to test for monophyly or non-monophyly of groups critical to alternative hypotheses about the evolution of bird-pollination.

Permutation tests only evaluate the significance of cladistic structure within a data set. They do not address underlying causes. Factors other than phylogeny may account for significant hierarchical structure in a data set, but a permutation test cannot detect the difference – it tests for hierarchical structure, whatever the cause (Trueman, 1993). In a case where characters are suspected of showing concerted evolution (above), either those characters should be discarded, or preferably their evidence should be tested for congruence with other data. As explained above, I suspected that concerted evolution might have occurred in morphological characters that were related to the bird-pollination function (characters marked with an asterisk in Appendix 1). Therefore I excluded these characters and re-analysed the data set to see whether the remaining characters supported a similar estimate of phylogeny.

Results

In all cases, both PAUP and Hennig86 found the same most parsimonious trees from the same data. This might be expected, because exact algorithms were used ('branch and bound' in PAUP, and 'ie' in Hennig86), although the character transformations had to be different for polymorphic taxa. Figure 5 shows the three most parsimonious cladograms obtained from the cladistic analysis of the primary data set. In all three cases, *Brachysema* is monophyletic and *Nemcia* is paraphyletic with respect to *Jansonia*. They differ only in the relative arrangement of the three basal clades: *Oxylobium lineare*, *Brachysema* and {*Nemcia* + *Jansonia*}. A strict consensus tree is identical to one of these (Fig. 5a). Re-analysis excluding the characters putatively related to bird-pollination, marked with an asterisk in Appendix 1, gave a single minimal cladogram (Fig. 6). This differs from Fig. 5a in being rather less resolved, and in showing different relationships among *Brachysema* species, but still includes the two clades *Brachysema* and {*Nemcia* + *Jansonia*}. The latter is unresolved, and therefore is uninformative with respect to the monophyly or otherwise of *Nemcia*. Analysis of the secondary data set yielded 10 most parsimonious cladograms. Only their strict consensus tree is shown (Fig. 7). All these contain the same three basal clades as in Fig. 5, and differ in the arrangement of these clades, as well as in the internal relationships of *Nemcia*. All show *Jansonia* as more closely related to the bird-pollinated species of *Nemcia* than either group is to the insect-pollinated species of *Nemcia*. In other words, *Nemcia* again appears to be paraphyletic with respect to *Jansonia*. Moreover, the {*Jansonia* + bird-pollinated *Nemcia*} clade is removed by two nodes from the base of *Nemcia*, even in the strict consensus tree (Fig. 7).

Results of permutation tests on the cladograms in Figs 5a and 7 are shown in Table 2. Both the primary and secondary data sets show significant cladistic structure (respectively $P < 0.01$ and $P < 0.05$). Topology-dependent testing showed that the monophyly of *Brachysema* was significant in the primary data set ($P < 0.01$). Therefore this clade was reduced to an estimated ancestral node for further testing (cf. Faith, 1991: 375). Paraphyly of *Nemcia* with respect to *Jansonia* in the primary analysis (Fig. 5) was unexpected, given the prior recognition of these as separate genera (Crisp & Weston, 1987). Therefore, an *a posteriori* test of monophyly was applied to both the {*Nemcia* + *Jansonia*} clade and the {*Jansonia* + bird-pollinated *Nemcia*} clade (cf. Faith, 1991: 369). Both these were non-significant (Table 2).

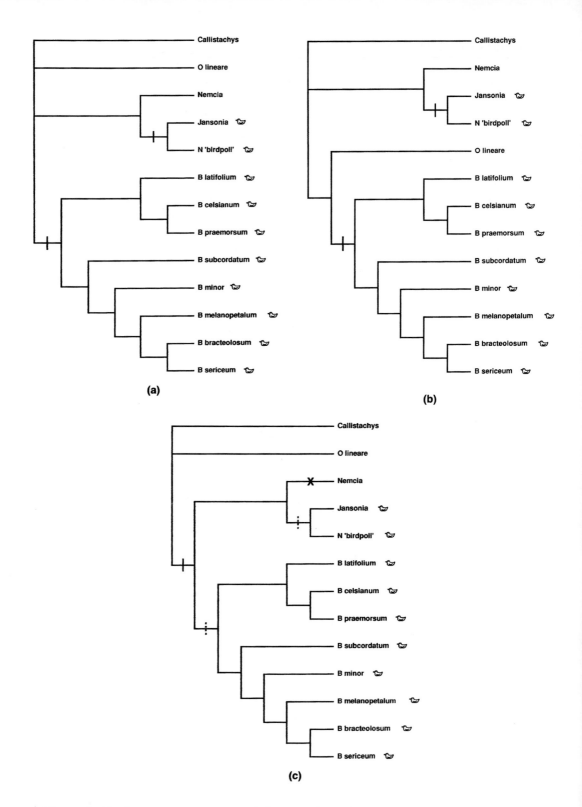

Figure 5 The three most parsimonious cladograms obtained from an analysis of the primary data set (including all rows in Table 1 except those marked with an asterisk). Tree 5a is identical with the strict consensus tree of all three. Bird-pollinated terminal taxa are marked with a bird symbol. Most parsimonious scenarios for origin and/or loss of bird-pollination are mapped as follows: a–b, two parallel gains (bars); c, two parallel gains (dotted bars) or gain (bar) followed by loss (cross).

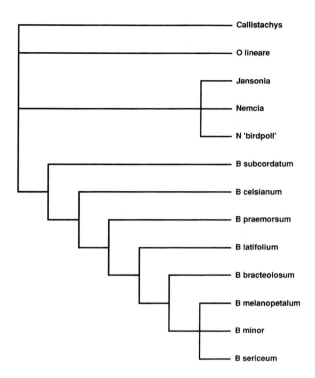

Figure 6 The single most parsimonious cladogram obtained from an analysis of the primary data set, excluding characters putatively related to bird-pollination (maked with an asterisk in Table 1).

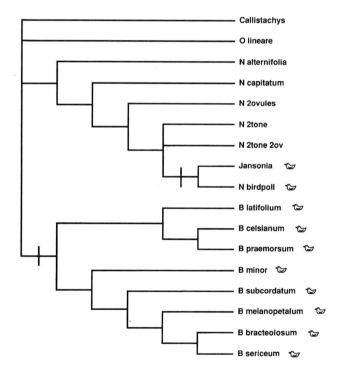

Figure 7 Strict consensus of ten trees obtained from an analysis of the secondary data set (the whole of Table 1 except the row '*Nemcia* bee-pollinated'). Scenarios for evolutionary gain or loss of bird-pollination are mapped as in Fig. 5: two parallel gains only are parsimonious (bars).

Table 2 Results of permutation tests on cladograms in Figs 5 (primary data set) and 7 (secondary data set). Asterisks indicate *a posteriori* tests, made because the clade was unexpected. All remaining tests were *a priori* (results expected). For the second set of tests, *Brachysema* was reduced to a single 'ancestral' node, because its monophyly had been supported by testing of the primary data set.

1. Primary data set (*Nemcia* as two taxa)	
Whole data set	$P<0.01$
Monophyly of:	
Brachysema	$P<0.01$
Nemcia + Jansonia*	n.s.
Jansonia + N. bird-pollinated*	n.s.
Non-monophyly of:	
All bird-pollinated taxa	n.s.
Nemcia	n.s.
2. Secondary data set (*Nemcia* as six taxa)	
Whole data set	$P<0.05$
Monophyly of:	
Nemcia + Jansonia	n.s.
Jansonia + N. bird-pollinated	$P<0.05$
Non-monophyly of:	
All bird-pollinated taxa	n.s.
Nemcia	n.s.

However, *a priori* tests were justified when a similar result was obtained from the modified secondary data set, that is *Nemcia* was again shown to be paraphyletic with respect to Jansonia (Fig. 7). This time, the monophyly of {*Jansonia* + bird-pollinated *Nemcia*} was significant (*P*<0.05), although monophyly of {*Nemcia* + *Jansonia*} was again non-significant (Table 2). Finally tests of non-monophyly were made on groupings which did not occur in either result, but which might have been expected (cf. Faith 1991: 372). The first of these comprised all bird-pollinated taxa, that is *Brachysema, Jansonia* and bird-pollinated species of *Nemcia*. Although these taxa did not form a clade, they might have, with bird-pollination as a synapomorphy, along with associated morphological characters, such as red and green petal coloration. In both sets of trees (Figs 5 & 7), non-monophyly was non-significant (Table 2); in other words, the null hypothesis that all bird-pollinated taxa form a clade, with a single origin of this trait in their common ancestor, was not rejected. The second test of non-monophyly was made on *Nemcia*, which appears paraphyletic by inclusion of *Jansonia* (Figs. 5 & 7). In both cases, the result was non-significant (Table 2); therefore, the null hypothesis that *Nemcia* is monophyletic as previously circumscribed was not rejected.

DISCUSSION

Cladistic analyses

The primary and secondary data sets in this study both show significant cladistic structure (Table 2). This is important if these cladograms are to be used to evaluate evolutionary hypotheses (Faith & Cranston, 1990), as I do below. Moreover, previously assumed monophyly of *Brachysema* is supported (Fig. 5), and this is significant (Table 2). It is worth noting that this result corroborates Crisp & Weston's (1987) inclusion of the monotypic genus *Cupulanthus* within *Brachysema* (as *B. bracteolosum*), even though those authors did not test their hypothesis by analysing relationships of these taxa at species level.

A surprising result of this study is the paraphyly of *Nemcia* with respect to its inclusion of *Jansonia* (Figs 5–7). The cladograms of Crisp & Weston (1987, 1994; see also Fig. 3) suggest that *Jansonia* is more closely related to *Brachysema* than either is to *Nemcia*, with bird-pollination as a synapomorphy of the first two, while appearing as a homoplasy in the third. However, their analyses did not include species or species-groups of *Nemcia*, and therefore the assumed monophyly of this genus was not tested. Support for the present result is equivocal. On the one hand, all three analyses using different data sets showed *Jansonia* to be monophyletic with *Nemcia*, and two of these (Figs 5 & 7) suggested that it is more closely related to bird-pollinated species of *Nemcia* than to any other group, including the remaining species of *Nemcia*. Given that these data sets are partially different, this amounts to partial corroboration of the result. The third analysis (Fig. 6) did not resolve the {*Nemcia* + *Jansonia*} node. On the other hand, statistical support for this result is equivocal. Only in the third analysis (Fig. 7) was monophyly of {*Jansonia* + bird-pollinated *Nemcia*} significant (Table 2). Moreover, non-monophyly of *Nemcia* as previously circumscribed, that is excluding *Jansonia*, was not significant. Thus, while it may be safe to conclude that {*Jansonia* + bird–pollinated *Nemcia*} is monophyletic, it is less certain that this clade is monophyletic with the insect-pollinated species of *Nemcia*.

Finally, it should be noted that the relationship of *Oxylobium lineare* to the other taxa in this study is uncertain, thus providing no advance from Crisp & Weston (1987).

Evolution of bird-pollination

It is clear that bird-pollination has arisen several times independently within the legume tribe Mirbelieae (Fig. 3). Such interpretation is simple, using the principle of parsimony, when bird-pollinated taxa are phylogenetically isolated among insect-pollinated relatives, as is the case in *Gompholobium*, *Daviesia*, *Leptosema* and *Gastrolobium*. Even in these examples, further investigation would be needed to estimate whether, say, bird-pollination in the two species of *Daviesia* or *Gompholobium* arose once or twice independently. If each of these events were independent, then they are homoplasies (Coddington, 1988), and upon close examination, their similar morphologies may prove to be different (cf. Patterson, 1982). Thus, *Leptosema* and *Brachysema*, which once were combined in a single genus on the basis of their similarity in floral morphology, as related to bird-pollination, proved to have differences in the shapes of the standard-petal, calyx and nectary, as well as in many other characters, and their relationship is now estimated to be rather distant (Crisp, 1982; Crisp & Weston, 1987; see also Fig. 3).

While it is clear that bird-pollination originated in the clade comprising *Brachysema*, *Jansonia* and *Nemcia* independently from all the above taxa (Fig. 3), events within this group are unclear because some species of *Nemcia* are pollinated by insects. Only by examining lower-level relationships within these taxa, as in Figs 5–7, can this question be pursued further. In the three equally parsimonious cladograms derived from the primary data set (Fig. 5), two scenarios for the evolutionary gain and/or loss of bird-pollination are equally likely. The first is two parallel origins, seen in all three cladograms: once in each of the hypothetical ancestors of *Brachysema* and the {*Nemcia* + *Jansonia*} clade. The second scenario, seen only in the third cladogram (Fig. 5c), is an origin in the ancestor of the clade {*Brachysema* + *Nemcia* + *Jansonia*}, followed by loss (reversal) in the ancestor of the insect-pollinated species of *Nemcia*. Of course, this second scenario presumes that the insect-pollinated species of *Nemcia* actually had a single common ancestor, in other words, that they are monophyletic. However, this assumption is questionable, for when the group is subdivided into lower-level terminal taxa (secondary data set) and re-analysed, the insect-pollinated species of *Nemcia* are paraphyletic (Fig. 7). In this result, ten minimum-length cladograms are obtained, but only one scenario is parsimonious for the evolutionary origination of bird-pollination: parallel gains in the ancestors of *Brachysema* and {*Jansonia* + bird-pollinated *Nemcia*}. Support for this scheme is provided by significant permutation tests for both these clades (Table 2).

Can one rule out the intuitively appealing hypothesis that bird-pollination originated only once in *Brachysema–Nemcia–Jansonia*, in a single hypothetical ancestor of all three genera? One of the equally parsimonious optimisations in Fig. 5c suggests this possibility, but it requires a reversal within *Nemcia*; that is, the regain of a fully functional insect-pollinated flower in essentially its original (plesiomorphic) form, as seen in the outgroups. Some authors reject on principle the regain of complex functional structures, once they have been lost. This is equivalent to constraining characters to be uniquely derived, while allowing reversals, or in other words, rejecting convergence (Dollo's 'Law': Le Quesne, 1974; Farris, 1977; Futuyma, 1986: 297; Swofford, 1991). Interestingly some authors argue the opposite (the Camin-Sokal model): that complex functional structures, once derived, cannot be lost; in other words, convergence is possible but reversal is not (Johnson & Briggs, 1985; Campbell & Barwick, 1990). Both viewpoints are similar in that they argue *a priori* for constraint on transformations of complex characters. Against both is evidence that at least some complex structures are under the control of homeotic genes, and are easily changed by simple mutations (above). Moreover, both models are methodologically unnecessary. Homoplasy, whether reversal or convergence, may be mistaken assessment of homology, and as such each is equally likely (Crisp & Weston, 1987: 78). In the primary analysis (Fig. 5), parsimony does not rule out either scenario: a single origin of bird-pollination, followed by a reversal, or two convergent origins. However, the secondary analysis (Fig. 7) favours only two convergent origins. This scenario is supported by the permutation tests of monophyly for both clades in whose ancestors bird-pollination is postulated to have originated independently: both are significant (Table 2). A single, non-homoplasious origin of bird-pollination would be supported if all bird-pollinated taxa (*Brachysema*, *Jansonia* and bird-pollinated *Nemcia* species) were monophyletic. In fact, non-monophyly of these taxa is not significant in either data set (Table 2), so this alternative cannot be ruled out. However, it seems unlikely, given that none of the three analyses using different data sets showed monophyly of all bird-pollinated taxa (Figs 5–7). Rather, the weight of evidence from all analyses favours two independent origins of bird-pollination.

Assuming that bird-pollination originated independently in *Brachysema* and the {*Jansonia* + bird-pollinated *Nemcia*} clade, several things can be inferred about evolution of morphological features related to pollination. First, each clade should have autapomorphies, which by definition are unique and therefore different. In the present data set, *Brachysema* has two autapomorphies: inflorescences axillary and a raised disc (nectary), and {*Jansonia* + bird-pollinated *Nemcia*} has one: valvate inflorescence bracts (Fig. 4c). Second, both clades should have structural similarities reflecting their convergent adaptation to bird-pollination. These include red petals, standard reduced relative to the keel, wings shorter than the keel and a sprawling, scandent growth habit (Figs 5b–d, 8). This last feature is related to bird-pollination in that an ground-hugging plant may be pollinated by birds standing on the ground (Keighery, 1982a). Convergent features may have only superficial similarity, and where their lack of homology is obvious, should not be used as characters in systematic analysis. Such features excluded from the present analysis, but seen in both bird-pollinated clades, are enlarged flowers and elongated stamens and style. More strongly convergent features can only be inferred to be homoplasious from a cladogram, such as those mapped on to Fig. 8 (cf. Nelson, in press). The third inference about character evolution is more subtle and relates to the problem of concerted evolution (above). If the characters interpreted as being synapomorphies for the species of *Brachysema* were in reality congruent convergences, so that the

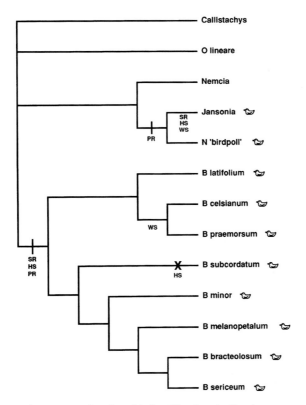

Figure 8 Convergent characters related to bird-pollination in Brachysema and the {*Jansonia* + bird-pollinated *Nemcia*} clade. HS = habit scandent; PR = petals red; SR = standard reduced (shorter than keel); WS = wings shorter than keel. Cross indicates a reversal.

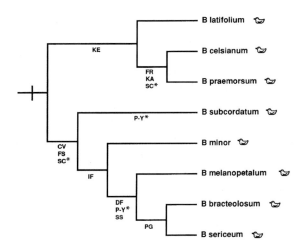

Figure 9 Elaboration (further transformation) of floral morphology within *Brachysema* beyond the autapomorphies inferred in the hypothetical ancestor of the genus. Synapomorphies for the species in the genus and convergences with *Jansonia* and *Nemcia* are not shown here; see instead text and Fig. 8. CV = calyx ventricose; DF = peduncles filiform; FR = flower resupinate; FS = flower 'sericeum' type; IF = inflorescence few-flowered; KA = keel acute; KE = keel elongate; PG = petals tending to green; P–Y = loss of yellow petal markings; SC = standard-lamina constricted; SS = standard-lamina subulate. An asterisk indicates homoplasy, shown as convergence; alternatively, earlier transformation followed by reversal is equally parsimonious, but not shown.

group supported by them were in fact polyphyletic, then one might expect conflict among characters within this group, and poor resolution of relationships among the eight species. Against this, relationships between *Brachysema* species were fully resolved (Figs 5 & 7), with more than one synapomorphy at each node. Furthermore, mapping floral characters onto the cladogram shows elaboration of features related to bird-pollination, with little homoplasy (Fig. 9). For example, in the clade {*B. latifolium* + *B. celsianum* + *B. praemosum*}, flowers invert (resupinate) and become very large with a long curving acute keel. In the sister-clade to this, the flower develops a very distinctive shape ('*sericeum*' type), the standard petal progressively constricts and narrows, the inflorescence reduces in number of flowers and becomes nutant, and in the terminal group {*B. melanopetalum* + *B. bracteolosum* + *B. sericeum*}, the petals lose their yellow markings then become green. All these features are acquired progressively, node by node (Fig. 9); in other words, they covary hierarchically, but are not correlated. This observation is significant, because one expects congruent convergence to result only in correlated characters. These would appear as a suite of 'synapomorphies', acquired simultaneously at a single node (the basal node of the false group).

Independence and corroboration

Permutation tests may be viewed as evaluating the corroboration of a phylogenetic hypothesis in the Popperian sense, in which the degree of corroboration is the severity of the various tests to which the hypothesis is subjected (Faith, 1992). However, if such tests are applied only to a single data set, then they have validity only within the universe of those data. Corroboration may also be viewed as tests provided by alternative sources of evidence (Ruse, 1979), although this is more usually known as congruence. Resort to alternative evidence becomes necessary if one source of data is suspect for a particular reason (e.g., Faith & Cranston, 1990; Doyle, 1992). For example, non-independence among morphological characters due to structural-functional correlation and convergence ('concerted evolution') may result in a cladogram that is positively misleading. The only protection against such a contingency is to estimate the phylogeny from data independent of the adaptive traits being investigated; that is to seek corroboration from, for example, molecular data (Harvey & Purvis, 1991; Albert, Williams & Chase, 1992). Whether data from different sources should be analysed and evaluated separately, at least in the first instance (Patterson, 1987; Doyle, 1992), or simply combined in a 'total evidence' approach (Kluge, 1989; Donoghue & Sanderson, 1992; Trueman, 1993) is a contentious issue. A solution which could reconcile the two notions of corroboration is to combine the data sets and carry out permutations tests (Faith & Cranston, 1990). Otherwise, two data sets could be congruent but without significant cladistic structure, in which case their congruence is trivial. Even though corroboration (in the sense of Ruse) is sought by compiling new data, it is not quantified in the Popperian sense (Faith, 1992) until a permutation or similar test is done to indicate a probability of departure from randomness.

Alternative data sets were considered in the present study first by excluding characters that were putatively related to the bird-pollination function (Fig. 6), and then by adding taxa (Fig. 7). Given that there was considerable overlap between these data sets, the test of corroboration was weak. I am now proceeding to reconstruct the phylogeny of the *Brachysema* group using molecular data (ribosomal DNA sequences).

CONCLUSIONS

Bird-pollination apparently originated independently in *Brachysema* and the *Nemcia–Jansonia* lineage. Detailed phylogenetic analysis of this group at species-level has shown that the previous classification is probably incorrect. In particular, *Nemcia* appears to be paraphyletic with respect to the monotypic genus *Jansonia*. Previously, *Jansonia* was placed as sister-taxon to *Brachysema* using convergent morphological characters which function in adaptation to bird-pollination. Permutation testing supports a sister-group relationship between *Jansonia* and the bird-pollinated species of *Nemcia*, and these appear to have a more distant relationship to *Brachysema*. Detailed mapping of floral characters on the cladogram suggests that these two bird-pollinated taxa have different autapomorphies (not counting convergent characters such as red petals), and that within *Brachysema*, floral morphology has progressively elaborated along diverging pathways. This last observation suggests that correlated convergence is not a problem in this group. However, independent evidence is needed to test this conclusion. It is clear from this study that the degree of corroboration of a phylogeny should be considered before evaluating adaptive hypotheses.

ACKNOWLEDGEMENTS

I am grateful to David Swofford for modifying PAUP to facilitate T-PTP testing. The curators of herbaria CBG, CANB and PERTH gave access to their collections. Dan Faith, Peter Linder, Duncan McLaughlin, John Trueman, Dick Vane-Wright, Peter Weston and an anonymous referee made helpful comments on the manuscript.

REFERENCES

ALBERCH, P. & GALE, E.A., 1985. A developmental analysis of an evolutionary trend: digital reduction in amphibians. *Evolution, 39*: 8–23.

ALBERCH, P., GOULD, S.J., OSTER, G.F. & WAKE, D.B., 1979. Size and shape in ontogeny and phylogeny. *Paleobiology, 5*: 296–317.

ALBERT, V.A., WILLIAMS, S.E. & CHASE, M.W., 1992. Carnivorous plants: phylogeny and structural evolution. *Science, 27*: 1491–1495.

AMBRIOS, V., 1988. Genetic basis for heterochronic variation. In M.L. McKinney (Ed.), *Heterochrony in Evolution*, 269–285. New York: Plenum Press.

ARCHIE, J.W., 1989. A randomization test for phylogenetic information in systematic data. *Systematic Zoology, 38*: 239–252.

ARMSTRONG, J.A., 1979. Biotic pollination mechanisms in the Australian flora – a review. *New Zealand Journal of Botany, 17*: 467–508.

ARMSTRONG, J.A., 1987. Pollination syndromes as generic determinants. *Australian Systematic Botany Society Newsletter, 53*: 54–59.

ARROYO, M.T.K., 1981. Breeding systems and pollination biology in Leguminosae. In R.M. Polhill & P.H. Raven (Eds), *Advances in Legume Systematics, (2)*, 723–769. Kew: Royal Botanic Gardens.

BENTHAM, G., 1864. *Flora Australiensis, 2*. London: Reeve & Co.

BMR PALAEOGEOGRAPHIC GROUP, 1990. *Australia: Evolution of a Continent*. Canberra: Bureau of Mineral Resources.

BROOKS, D.R. & McLENNAN, D.A., 1994. Historical ecology as a research programme: scope limitations and future. In P. Eggleton & R.I. Vane-Wright (Eds), *Phylogenetics and Ecology*, 1–27. London: Academic Press.

BOUGHTON, D.A., COLLETTE, B.B. & McCUNE, A.R., 1991. Heterochrony in jaw morphology of Needlefishes (Teleostei: Belonidae). *Systematic Zoology, 40*: 329–354.

BRUNDIN, L., 1976. A Neocomian chironomid and Podonominae-Aphroteniinae (Diptera) in the light of phylogenetics and biogeography. *Zoologica Scripta, 5*: 139–160.

CAMPBELL, K.S.W. & BARWICK, R.E., 1990. Paleozoic dipnoan phylogeny: functional complexes and evolution without parsimony. *Paleobiology, 16*: 143–169.

CHRISTIDIS, L. & SCHODDE, R., 1991. Relationships of Australo–Papuan songbirds – protein evidence. *Ibis, 133*: 277–285.

CODDINGTON, J.A., 1988. Cladistic tests of adaptational hypotheses. *Cladistics, 4*: 3–22.

CODDINGTON, J.A., 1992. Avoiding phylogenetic bias. *Trends in Ecology and Evolution, 7*: 68–69.

CODDINGTON, J.A., 1994. The roles of homology and convergence in studies of adaptation. In P. Eggleton & R.I. Vane-Wright (Eds), *Phylogenetics and Ecology*, 53–78. London: Academic Press.

COEN, E.S., 1991. The role of homeotic genes in flower development and evolution. *Annual Review of Ecology and Systematics, 42*: 241–279.

COEN, E.S. & MEYEROWITZ, E.M., 1991. The war of the whorls: genetic interactions controlling flower development. *Nature, 353*: 31–37.

CRANSTON, P.S. & HUMPHRIES, C.J., 1988. Cladistics and computers: a chironomid conundrum. *Cladistics, 4*: 72–92.

CREPET, W.L. & FRIIS, E.M., 1987. The evolution of insect pollination in angiosperms. In E.M. Friis, W.G. Chaloner & P.R. Crane (Eds), *The Origins of Angiosperms and their Biological Consequences*, 181–201. Cambridge: Cambridge University Press.

CRISP, M.D., 1982. Evolution and biogeography of *Leptosema* (Leguminosae: Papilionoideae). In W.R. Barker & P.J.M. Greenslade (Eds), *Evolution of the Flora and Fauna of Arid Australia*, 317–322. Adelaide: Peacock Publications.

CRISP, M.D. & WESTON, P.H, 1987. Cladistics and legume systematics, with an analysis of the Bossiaeeae, Brongniartieae and Mirbelieae. In C. Stirton (Ed.), *Advances in Legume Systematics, (3)*, 65–130. Kew: Royal Botanic Gardens.

CRISP, M.D. & WESTON, P.H. 1994. Mirbelieae. In M.D. Crisp & J.J. Doyle (Eds), *Advances in Legume Systematics, (7), Phylogeny*. Kew: Royal Botanic Gardens.

DETTMAN, M.E. & JARZEN, D.M., 1991. Pollen evidence for late Cretaceous differentiation of Proteaceae in southern polar forests. *Canadian Journal of Botany, 69*: 901–906.

DONOGHUE, M.J., 1989. Phylogenies and the analysis of evolutionary sequences, with examples from seed plants. *Evolution, 43*: 1137–1156.

DONOGHUE, M.J. & SANDERSON, M.J., 1992. The suitability of molecular and morphological evidence in reconstructing plant phylogeny. In P.S. Soltis, D.E. Soltis & J.J. Doyle (Eds), *Molecular Systematics of Plants*, 340–368. New York: Chapman & Hall.

DOYLE, J.J., 1992. Gene trees and species trees: molecular systematics as one-character taxonomy. *Systematic Botany, 17*: 144–163.

DOYLE, J.A. & DONOGHUE, M.J, 1992. Fossils and seed plant phylogeny reanalyzed. *Brittonia, 44*: 89–106.

EDWARDS, S.V., ARCTANDER, P. & WILSON, A.C., 1990. Mitochondrial resolution of a deep branch in the genealogical tree for perching birds. *Proceedings of the Royal Society of London, Series B, 243*: 99–107.

FAEGRI, K. & VAN DER PIJL, L., 1978. *The Principles of Pollination Ecology* (3rd edition). Oxford: Pergamon Press.

FAITH, D.P., 1989. Homoplasy as pattern: multivariate analysis of morphological convergence in Anseriformes. *Cladistics, 5*: 235–258.

FAITH, D.P., 1991. Cladistic permutation tests for monophyly and nonmonophyly. *Systematic Zoology, 40*: 366–375.

FAITH, D.P., 1992. On corroboration: a reply to Carpenter. *Cladistics, 8*: 265–273.

FAITH, D.P. & BELBIN, L., 1994. Distinguishing phylogenetic effects in multivariate models relating *Eucalyptus* convergent morphology to environment. In P. Eggleton & R.I. Vane-Wright (Eds), *Phylogenetics and Ecology*, 169–188. London: Academic Press.

FAITH, D.P. & CRANSTON, P.S., 1990. Could a cladogram this short have arisen by chance alone?: on permutation tests for cladistic structure. *Cladistics, 7*: 1–28.

FARRIS, J.S., 1977. Phylogenetic analysis under Dollo's Law. *Systematic Zoology, 26*: 77–88.

FARRIS, J.S., 1989a. Hennig86: a PC-DOS program for phylogenetic analysis. *Cladistics, 5*: 163.

FARRIS, J.S., 1989b. The retention index and the rescaled consistency index. *Cladistics, 5*: 417–419.

FITZHUGH, K., 1989. Cladistics in the fast lane. *Journal of the New York Entomological Society, 97*: 234–241.

FORD, H.A. & PATON, D.C. (Eds), 1986. *The Dynamic Partnership: Birds and Plants in Southern Australia*. Adelaide: Government Printer.

FORD, H.A., PATON, D.C. & FORDE, N., 1979. Birds as pollinators of Australian plants. *New Zealand Journal of Botany, 17*: 509–519.

FRITH, H.J. (Ed.), 1976. *The Reader's Digest Complete Book of Australian Birds*. Sydney: Reader's Digest.

FUTUYMA, D.J., 1986. Evolutionary Biology (2nd edition). Sunderland, MA: Sinauer.

GOLDBLATT, P., 1990. Status of the southern African *Anapalina* and *Antholyza* (Iridaceae) genera, based solely on characters for bird pollination, and a new species of *Tritoniopsis*. *South African Journal of Botany, 56*: 577–582.

GOTTLIEB, L.D., 1984. Genetics and morphological evolution in plants. *American Naturalist, 123*: 681–709.

GRANT, V., 1949. Pollination systems as isolating mechanisms in angiosperms. *Evolution, 3*: 82–97.

GROSS, C.L., 1992. Floral traits and pollinator constancy: foraging by native bees among three sympatric legumes. *Australian Journal of Ecology, 17*: 67–74.

GUERRANT, E.O., 1982. Neotenic evolution of *Delphinium nudicaule* (Ranunculaceae): a hummingbird-pollinated larkspur. *Evolution, 36*: 699–712.

HARVEY, P.H., 1991. Comparing uncertain relationships: the Swedes in revolt. *Trends in Ecology and Evolution, 6*: 38–39.

HARVEY, P.H. & PAGEL, M.D., 1991. *The Comparative Method in Evolutionary Biology.* Oxford: Oxford University Press.

HARVEY, P.H. & PURVIS, A., 1991. Comparative methods for explaining adaptations. *Nature, 351*: 619–623.

HERENDEEN, P.S., CREPET, W.L. & DILCHER, D.L., 1992. The fossil history of the Leguminosae: phylogenetic and biogeographic implications. In P.S. Herendeen & D.L. Dilcher (Eds), *Advances in Legume Systematics, (4), The Fossil Record*. Kew: Royal Botanic Gardens.

HOPPER, S.D. & BURBIDGE, A.H., 1986. Speciation of bird-pollinated plants in south-western Australia. In H.A. Ford & D.C. Paton (Eds), *The Dynamic Partnership: Birds and Plants in Southern Australia*, 20–31. Adelaide: Government Printer.

JOHNSON, L.A.S. & BRIGGS, B.G., 1985. Myrtales and Myrtaceae – a phylogenetic analysis. *Annals of the Missouri Botanical Garden, 71*: 700–756.

KÄLLERSJÖ, M., FARRIS, J.S., KLUGE, A.G. & BULT, C., 1992. Skewness and permutation. *Cladistics, 8*: 275–287.

KEAST, J.A., 1976. The origins of adaptive zone utilizations and adaptive radiations, as illustrated by the Australian Meliphagidae. *Proceedings of the 16th International Ornithological Congress, Canberra*, Australian Academy of Science, 71–82.

KEIGHERY, G.J., 1980. Bird pollination in south Western Australia: a checklist. *Plant Systematics and Evolution, 135*: 171–176.

KEIGHERY, G.J., 1982a. Bird-pollinated plants in Western Australia. In J.A. Armstrong, J.M. Powell and A.J. Richards (Eds), *Pollination and Evolution*, 77–89. Sydney: Royal Botanic Gardens.

KEIGHERY, G.J., 1982b. Notes on the biology and phytogeography of Western Australian plants, (13): ecological notes on the Fabaceae. Mimeographed report, 54 pp. Perth: Kings Park Board.

KEIGHERY, G.J., 1984. Pollination of Jansonia formosa Kipp. ex Lindl. (Papilionaceae). *Western Australian Naturalist, 16*: 21.

KLUGE, A.G., 1989. A concern for evidence and a phylogenetic hypothesis of relationships among *Epicrates* (Boidae, Serpentes). *Systematic Zoology, 38*: 7–25.

LANGE, R.T., 1981. Australian Tertiary vegetation. In J.M.B. Smith (Ed.), *A History of Australasian Vegetation*, 44–89. Sydney: McGraw Hill.

LE QUESNE, W., 1974. The uniquely evolved character concept and its cladistic application. *Systematic Zoology, 23*: 513–517.

LEPPIK, E.E., 1966. Floral evolution and pollination in Leguminosae. *Annales Botanici Fennici, 3*: 299–308.

LINDER, H.P., 1991. A review of the southern African Restionaceae. *Contributions from the Bolus Herbarium, 13*: 209–264.

LLOYD, D.G. & WELLS, M.S., 1992. Reproductive biology of a primitive angiosperm, Pseudowintera colorata (Winteraceae), and the evolution of pollination systems in the Anthophyta. *Plant Systematics and Evolution, 181*: 77–95.

LOCONTE, H. & STEVENSON, D.W., 1990. Cladistics of the spermatophyta. *Brittonia, 42*: 197–211.

McLOUGHLIN, S., 1993. Preliminary report on the late Eocene flora of the Kojonup Sandstone, southwestern Australia. In R.S. Hill (Ed.), *Abstracts, Southern Temperate Ecosystems: Origin and Diversification*, p. 68. Hobart: Southern Connection, Australian Systematic Botany Society, Ecological Society of Australia.

MINDELL, D.P., 1992. DNA–DNA hybridization and avian phylogeny. *Systematic Biology, 41*: 126–143.

NELSON, G., in press. Homology and systematics. In B.K. Hall (Ed.), *Homology: the Hierarchical Basis of Comparative Biology*. San Diego: Academic Press.

NELSON, G. & PLATNICK, N., 1981. *Systematics and Biogeography: Cladistics and Vicariance*. New York: Columbia University Press.

OLSON, S.L., 1988. Aspects of global avifaunal dynamics during the Cenozoic. In H. Ouellet (Ed.), *Acta XIX Congressus Internationalis Ornithologici, 2*, 2023–2029. Ottawa: University of Ottawa Press.

PAGEL, M., 1994. The adaptationist's wager. In P. Eggleton & R.I. Vane-Wright (Eds), *Phylogenetics and Ecology*, 29–52. London: Academic Press.

PAGEL, M.D. & HARVEY, P.H., 1988. Recent developments in the analysis of comparative data. *The Quarterly Review of Biology, 63*: 413–440.

PATON, D.C., 1986a. Honeyeaters and their plants in south-eastern Australia. In H.A. Ford & D.C. Paton (Eds), *The Dynamic Partnership: Birds and Plants in Southern Australia*, 9–19. Adelaide: Government Printer.

PATON, D.C., 1986b. Evolution of bird-pollination in Australia. In H.A. Ford & D.C. Paton (Eds), *The Dynamic Partnership: Birds and Plants in Southern Australia*, 32–41. Adelaide: Government Printer.

PATTERSON, C., 1982. Morphological characters and homology. In K.A. Joysey & A.E. Friday (Eds), *Problems of Phylogenetic Reconstruction*, 21–74. London: Academic Press.

PATTERSON, C., 1987. Introduction. In C. Patterson (Ed.), *Molecules and Morphology in Evolution: Conflict or Compromise?*, 1–22. Cambridge: Cambridge University Press.

PORSCH, O., 1927. Kritische Quellenstudien uber Blumenbesuch durch Vogel. III. *Biologia Generalis, 3*: 475–548.

RAVEN, P.H. & POLHILL, R.M., 1981. Biogeography of the Leguminosae. In R.M. Polhill & P.H. Raven (Eds), *Advances in Legume Systematics, (1)*, 27–34. Kew: Royal Botanic Gardens.

RIEDL, R., 1978. *Order in Living Organisms*. New York: Wiley.

RUSE, M., 1979. Falsifiability, consilience, and systematics. *Systematic Zoology, 28*: 530–536.

SARICH, V.M., SCHMID, C.W. & MARKS, J., 1989. DNA hybridization as a guide to phylogenies: a critical analysis. *Cladistics, 5*: 3–32.

SCHRIRE, B., 1989. A multidisciplinary approach to pollination biology in the Leguminosae. In C.H. Stirton & J.L. Zarucchi (Eds), *Advances in Legume Biology. Monographs in Systematic Botany from the Missouri Botanical Garden, 29*: 183–242.

SCHODDE, R. & TIDEMANN, S.C. (Eds), 1986. *The Reader's Digest Complete Book of Australian Birds* (2nd edition). Sydney: Reader's Digest.

SIBLEY, C.G. & AHLQUIST, J.E., 1985. The phylogeny and classification of the Australo-Papuan passerine birds. *Emu, 85*: 1–14.

SIBLEY, C.G. & AHLQUIST, J.E., 1990. *Phylogeny and Classification of Birds: a Study in Molecular Evolution*. New Haven, Conn.: Yale University Press.

SWOFFORD, D.L., 1991. PAUP: Phylogenetic Analysis Using Parsimony, Version 3.0s.

Computer program distributed by the Illinois State Natural History Survey, Champaign, Illinois 61820, U.S.A.

THOMPSON, J., 1990. New species and new combinations in the genus *Swainsona* (Faba-ceae) in New South Wales. *Telopea, 4*: 1–5.

TRUEMAN, J.W.H., 1993. Randomisation confounded: a response to Carpenter. *Cladistics, 9*: 101–109.

TUCKER, S.C., 1989. Evolutionary implications of floral ontogeny in legumes. In C.H. Stirton & J.L. Zarucchi (Eds), *Advances in Legume Biology. Monographs in Systematic Botany from the Missouri Botanical Garden, 29*: 59–75.

WANNTORP, H-E., BROOKS, D.R., NILSSON, T., NYLIN, S., RONQUIST, F., STEARNS, S.C. & WEDELL, N., 1990. Phylogenetic approaches in ecology. *Oikos, 57*: 119–132.

WENZEL, J. & CARPENTER, J., 1994. Comparing methods: adaptive traits and tests of adaptation. In P. Eggleton & R.I. Vane-Wright (Eds), *Phylogenetics and Ecology*, 79–101. London: Academic Press.

APPENDIX 1

List of characters and their states as used in the cladistic analysis. Those marked with an asterisk are putatively related to bird-pollination, and were deleted for further analysis (Fig. 6).

1   Habit: scandent or semi-scandent (1), shrubby (0)
2   Phyllotaxis: opposite or whorled (1), alternate (0)
3   Stipules: terete or angular (1), flat (0)
4   Accessory inflorescence shoots: present (1), absent (0)
5   Inflorescence: condensed, *Nemcia*-type (1), otherwise (0)
6* Unit inflorescence, no. of flowers: 1–3, rarely more (1), 4 or more, rarely 2–3 (0)
7* Inflorescence capitate with an involucre: present (1), absent (0)
8   Inflorescence: predominantly axillary (1), terminal and anauxotelic (0)
9* Flower nutant on a filiform peduncle: present (1), absent (0)
10* Flower orientation: resupinate (1), normal (0)
11* Bract aestivation: valvate in bud (1), imbricate or scattered (0)
12 Bract persistence: caducous (1), persistent (0)
13 Basal bracts: bract-like no lamina (1), leaf-like with lamina (0)
14* Floral morphology: '*sericeum*' type (1), otherwise (0)
15* Calyx dorsally ventricose: present (1), absent (0)
16 Hairs of calyx-lobes: golden or rusty, usually silver on tube (1), uniformly silver to brown (0)
17* Standard lamina: subulate constricted and channelled (3), constricted to a waist above auricular base (2), narrowed and channelled (1), orbicular and not channelled (0), subulate and not channelled nor constricted {03}
18* Standard/keel, relative length: <0.85 (1), *c.* 1.0 (0)
19* Wings/keel, relative length: shorter (1), equal or longer (0)
20* Wing auricles: elongated (1), not elongated (0)
21* Keel length: elongate, 30 mm or longer (1), shorter, <25 mm (0)
22* Keel shape: half narrow-elliptic, acute (1), oblique, *c.* obtuse (0)
23* Petal colours: reddish tending to green, no yellow (3), all red with no yellow

(2), predominantly red with yellow markings (1), predominantly yellow with red markings (0)

24* Ovule number: 2–9 (2), 10–14 (1), >14 (0)

25* Disc: raised (1), not raised or absent (0)

26 Pod indumentum: sparse (1), moderate to dense, sericeous to villous (0)

27 Aril shape: horseshoe-shaped (1), downwardly lobed or absent (0)

APPENDIX 2

Terminal taxa included in the cladistic analysis and (where relevant) their subordinate species, with author citations; * = terminal taxa in primary analysis, † = informal species groups into which 'Nemcia insect-pollinated' group was subdivided for further analysis.

Brachysema bracteolosum F. Muell., *B. celsianum* Lemaire, *B. latifolium* R. Br., *B. melanopetalum* F. Muell., *B. minor* Crisp ined., *B. praemorsum* Meissner, *B. sericeum* (Smith) Domin, *B. subcordatum* Benth.

Callistachys lanceolata Vent.

Jansonia formosa Kippist

Nemcia 'bird-pollinated' group: *N. leakeana* (J. Drumm.) Crisp, *N. rubra* Crisp, *N. vestita* Domin

Nemcia 'bee-pollinated' group (this was subdivided into the following groups for further analysis)

  †*N.* '2-ovules' group (2 ovules per ovary and uniform calyx-hairs): *N. acuta* (Benth.) Domin, *N. carinata* Crisp, *N. epacridoides* (Meissner) Crisp, *N. hookeri* (Meissner) Crisp, *N. ilicifolia* (Meissner) Crisp, *N. lehmannii* (Meissner) Crisp, *N. obovata* (Benth.) Crisp, *N. plicata* (Turcz.) Crisp, *N. pulchella* (Turcz.) Crisp, *N. punctata* (Turcz.) Crisp, *N. spathulata* (Benth.) Crisp, *N. stipularis* (Meissner) Crisp

  †*N. alternifolia* Crisp ined. (alternate leaves)

  †*N.* '2-tone' group (4 or more ovules per ovary and two-tone calyx-hairs): *N. coriacea* (Smith) Domin, *N. dilatata* (Benth.) Crisp, *N. emarginata* (S. Moore) Crisp, *N. retusa* (Lindley) Domin

  †*N.* '2-tone/2-ovules' group (2 ovules and two-tone calyx-hairs): *N. crenulata* (Turcz.) Crisp, *N. pyramidalis* (T. Moore) Crisp

  †*N.* capitata group (accessory inflorescence shoots present): *N. axillaris* (Meissner) Crisp, *N. capitata* (Benth.) Domin, *N. reticulata* (Meissner) Domin

Oxylobium lineare (Benth.) Benth.

On the use of discrete characters in phylogenetic trees with special reference to the evolution of avian mating systems

BIRGITTA SILLÉN-TULLBERG & HANS TEMRIN

CONTENTS

Keywords: Bird young – feeding – mating pattern – bond length.

Abstract

A convergence approach was used to study the evolution of avian mating systems, components of which were treated as discrete characters in an extensive phylogeny (227 taxa). The sample of taxa was unbiased with respect to mating systems apart from the passerines being excluded, and we used a statistical method to investigate whether transitions to polygamy and short pair bonds are independent of the ability of young to feed by themselves. Both female polygamy and short pair bond were more likely to originate on branches where young were inferred to be self-feeding, than on branches where they were inferred to be fed, and these relationships would be even stronger had the passerines been included. It is difficult, however, to draw conclusions concerning the relationship between origins of male polygamy and the feeding ability of the young until the passerines can be included. In general, major alterations to the phylogeny are not likely to affect our main conclusions because, in general, transitions in the young-feeding character are usually inferred for basal branches, and transitions in mating system in apical (often terminal) branches.

INTRODUCTION

The evolution of avian mating systems has received considerable attention (e.g. Selander, 1972; Emlen & Oring, 1977; Wittenberger & Tilson, 1980; Oring, 1983; Clutton-Brock, 1991; Ligon, 1993), but few attempts have actually been made to extensively analyse the phylogenetic history of mating systems (but see McKitrick, 1992). Birds are an unusually 'balanced' group of animals with respect to mating systems (Mock, 1983) and an extensive phylogenetic analysis could greatly improve our understanding of the circumstances under which different mating systems evolve. One factor that is likely to be important in the evolution of mating systems is the characteristics of the young, because the degree of care necessary for the young will influence the possibility for parents to invest in mating instead of parental care (Orians, 1969; Clutton-Brock, 1991). Thus, if the young are relatively independent, there is a greater possibility for one of the parents to solely care for them, which would give the other an opportunity to invest in other activities, such as mating with several partners.

There are various ways in which phylogenetic information can be used for an increased understanding of the evolution of various traits (see Brooks & McLennan, 1991; Coddington, Chapter 3; Nylin and Wedell, Chapter 12; Pagel, Chapter 2). We use the investigation of convergence, the 'homoplasy approach' (Coddington, Chapter 3), and hold that directional methods (see Nylin & Wedell, Chapter 12) are necessary for inferences about causation, especially for the possibility to refute hypotheses about causation (e.g. Sillén-Tullberg, 1988; Sillén-Tullberg & Möller, 1993), because the relative timing of events can be inferred using such methods. Non-directional methods are more fruitful for studying correlation between coevolving traits (i.e. where two traits influence each other), or between traits for which the question of causation may be solved on other grounds.

The analysis in this paper is part of a larger investigation of the evolution of avian mating systems (Temrin & Sillén-Tullberg, in press; Temrin & Sillén-Tullberg, unpublished), and here we specifically address the question of whether mating systems are related to the need of the young to be fed or not. Thus, the first thing to consider is the relative timing of transitions in characters related to mating systems, that is to polygamy and to a short bond between the sexes, on one hand, and transitions in the requirement of the young on the other. As it turns out, changes in the latter character generally takes place in more basal branches, that is before changes in characters related to mating systems. We may then ask whether transitions to polygamy and short bonds are associated with a particular state of the young, and this question requires a statistical test and an appropriate sample (see Sillén-Tullberg, 1993) for such a test.

For various reasons the literature on statistical treatment of continuous characters, usually non-directional methods, is far more extensive (e.g. Huey & Bennett, 1987; Felsenstein, 1988; Grafen, 1989, 1992; Losos, 1990; Harvey & Pagel, 1991; Martins & Garland, 1991; Garland, Harvey & Ives, 1992; Björklund, unpublished), than that of discrete characters, for which three methods have been presented (Ridley, 1983; Maddison, 1990; Sillén-Tullberg, 1993). Ridley's method deals with independent cases of associations between two traits, without considering the temporal sequence of events, whereas Maddison (1990) and Sillén-Tullberg (1993) start from considering optimizations of two characters and the relative timing of transitions in these characters. Thus, both of these methods are appropriate for dealing with a question of whether transitions to polygamy are more likely on branches for which one of the feeding states of the young is inferred. Although these two methods, when applied to the same real (Sillén-Tullberg, 1993) or simulated (Werdelin & Sillén-Tullberg, unpublished) data sets, give similar results under certain conditions, the concentrated changes test of Maddison (1990) is sensitive to tree topology and, through considering character reconstructions not allowed by parsimony, may increase type I error rate (Werdelin & Sillén-Tullberg, unpublished). Thus, in this paper we apply the contingent states test (Sillén-Tullberg, 1993) to the bird data.

MATERIAL AND METHODS

The characters

The classical definitions of mating systems often include many aspects of behaviour, such as the number of sexual partners, the way these are acquired, the characteristics of the pair bond and care of young etc. Some definitions include functional aspects as well as implications about how the mating system has evolved (Emlen & Oring, 1977). One can easily see the difficulty with using this large array of mating systems in a cladistic study – one would end up with a character containing a great number of states, each of which really is a mixture of several traits.

In an extensive phylogenetic analysis of different mating systems in non-passerine birds (Temrin & Sillén-Tullberg, in press), we have differentiated between male and female mating pattern, defined as the frequency of individuals having several partners, and the length of the pair bond between the sexes. In that study, mating pattern was treated as a four-state character and pair bond as having five states. In the present study we regard each of these characters as having two states. For female and male mating pattern, respectively, these states are monogamy and polygamy, with polygamy being defined as 5% or more of mated individuals having more than one partner (e.g. Möller, 1986). For pair bond, the two states are short bond, defined as a bond that dissolves before hatching, and long bond, that involves at least some care of the young. All three characters are related to one breeding season, which means that, in this paper, no distinction has been made between seasonal monogamy and monogamy that lasts for more than one breeding season.

The fourth character is related to a characteristic of the young, namely whether they are being fed or are self-feeding. The three states in this character are defined as follows, with day 1 being the day of hatching: (1) fed day 7 or later; (2) fed day 2–6; and (3) self-feeding from start.

The phylogeny and sampling of taxa

The most extensive phylogenetic study of birds is that by Sibley & Ahlquist (1990), based on DNA hybridization, which we have used as a basis for a cladistic analysis of characters related to mating systems. The methods used in Sibley & Ahlquist (1990) have been subject to much criticism (e.g. Cracraft, 1987; Sarich, Schmid & Marks, 1989; O'Hara, 1991; Mindell, 1992). However, we will discuss whether and how the use of alternative phylogenies would affect our conclusions.

The detailed description of the working procedure, the taxa included and references are found in Temrin & Sillén-Tullberg (in press). We have made an extensive literature search for information on mating systems; and we have sampled all taxa for which we have obtained detailed information on the components of mating systems and which are also included in the Sibley & Ahlquist (1990) phylogeny. This is likely to give an unbiased sample with regard to mating systems, since Sibley & Ahlquist (1990) have not chosen their taxa with respect to this behavioural feature. As a next step we have gathered data on the characteristics of the young for these taxa.

Generally, the terminal taxa in our study have been genera and we have assumed that genera are monophyletic groups. However, in cases where there is variation within a genus in one or more of the characters of interest, we have treated this genus as a soft polytomy (dichotomous branching assumed but unknown).

The passerines have not been subject to an extensive analysis, mainly due to a lack of data on a broad range of groups. However, we have made a tentative mapping of the occurrence of different mating patterns in the passerine families and in the Discussion we evaluate whether our conclusions concerning the non-passerines are likely to hold were the passerines to be included as well.

Character optimizations

The characters were optimized using MacClade 3.0. (Maddison & Maddison, 1992). For the three characters related to mating system it was difficult to establish the outgroup conditions, because not enough is known about the mating systems in Crocodylia (J.P. Ross, personal communication). Consequently these characters (but not the characters related to the young) were optimized without using an outgroup. Although the characters related to the young may be regarded as ordered, we have optimized them in an unordered fashion. This is so because we were interested in transition sequences *per se*, so that it would be inappropriate to make assumptions concerning order. After optimizations the characters have been compared to each other with regard to transitions to various states.

The contingent states test

The purpose of this test is to see whether a certain transition in one (dependent) character is related to a certain state of another (independent) character, and is fully described in Sillén-Tullberg (1993). The test differs from the concentrated changes test (Maddison, 1990) in that origins (gains) and reversals (losses) in a character are tested separately instead of being mixed in one test. The test can also be extended to investigate contrasts in the independent character (Sillén-Tullberg & Hunter, unpublished).

Consider two characters, each of which has two states. These may be young-feeding and pair bond, as in this chapter. When both characters have been optimized we may ask whether the presence of a certain state, self-feeding, in the character regarded as independent, increases the probability of a certain state transition, from long to short bonds, in the character regarded as dependent. In the dependent character which is to be traced 'over' the independent character, there are four things that can happen on a branch or internode: maintenance of the state long bond; transition from long to short bond; maintenance of short bond; and transition from short to long bond. We need to keep track of all events in the dependent character over the whole tree, and then specifically with reference to the two-character states of the independent character. The total number of events should sum up to the number of internodes or branches.

Since we ask about the probability of a transition from long to short pair bond we are especially interested in all the branches that start with a long bond. This is so because only these branches (that is branches with the maintenance of the long bond and with a transition from long to short bond) carry a potential for such a transition. The aim with the method is to compare the two states of the independent character, fed and self-feeding, with regard to the proportion of branches with a transition from long to short bond. This proportion, the conditional probability of state transition long→short, is simply calculated as long→short/[(long→long) + (long→short)]. The number of origins, long→short, and number of branches with no change, long→long, for branches with fed and self-feeding, are compared in a 2×2 table under the null hypothesis that a short pair bond is equally likely to originate on these two states of the independent character.

The internodes, with a maintenance of the short bond and with a reversal from short to long bond, are regarded as occupied, that is there is no possibility for a transition from long to short bond. These events are left out of an analysis dealing with origins of a short bond, but can be used, for instance, when dealing with a problem related to maintenance and stability of this state.

In this paper there are three analyses, one for each of the characters related to mating system as the dependent character, and each with young-feeding as the independent character. For simplicity, the branches carrying the state 'fed day 2–6' were pooled with those carrying the state 'self-feeding', and this pooling has little effect on the results since there are few branches with this 'intermediate' state. Moreover, we used one of two extreme optimizations, where equivocal branches are assumed to carry the state self-feeding. Maximizing the number of such branches bias against the hypothesis of polygamy and short bonds originating more frequently in lineages where the young are self-feeding. For each of the three dependent characters, female mating pattern, male mating pattern and pair bond, both extreme optimizations were used with regard to equivocal branches.

RESULTS

Mating pattern and pair bond optimizations

For illustrative purposes we have compressed the avian phylogeny to 13 monophyletic groups and shown the number of taxa included in the present study (Fig. 1). A more extensive phylogeny is shown in Temrin and Sillén-Tullberg (in press). The mating pattern inferred for the overwhelming majority of branches in our phylogeny

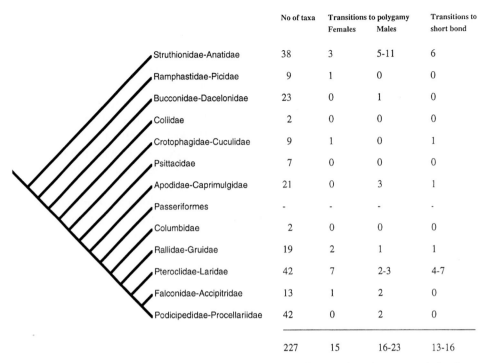

	No of taxa	Transitions to polygamy		Transitions to short bond
		Females	Males	
Struthionidae-Anatidae	38	3	5-11	6
Ramphastidae-Picidae	9	1	0	0
Bucconidae-Dacelonidae	23	0	1	0
Coliidae	2	0	0	0
Crotophagidae-Cuculidae	9	1	0	1
Psittacidae	7	0	0	0
Apodidae-Caprimulgidae	21	0	3	1
Passeriformes	-	-	-	-
Columbidae	2	0	0	0
Rallidae-Gruidae	19	2	1	1
Pteroclidae-Laridae	42	7	2-3	4-7
Falconidae-Accipitridae	13	1	2	0
Podicipedidae-Procellariidae	42	0	2	0
	227	15	16-23	13-16

Figure 1 Overview of 13 monophyletic groups of birds in the tree based on DNA hybridization data (Sibley & Ahlquist, 1990). For information concerning the taxa that belong to each group, we refer to Temrin & Sillén-Tullberg (in press). The figure shows the number of taxa included in the present study, the number of transitions from monogamy to polygamy inferred for females and males, respectively, and the number of inferred transitions from long to short pair bond. See text for definition of character states.

is monogamy, both for females and males. Likewise, the long bond is expressed by most taxa and inferred for the majority of branches in the tree. However, transitions from monogamy to polygamy for females and males, respectively, as well as transitions to short bond have taken place a number of times (Fig. 1). As can be seen such transitions are concentrated within certain monophyletic groups. Thus, there is a preponderance of male polygamy in the group Struthionidae–Anatidae, which includes the ratites, the gallinaceous birds and the anserine birds. Moreover, there is a concentration of inferred origins of female polygamy in the group Pteroclidae–Laridae, which consists of the sandgrouse and shorebirds. Also, there is a concentration of origins of short bonds in these two groups. Generally, both male and female polygamy originates on branches where the other sex is inferred to be monogamous, the two exceptions being the ratites and *Alectura* in the family Megapodidae (Temrin & Sillén-Tullberg, in press).

Young-feeding optimizations

Figure 2 shows the distribution of various states in the character young-feeding. For illustrative purpose the bird phylogeny has been resolved so that all transitions between states can be observed at lower levels than the 13 monophyletic groups depicted in Fig. 1.

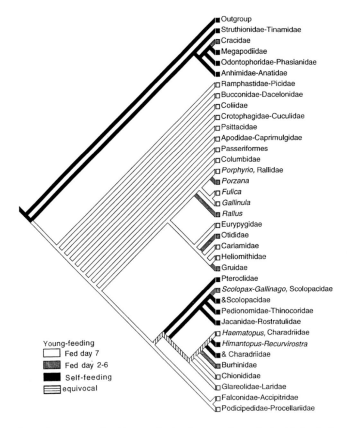

Figure 2 The character young-feeding optimized over the whole bird phylogeny, which has been resolved (from Fig. 1) to show transitions between various states.

There are 12 steps required to explain the present-day distribution of states of the character young feeding, with the outgroup condition being 'self-feeding' (Crocodylia). Depending on equivocal branches there are one to four transitions to 'self-feeding', and one to four transitions to the state 'fed day 7' or later. There are seven transitions to the intermediate feeding state.

The relationship between mating system characters and young-feeding

The proportion of branches with transitions to female polygamy, male polygamy and short bonds is higher for branches labelled 'self-feeding' (including the intermediate state, see Methods) than those labelled 'fed' (Table 1). The relationships in Table 1 are significant for all character optimizations in the analyses using young-feeding as the independent character.

Two types of female polygamy in relation to young-feeding

There is an expectation that there should be more cases of male than female polygamy, which stems from a general expectation that males should try to maximize the number of fertilized eggs (e.g. Krebs & Davies, 1981). Why, then, does

Table 1 The proportion of branches with transitions from monogamy to polygamy for females and males, and from long to short pair bond. Proportions are based on all branches where a transition could potentially take place (see text for explanation). Proportions are related to the state of the character young-feeding, where branches with the intermediate state 'fed day 2–6' are included under 'self-feeding'. P-values (two-tailed) from Contingent states test (see text). Ranges in proportions and P-values due to alternative character optimizations. Sample sizes (number of branches) within parenthesis.

Young-feeding	female polygamy	male polygamy	short pair bond
		Proportion of branches with transitions to:	
Fed	0.02 (289)	0.03 (283)	0.01 (277)
	$P = 0.000$	$P = 0.001 - 0.037$	$P = 0.000$
Self-feeding	0.09 (108)	0.08–0.12 (98–126)	0.13–0.14 (88–102)

female polygamy evolve relatively frequently in our sample of birds? More specifically, what does a female bird gain from being polygamous?

Since female polygamy invariably involves male care of the young, we can see two different benefits: (1) the female can lay more eggs; or (2) the offspring receive more parental care. The first benefit is likely to be realized when a female lays several clutches (either in succession or 'simultaneously'), one clutch for each male. The second benefit is likely to be realized when polygamy takes the form of cooperation, that is when several males help the female with the same clutch (see e.g. Davies & Houston, 1986). Since the second benefit is likely to be more important for young that need to be fed, and hence which need more parental care, we tested whether transitions to cooperative female polygamy are equally frequent when the young need to be fed as when they are self-feeding. In this analysis we included the only case of female polygamy that we have found in passerine taxa that is also included in Sibley & Ahlquist (1990). The frequency of transitions to cooperative polygamy was significantly higher among taxa where the young are being fed than where they are self-feeding ($P = 0.001$, $N = 16$, Fisher Exact Test). It should be mentioned that we have found only five origins of cooperative female polygamy, all with the young being fed, using taxa included by Sibley & Ahlquist (1990) (*Porphyrio porphyrio* and *Gallinula tenebrosa*, both in Rallidae, *Buteo galapagoensis*, *Melanerpes foramicivorus* and one case in the passerines, *Prunella modularis*), and that the majority of cases of female polygamy is of a non-cooperative kind, where the male alone may take care of the young. Thus, it is likely that most cases of female polygamy involve the benefit of higher egg production, not of increased care per young.

DISCUSSION

This study gives support to the general notion of polygamy being related to the young characteristics. Thus, we found that origins of both polygamy and short bonds were significantly concentrated on branches where the young are inferred to be self-feeding. Moreover, we found that transitions to short pair bonds, implying that one parent takes care of the young, were also significantly concentrated on such branches.

We may then ask whether these relationships would hold if we could include the passerines, all of which feed their young, in an extensive analysis of all birds. Given the great number of branches in this group, in addition to the few instances of female polygamy, we would guess that the relationship between origin of female polygamy and young-feeding would be even stronger. The same may apply to transitions to short bonds, since short bonds appear to be rare among the passerines (Lack, 1968). The question then concerns male polygamy which appears relatively frequently in the passerines (Lack, 1968; Möller, 1986). Although there are many branches in this group on which male polygamy could potentially evolve, the question is whether the effect of the amount of branches outweighs the effect of the number of polygamy origins. The question cannot be resolved until an extensive study is made on this group of birds. Thus, the relationship between male polygamy and feeding the young remains obscure.

As mentioned in Methods (p. 314), the Sibley & Ahlquist (1990) bird phylogeny has been subject to much criticism, and an important question is how alternative phylogenies would affect our results. We have optimized our characters related to mating system and found that replacing the original (Sibley & Ahlquist, 1990) with alternative trees (for instance the ratites; Cracraft, 1974; Bledsoe, 1988; Mindell, 1992) will affect the inference concerning some of the ancestral states for this clade, and that using alternative reconstructions concerning the basal branches for all birds (see Cracraft and Mindell, 1989) will affect inferences concerning ancestral states for all birds (Sillén-Tullberg & Temrin, unpublished). However, the question of ancestral states cannot be resolved until we have better knowledge of the outgroup states in Crocodylia for the mating characters. The importance of the outgroup is also obvious from the fact that optimizing the young-feeding character using alternative reconstructions of the basal branches will not affect the ancestral state (self-feeding).

The main question is, however, how our results with regard to the relationship between mating system and young-feeding would be affected, should Sibley & Ahlquist (1990) be wrong concerning the relationships among higher groups such as families. In a reconstruction highly deviating from that of Sibley & Ahlquist (1990), transitions would still take place in basal branches relative to character transitions for mating systems which generally take place within families, often in terminal taxa. Thus, we contend that our main conclusions concerning the number and relative position of transitions in these characters would not be affected with alternative phylogeny reconstructions.

Today there is no alternative phylogeny to Sibley & Ahlquist (1990) that includes all the major taxa, but there are several alternative phylogenies for various monophyletic groups within the large phylogeny, for instance anserine birds (Livezey, 1986; Kessler & Avise, 1984), and parrots (Christidis et al., 1991). How would a replacement of Sibley & Ahlquist's (1990) resolutions for such groups affect our results? One likely effect would be an increase in the total number of transitions in mating system characters because detailed phylogenies for smaller groups include more taxa. However, a replacement of certain monophyletic groups would confer a bias toward such groups, and one of the main purposes using the Sibley & Ahlquist (1990) phylogeny is to have an unbiased sample of taxa with respect to all birds.

At last, as behavioural ecologists appreciating the great utility of phylogenetic information in our research, we want to express our concern about using the work of systematists. We would of course prefer a situation where we could pick phylogenies for those groups of organisms that we are interested in, choosing the ones that according to us used the best methods and were substantiated by the best data. One problem is that for many groups there simply are no existing phylogenies at all, and when there are published phylogenies, there is often disagreement among systematists about their value. In our view behavioural ecologists should, rather than disregard existing phylogenies because they are controversial, use them with care and check how alterations would affect conclusions. Thus, the question of whether a phylogeny is 'good enough' for comparative studies (Coddington, 1992) must be related to the characters that we study and to the questions we ask.

ACKNOWLEDGEMENTS

We thank two anonymous referees for comments on the manuscript. This study was supported by the Swedish Natural Science Research Council.

REFERENCES

BLEDSOE, A.H., 1988. A phylogenetic analysis of postcranial skeletal characters of the ratite birds. *Annals of Carnegie Museum, 57*: 73–90.

BROOKS, D.R., & McLENNAN, D.A., 1991., *Phylogeny, Ecology, and Behavior.* Chicago: The University of Chicago Press.

CHRISTIDIS, L., SCHODDE, R., SHAW, D.D., & MAYNES, S.F., 1991. Relationships among the Australo-Papuan parrots, lorikeets, and cockatoos (Aves: Psittaciformes): protein evidence. *The Condor, 93*: 302–317.

CLUTTON-BROCK, T.H., 1991. *The Evolution of Parental Care.* Princeton, NJ: Princeton University Press.

CODDINGTON, J.A., 1992. Avoiding phylogenetic bias. *Trends in Ecology and Evolution, 7*: 68–69.

CODDINGTON, J.A., 1994. The roles of homology and convergence in the studies of adaptation. In P. Eggleton & R.I. Vane-Wright (Eds), *Phylogenetics and Ecology*, 53–78. London: Academic Press.

CRACRAFT, J., 1974. Phylogeny and evolution of the ratite birds. *Ibis 116*: 494–521.

CRACRAFT, J., 1987. DNA hybridization and avian phylogenetics. *Evolutionary Biology 21*: 47–96.

CRACRAFT, J. & MINDELL, D.P., 1989. The early history of modern birds: a comparison of molecular and morphological evidence. In B. Fernholm, K. Bremer & H. Jörnvall (Eds), *The Hierarchy of Life*, 389–403. Amsterdam: Elsevier.

DAVIES, N.B. & HOUSTON, A.I., 1986. Reproductive success of dunnocks, *Prunella modularis*, in a variable mating system. II. Conflicts of interest among breeding adults. *Journal of Animal Ecology, 55*: 139–154.

EMLEN, S.T. & ORING, L.W., 1977. Ecology, sexual selection, and the evolution of mating systems. *Science, 197*: 215–223.

FELSENSTEIN, J., 1988. Phylogenies and quantative characters. *Annual Review of Ecology and Systematics, 19*: 445–471.

GARLAND, T. JR., HARVEY, P.H., & IVES, A.R., 1992. Procedures for the analysis of comparative data using phylogenetically independent contrasts. *Systematic Biology, 41*: 18–32.

GRAFEN, A., 1989. The phylogenetic regression. *Philosophical transactions of the Royal Society of London, Series B, 326*: 119–157.

GRAFEN, A., 1992 The uniqueness of the phylogenetic regression. *Journal of Theoretical Biology, 156*: 405–423.

HARVEY, P.H. & PAGEL, M.D., 1991., *The Comparative Method in Evolutionary Biology*. Oxford: Oxford University Press.

HUEY, R.B. & BENETT, A.F., 1987. Phylogenetic studies of coadaptation: preferred temperatures versus optimal performance temperatures of lizards. *Evolution, 41*: 1098–1115.

KESSLER, L.G. & AVISE, J.C., 1984., Systematic relationships among waterfowl (Anatidae) inferred from restriction endonuclease analyse of mitochondrial DNA. *Systematic Zoology, 33*: 370–380.

KREBS, J.R. & DAVIES, N.B., 1981. *An Introduction to Behavioural Ecology*. Oxford: Blackwell Scientific.

LACK, D., 1968. *Ecological Adaptations for Breeding in Birds*. London: Chapman and Hall.

LIGON, J.D., 1993. The role of phylogenetic history in the evolution of contemporary avian mating and parental care systems. In *Current Ornithology*, D.M. Power (Ed.), 1–46. Plenum Press, New York.

LIVEZEY, B.C., 1986. A phylogenetic analysis of recent anseriform genera using morphological characters. *Auk, 103*: 737–754.

LOSOS, J.B., 1990. Ecomorphology, performance capability, and scaling of West Indian *Anolis* lizards: an evolutionary analysis. *Ecological Monographs 60*: 369–388.

MADDISON, W.P., 1990. A method for testing the correlated evolution of two binary characters: are gains or losses concentrated on certain branches of a phylogenetic tree? *Evolution, 44*: 539–557.

MADDISON, W.P. & MADDISON, D.R., 1992. MacClade: Analysis of phylogeny and character evolution. Version 3.0. Sunderland, Massachusetts: Sinauer.

MARTINS, E.P. & GARLAND, T. JR., 1991. Phylogenetic analyses of the correlated evidence of continuous characters: a simulation study. *Evolution, 45*: 534–557.

McKITRICK, M.C., 1992. Phylogenetic analysis of avian parental care. *Auk 109*: 828–846.

MINDELL, D.P., 1992. DNA–DNA hybridization and avian phylogeny. *Systematic Biology 41*: 126–134.

MOCK, D.W., 1983. On the study of avian mating systems. In A.H. Brush & G.A. Clark, Jr. (Eds), *Perspectives in Ornithology*, 55–91. Cambridge: Cambridge University Press.

MÖLLER, A.P., 1986. Mating systems among European passerines: a review. *Ibis, 128*: 234–250.

NYLIN, S. & WEDELL, N., 1994. Sexual size dimorphism and comparative methods. In P. Eggleton & R.I. Vane-Wright (Eds.) *Phylogenetics and Ecology*, 253–280. London: Academic Press.

O'HARA, R.J., 1991., Phylogeny and classification of birds: a study in molecular evolution. *Auk, 108*: 990–994.

ORIANS, G.H., 1969. On the evolution of mating systems in birds and mammals. *American Naturalist 103*: 589–603.

ORING, L.W., 1983. Avian polyandry. In R.F. Johnston (Ed.), *Current Ornithology, 3*: 309–351.

PAGEL, M., 1994. The adaptionist wager. In P. Eggleton & R.I. Vane-Wright (Eds), *Phylogenetics and Ecology*, 29–51. London: Academic Press.

RIDLEY, M., 1983. *The Explanation of Organic Diversity: The Comparative Method and Adaptations for Mating*. Oxford: Clarendon Press.

SARICH, V.M., SCHMID, C.W. & MARKS, J., 1989. DNA hybridization as a guide to phylogenies: A critical analysis. *Cladistics, 5*: 3–32.

SELANDER, R.K., 1972. Sexual selection and dimorphism in birds. In B. Campbell (Ed.), *Sexual Selection and the Descent of Man*, 180–230. London: Heinemann.

SIBLEY, C.G. & AHLQUIST, J.E., 1990. *Phylogeny and Classification of Birds.* New Haven, CT: Yale University Press.

SILLÉN-TULLBERG, B., 1988. The evolution of gregariousness in aposematic butterfly larvae: a phylogenetic analysis. *Evolution, 42*: 293–305.

SILLÉN-TULLBERG, B., 1993. The effect of biased inclusion of taxa on the correlation between discrete characters in phylogenetic trees. *Evolution, 47*: 1182–1191.

SILLÉN-TULLBERG, B. & MÖLLER, A.P., 1993. The relationship between concealed ovulation and mating systems in anthropoid primates: a phylogenetic analysis. *The American Naturalist, 141*: 1–25.

TEMRIN, H. & SILLÉN-TULLBERG, B., in press. The evolution of avian mating systems: A phylogenetic analysis of male and female polygamy and length of pair bonds. *The Biological Journal of Linnean Society.*

WITTENBERGER, J.F. & TILSON, R.L., 1980. The evolution of monogamy: hypotheses and evidence. *Annual Review of Ecology and Systematics, 11*: 197–232.

FRANCIS GILBERT, GRAHAM ROTHERAY, PAUL EMERSON
& REHENA ZAFAR

CONTENTS

Keywords: Morphology – specialization – predators – predatory behaviour – Syrphidae.

Abstract

We analyse the evolution of feeding strategies in the Syrphidae (Diptera), an insect group containing phytophages, saprophages, carnivores and even ectoparasitoids. To consider their evolution, we first develop a generic phylogeny for the family based on larval characters, and discuss the evolution of larval ecomorphology. We then focus on the predators, choosing the aphidophagous Syrphinae for detailed study. From their generic phylogeny we see a trend from feeding in the herb layer to feeding on arboreal aphid colonies. A trend from plesiomorphic generalists to apomorphic specialists is suggested, but not supported by the admittedly poor-quality literature data.

Finally, we consider the evolution of species in two closely related plesiomorphic genera of predators. After estimating the phylogeny from adult characters, we use

Phylogenetics and Ecology
ISBN 0-12-232990-2

Lynch's comparative method to analyse the evolution of behavioural, ecological and morphological characters. There is evidence for a small phylogenetic component to ecological measures (prey range, number of generations per year, growth rates and abundance). Predictions about the directions of evolutionary change in behavioural characters were all supported by the evidence (for measures of casting rate, capture rate, capture efficiency, handling time, and response to starvation). Morphology is strongly constrained by phylogenetic relatedness: most evolutionary change is in size and its associated allometry.

INTRODUCTION

Much of the work of community ecologists during the last 40 years has involved inferring causes of ecological differences between related species from morphological, ecological or behavioural data. This has been an attempt to explain the so-called 'problem of species coexistence'. The whole field of ecomorphology involves the mapping of morphological onto ecological differences (Karr & James, 1975), often for single variables, but increasingly using multivariate techniques (e.g. Miles & Ricklefs, 1984; Gilbert, 1985a). Implicit in most of these studies is the assumption of equilibrium, (that is that the observed patterns reflect the product of current measurable forces. The morphology of each species is assumed to be moulded independently so as to reflect the forces of natural selection, acting via ecological and behavioural features, with the main result of producing an integrated community. Throughout the early part of this research programme, explanations were dominated by the resource-based competition model of species packing, which provided an all-embracing theoretical framework (e.g. Pianka, 1975).

The collapse of this paradigm in the 1980s was in large part due to a re-emphasis of two factors: the advent and use of null hypotheses in ecology (Strong, 1980), and the insistent cry that evolutionary history cannot be ignored. Species are not ecologically less similar that we would expect from random expectation, but are more similar (Kikkawa, 1977), not necessarily because competition is a weak force in communities, but because species are related through their evolutionary history.

The historical perspective on these questions now adds a new dimension to Hutchinson's famous question "Why are there so many kinds of animals?", first posed while he gazed into the pool at the shrine of Santa Rosalia, looking for water boatmen. Hutchinson (1959) considered single morphological characters; we are now able to combine multivariate with phylogenetic techniques (e.g. Miles & Dunham, 1992). This is clearly a fruitful area of future research, as Losos' (1990a,b) elegant studies of ecomorphology and performance in lizards demonstrate. Valuable insights can also be gained into the process of diversification. In a beautiful study of *Phylloscopus* warblers, Richman & Price (1992) showed that there was rapid and extensive morphological change early in the diversification of the group, associated with differences in prey size and feeding method. Later, smaller morphological changes leading to differences between sister species were mainly associated with habitat differences. This is interesting in the light of Schoener's (1974) classic paper on differences between coexisting species with similar niches: his review indicated that such species differed mainly along habitat dimensions, less often on food dimensions, and rarely on a temporal dimension to the niche.

With the advent of all this renewed interest in the role of phylogeny have come new tools for the analysis of this role. For quantitative data, we can divide these methods according to the type of ecological question being asked. First, one can ask whether one trait is *influenced* by the phylogenetic relationships between species. The appropriate methods here are either the Mantel test for association between phylogenetic and ecological distance matrices (J.M. Cheverud, personal communication), or the phylogenetic autocorrelation (Cheverud, Dow & Leutenegger, 1985; Gittleman & Kot, 1990). A different sort of question is whether two traits *covary*, after having allowed for the fact that data are not independent, by *removing* the effect of phylogenetic relationships: the most popular method here is 'independent comparisons' (Harvey & Pagel, 1991; Harvey & Purvis, 1991).

Several authors have viewed the autocorrelation and independent comparisons methods as alternatives, only considering the second type of question. Following on from this, various workers have explored the accuracy with which these and other techniques remove phylogenetic effects (Martins & Garland, 1991; Gittleman & Luh, 1992, Chapter 5).

Using an extension of the maximum likelihood methods of quantitative genetics, both types of question are combined in the comparative method proposed by Lynch (1992). In this method, each quantitative trait is split into a grand mean value (shared by all members of the phylogeny), a phylogenetically inherited additive value (a deviation from the grand mean), and a residual. The residual contains all non-additive genetic components, environmental effects and sampling errors. The output consists of parameter estimates and their standard errors (s.e.), the variance–covariance matrices of both additive and residual values, estimates (with standard errors) of hypothetical ancestral values, and log-likelihood significance tests of a variety of hypotheses. Lynch argues that his method is an advance for three reasons: it uses the multivariate intercorrelated data in an efficient way to estimate the parameters; log-likelihood tests form a sound basis for hypothesis testing; and it uses only the additive evolutionary values to estimate ancestral phenotypes, minimizing bias.

In this paper, we use phylogeny to study the evolutionary ecology of insects, applying Lynch's technique where appropriate. We are principally interested in the evolution of morphology and ecology, and especially in the way in which larval and adult characters covary.

The study group

Our chosen study group are the hoverflies (Syrphidae), one of the largest families of the Diptera with more than 5500 species already described world-wide. The adults are well known for their habit of hovering and of visiting flowers for nectar and pollen; these habits have resulted in natural selection for mimicry, since their other well-known feature is their wasp- and bee-like patterns, which have evolved independently several times within the family (Gilbert, unpublished data).

In contrast to the fairly uniform feeding behaviour of adults, larvae have amazingly diverse feeding habits (Gilbert, 1986, 1990), with substantial proportion species being saprophages (44%), zoophages (40%) and phytophages (16%). Saprophages are exceedingly diverse, from aquatic filter-feeders to wood-borers. Whilst many of the zoophages are aphid (Rotheray & Gilbert, 1989) or ant-brood predators (Garnett,

Akre & Sehlke, 1985), at least one species group consists of ectoparasitoids on wasp larvae (Rupp, 1989). This great diversity makes the family very suitable for testing ideas about the evolution of different feeding modes.

Some work has been carried out on the multivariate ecomorphology of adult syrphids (Gilbert, 1985a,b,c,d, 1990; Gilbert *et al.*, 1985; Owen & Gilbert, 1989; Gilbert & Owen, 1990). Synthesizing these results, and despite some dissenting voices (see Gilbert, 1991), the conclusion is that the quantitative relationships between feeding ecology, activity patterns, egg size/number and morphology are very strong. The strongest pattern, established by canonical correlation analysis, is the relationship between proboscis shape and the time spent feeding on nectar or on pollen (Gilbert, 1985a). These mappings between morphology and ecology do not, however, extend to detectable competitive effects between species close in morphological or ecological space (Gilbert & Owen, 1990).

Virtually none of this work was carried out in a phylogenetic context, however, because at the time there were no phylogenies of syrphids except for some very sketchy ideas based on rather crude data. Inspired by the comparative larval morphological work of the Czech dipterologist Pavel Laska, in 1986 we decided to develop phylogenies of syrphids (Rotheray & Gilbert, 1989, unpublished manuscript).

THE PHYLOGENY OF THE FAMILY

We now have a fairly clear picture of the main features of the generic phylogeny of hoverflies. Full details will be published elsewhere (Rotheray & Gilbert, unpublished manuscript); briefly, more than 160 morphological characters were scored for more than 110 Holarctic genera, and these data subjected to analysis using the Hennig86 phylogenetic program (Farris, 1988). To obtain the final trees, we frequently used Farris' method of successive weighting to reduce the number of equally parsimonious trees. We used only larval structural features, and worked at a generic level. This is because of an interesting contrast that we believe has rarely been noted before, namely that although larval ecological differences are much wider than adult ecological differences, larval structure is much more conservative than adult morphology in the sense that it retains more phylogenetic information. We suggest this is a general feature of insect evolution.

Syrphids belong to the group Aschiza, assumed to be the sister-group of the higher flies, the Cyclorrhapha. One of us (G.E.R.) studied larvae of the other aschizan groups in order to establish the root of the tree. Over the last 100 years there has been a great deal of controversy over the exact relationships between aschizan taxa (see McAlpine, 1989), perhaps because assessments have always been made on the basis of adult characters. Adult characteristics may evolve too quickly to retain enough higher-level phylogenetic information: in larvae, in contrast, the result is very clear. The Lonchopteridae are the most plesiomorphic of the Aschiza, and the Syrphidae the most apomorphic. The sister-group of the syrphids is the Pipunculidae, and of the Syrphidae + Pipunculidae, the Phoridae (cf. Wada's, 1991 conclusions).

Lonchopterids and platypezids feed as larvae on fungi and other microorganisms, and the phorids also contain fungal-feeding larvae (although as larvae they are, like the syrphids, highly diverse: see Disney, 1990). Although the pipunculids are clearly the sister-group of the syrphids, as far as we know as larvae they are all endo-parasitoids of Homoptera. According to our analysis, the most plesiomorphic syrphid is *Eumerus*, which feeds in wet semi-liquid rot pockets of plant roots infected with basal rot *Fusarium* (Creager & Spruijt, 1935). Feeding on micro-organisms must have bridged the gap between platypezids and syrphids, and therefore the original pipunculids must also have been microphagous. A further possibility is that predatory phorids gave rise to the parasitic pipunculids, and syrphids evolved from these via a return to saprophagy.

THE EVOLUTION OF LARVAL ECOMORPHOLOGY

From the analysis of aschizan larvae, the key innovations of larval structure in the first syrphid were probably the following: large size; the development of 'new' structures around the mouth that help in gathering and filtering the food, as well as guiding it to the mouth; an enlarged prothorax with longitudinal grooves, increasing its contractile abilities and resulting in an increased food-gathering capability; a modified anal segment, especially in the fusion of the two posterior spiracles into the posterior respiratory process; and the development of a longer ventral than the dorsal surface, leading to a tilting upwards of the posterior respiratory process.

Most of these innovations are associated with living in a semi-liquid environment. An increased food-gathering ability was probably necessary because syrphid larvae are so much larger than their aschizan relatives, although comparable modifications are found in the small aquatic Phoridae (Disney, 1991).

Figure 1 shows our current best estimate for the phylogeny of the Syrphidae. Note that while all the predatory Syrphinae are lumped together (for clarity and because of the unity of their larval form), this group in fact represents one third of all species and many different genera. Five major features are notable from this tree: we sketch these features here, but they are described and analysed fully elsewhere (Rotheray & Gilbert, unpublished manuscript).

The greatest diversification of feeding habit occurs low down on the tree, and hence presumably early in the evolutionary history of the group. Above *Brachyopa* all genera are saprophages of one sort or another, the differences probably being caused by repeated transitions between the habitat sets of decaying vegetation *versus* tree sap or heartwood.

The transition between the ancestral form and phytophagous genera (mainly *Cheilosia*) is fairly easy to imagine, and few structural modifications were probably necessary. We do not know what the 'phytophages' actually digest; they do ingest plant tissue, but it is possible that they remain feeders on the bacteria and fungi of decay, vastly increased by comminution of the plant substrate during feeding. Diversification of these phytophages has given rise to leaf miners, stem/root tunnellers and the bizarre cambial feeders on conifers (Rotheray, 1988, 1990).

We can follow genera feeding on the bacteria and fungi of decay right up the spine of the tree, from *Cheilosia* (fungi) to *Rhingia* (dung) to sap runs on trees (*Ferdinandea*, *Brachyopa*) and on into the rest of the saprophages (see Fig. 1).

The evolution of the predatory habit probably occurred twice, possibly three times. One pathway gave rise to the sister-taxa (currently tribes) of the Syrphinae and Microdontinae. The Syrphinae are nearly all aphid predators of great agricultural

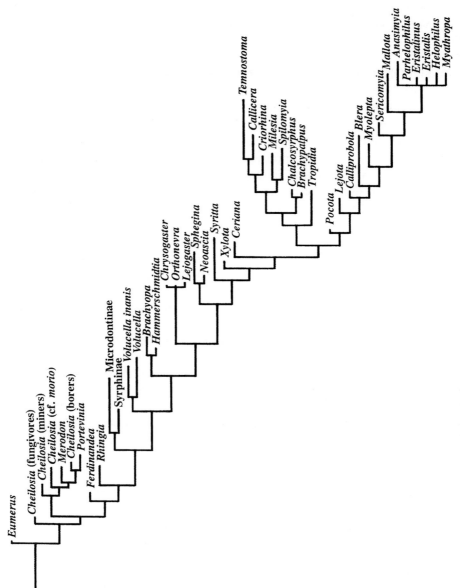

Figure 1 Phylogeny of the genera of hoverflies (Syrphidae). This consensus tree was produced from larval and puparial characters by the Hennig86 program. Full details of the analysis are in Rotheray & Gilbert (unpublished manuscript).

importance, while microdons are predators of ant brood (although early and rather sketchy European work suggests that they might be saprophages). *Microdon* larvae are particularly odd, and were first described as a new taxon of mollusc, so different are they from the normal syrphid or even dipteran larval form (Donisthorpe, 1927). The differences are certainly connected with their myrmecophilous habits, since they resemble other insect myrmecophiles. Do these two taxa represent independently evolved predators? There is an intermediate form, currently classified as a *Platycheirus* (Syrphinae) on the basis of its adult characters, but having a very microdon-like larva. This is *P. milleri* from New Zealand, whose larva lives in ant nests (Thompson, 1972). It is almost certain that this apparent 'intermediate' is actually the product of convergent evolution (see Rotheray & Gilbert, unpublished manuscript).

The last feature we highlight here is the great diversity of saprophages. Three taxa have evolved independently to exploit more aquatic environments (chrysogasterines, spheginines and eristalines). Alternatively, there are those living in decaying cambium and heartwood, consisting mainly of the line leading from *Chalcosyrphus* to *Temnostoma*.

Associated with these ecological changes, there are three main morphological transitions. Again we do not have the space to describe these in detail here (see Rotheray & Gilbert, unpublished manuscript), but they are illustrated in Figs 2–4.

The larval thorax (Fig. 2)

In *Eumerus* the thorax is broad and the head more or less absent, involuted as in the Cyclorrhapha. In the syrphine predators the thorax is strongly narrowed and elongated, probably in connection with their need to grasp small food items and suck out their contents. In *Volucella* there is a new development, the expansion of the anterior fold; this carries a new coating of spicules, and is possibly defensive in function.

The gradual enlargement of this anterior fold, and the appearance of additional longitudinal grooves, forms a transition series from *Syritta* to *Sericomyia* and the eristalines. We interpret this as allowing an enhanced rate of food-gathering and hence increased size. (The eristalines are the insect equivalents of baleen whales: giant, aquatic filter-feeders.) This transition series is associated with inhabiting more aquatic habitats, and a major problem is the ingress of water into the anterior spiracles. The chrysogasterines have lost their anterior spiracles completely in response to this problem, but the eristalines have developed an alternative: retractile spiracles, which are withdrawn into an invaginated pocket.

There has been an enormous development of the thorax along the other branch of the evolutionary tree, from *Tropidia* to *Temnostoma*, with the appearance of large hooks and spicules all over the thoracic surface. These were probably developed in response to the problem of preventing wear and of movement through particle-filled media, since these larvae occur under bark and in decaying heartwood. *Neoascia* species occur in similar material within decaying vegetation, and have independently evolved thoracic hooks, probably for much the same reason. From *Brachypalpus* to *Spilomyia* there is also an increase in the folding pattern because of the requirement for hook musculature. This culminates in the highly derived *Temnostoma*. This genus has all but lost the prothorax, and the meso- and metathorax are enlarged enormously and partly coalesced. On the anterior margin of the

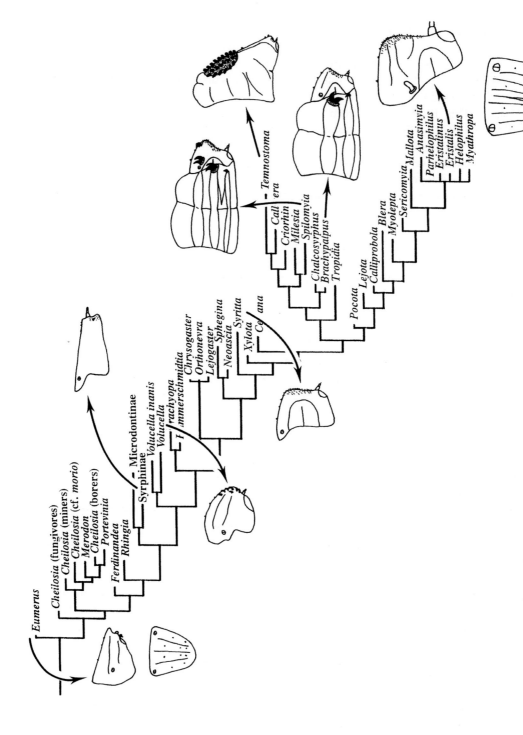

Figure 2 The evolution of the larval thorax in hoverflies.

mesothorax are huge plate-like structures with massive hooks, used in tunnelling in wet wood.

Locomotory organs (Fig. 3)

Locomotory organs are rudimentary in all plesiomorphic groups, consisting merely of pairs of bulges in the first six abdominal segments. These have no muscles, and work hydrostatically using the haemolymph.

The terrestrial predatory syrphines hunt aphids on plants, and this creates a severe problem of attachment to the plant (Rotheray, 1986, 1987). In response to this, they have developed a network of grooves on locomotory organs, around the anus and on the enlarged tip of the anal segment, all of which fill with saliva-like fluid from the anal segment. This sets up strong surface-tension forces that keep the larva on the plant. Different species have adapted to searching on different parts of plants (Rotheray, 1987): for example, *Epistrophe* is flattened and moves well on the leaf lamina, whereas *Scaeva* and *Eupeodes* have invented a U-shaped grasping organ that allows them to search much more efficiently on plant stems. All of this movement operates using haemolymph pressure, without musculature.

As in the case of the larval thorax, *Volucella* is a key taxon in the development of locomotory organs, for it is here that true prolegs, that is the locomotory organs with muscles, make their first appearance. In addition, the prolegs bear large hooked setae called crochets, that presumably help them to grip the wasp combs where they live.

The remaining evolution of the larval thorax merely involves increases or decreases in the relative sizes of the prolegs and/or crochets, both decreasing to almost complete absence in the line leading to *Temnostoma*, while increasing to very large sizes in the line leading to the eristalines.

The anal segment (Fig. 4)

It is in the development of the anal segment that we see the true potential of the syrphid design realized. In the plesiomorphic groups there is only a single fold to the segment. In *Eumerus* this develops a further ring, and each has a pair of lappets, that is, fleshy protuberances. In the predatory syrphines these lappets have moved to the rear of the segment, and the whole segment is enlarged for gripping the substrate.

Once again there is significant innovation in *Volucella*, with the appearance of a third ring to the anal segment with its associated pair of lappets. Each ring then evolves semi-independently during the subsequent evolution of the family. There are increases in relative length when invading more aquatic habitats, and decreases in relative length when invading semi-solid habitats. Thus elongated anal segments are seen in the chrysogasterines, the spheginines and the eristalines, with the length of each ring extended approximately equally. In the line leading from *Pocota* to *Eristalis*, however, the third ring elongates (*Calliprobola*), then the first and second together also elongate (*Myolepta*), and finally the first ring narrows and extends enormously (eristalines).

What happened to adult form during this diversification of larval structure and feeding habits? We know little as yet of these patterns, but data uncorrected for

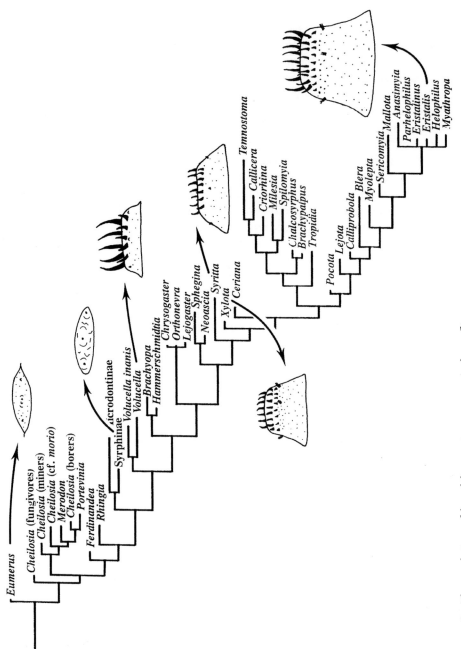

Figure 3 The evolution of larval locomotory organs in hoverflies.

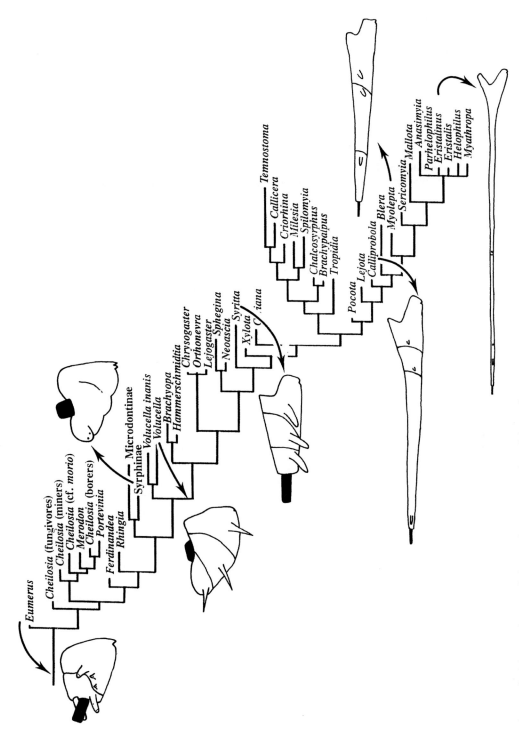

Figure 4 The evolution of the anal segment in larval hoverflies.

phylogenetic relatedness show that predators are relatively small, and saprophages generally relatively large. In addition, phytophages lay few large eggs, whereas saprophages lay many small eggs (Gilbert, 1990). Now with an estimate of the phylogeny of the family, we can begin to assess the way in which larval and adult evolution are interlinked, as we look at the predatory syrphids, and in more detail within the genus *Platycheirus.*

THE EVOLUTION OF PREDATORY GENERA

Our main current focus is on the evolution of predators, and in particular the evolution of specialized feeding habits (Owen & Gilbert, 1989; Rotheray & Gilbert, 1989; Gilbert, 1990; Gilbert & Owen, 1990). Very little is known about the selection pressures associated with predatory specialization (Bristowe, 1988; Gilbert, 1990).

Whilst there are three taxa of syrphid predators, we know very little about the microdons (mostly South American) or volucellines (relatively few species with rather poorly known behavioural ecology). Thus we concentrate on the aphid predators, the syrphines. They have an intrinsic interest since they are the only dipteran larvae with colours, used in crypsis (Rotheray, 1986), and they are economically important in aphid biocontrol (e.g. Chambers, 1986).

There are more than 1500 species of Syrphinae, covering very varied ecological types including feeding habits from extreme specialists to generalists. For example, *Triglyphus primus* larvae have only once been found feeding on any aphid other than in the galls of the aphid *Cryptosiphum artemisiae* on groundsel, *Artemisia vulgaris* (see Sedlag, 1966). In contrast, larvae of *Episyrphus balteatus* have been found on almost every aphid species that has been studied within its geographic range.

Again using larval characters, we have developed a generic phylogeny for this group (Rotheray & Gilbert, 1989: revised in Rotheray & Gilbert, unpublished manuscript). Figure 5 shows the resulting phylogeny. We used this in conjunction with two different types of data. The first is the type of aphid colony typically fed upon by members of the genus. There is a clear trend from genera feeding on aphids of the herb layer to genera feeding on arboreal aphids (see Gilbert, 1990). Interestingly the three genera furthest from the root of the tree (*Paragus, Scaeva, Eupeodes*) appear to have re-invaded herb-layer aphid colonies. In two of these, *Scaeva* and *Eupeodes*, a new U-shaped grasping organ allows greater efficiency in feeding on aphid colonies on stems: originally this organ may have evolved as an adaptation to feeding on aphids of conifer needles.

Is prey range influenced by phylogenetic relatedness? In an attempt to answer this we used the median number of prey species recorded for species in the genus, obtained from a comprehensive review of 20th-century literature on the family (Gilbert, unpublished). For these data to constitute an adequate test of whether there is an association between prey range and phylogeny, we would really need the prey range of the most plesiomorphic member of each genus. This we cannot estimate since we do not yet have species-level phylogenies of any but two to three syrphine genera.

In addition, these numbers may be especially biased for particular genera. For example, larvae of the plesiomorphic *Melanostoma* and some species of *Platycheirus* may be generalized predators in the leaf litter (Rotheray & Gilbert, 1989),

Figure 5   The phylogeny of genera of the predatory hoverflies (the Syrphinae), based upon larval and puparial characters, and produced by the Hennig86 program.

rather than specific aphid predators. While the adults are enormously abundant, larvae are rarely found at all in aphid colonies in the field. It was believed at one time that these larvae fed largely on very small aphid colonies on grass, but could subsist on rotting plant material while they moved between aphid colonies (Davidson, 1922; Goeldlin, 1974). This is incorrect, since larvae cannot feed on autoclaved rotting leaf material (Zafar, 1987); they are more likely to be general zoophages feeding on arthropods in the leaf litter, but we really do not know.

A further problem is the misleadingly high number of hosts found for the genus *Eupeodes*, particularly the subgenus *Metasyrphus*. Here it is very probably true that the more plesiomorphic species have restricted prey ranges, since we suspect that many of these species are specialized on conifer aphids.

Testing these data with Lynch's phylogenetic heritability gives a value close to zero. This clearly does not support what seems to us to be a reasonably clear pattern, namely an increased degree of specialization in the more apomorphic genera. However, data quality at this generic level may be problematic: data at the species level are more promising (see below). There is even a suggested mechanism for a putative general drift towards specialization, namely escape from competition. Elsewhere it is shown that there is evidence that generalists compete, but no evidence that generalists compete with specialists, nor specialists with each other (Gilbert, 1990).

THE GENERA *MELANOSTOMA* AND *PLATYCHEIRUS*

We now focus on one particular group of species in order to apply Lynch's methods to data at the species level. The genera we have chosen to use here are among the most plesiomorphic of the syrphines, and therefore may partly hold the key to the evolution of the predatory habit. *Melanostoma* and *Platycheirus* are overwhelmingly Holarctic genera with many species, especially *Platycheirus* (Vockeroth, 1990). All are small dark flies with a variable set of orange or white spots on the abdomen.

To generate the phylogeny, we need a set of characters that evolve reasonably quickly: larval characters may not be very useful because we suspect they evolve conservatively and, in any case, larvae are not available for many *Platycheirus* species. We therefore chose to use the remarkable forelegs of adult male *Platycheirus*. These have a great variety of extraordinarily modified hairs and tarsi (see Vockeroth, 1990); we suggest that these may be the result of a process of sexual selection similar to that which Eberhard (1985) hypothesized to be responsible for the remarkable variety of genitalic characters in animals. After careful examination of 83 species of *Platycheirus* (all the UK, three-quarters of the European and two-thirds of the Holarctic fauna), we settled upon 88 characters from the legs of the males. Male *Melanostoma* and *Xanthandrus* have unmodified legs, and we considered them to represent the outgroup. Full details of the character set and analysis will be published elsewhere (Emerson & Gilbert, in preparation).

The phylogeny of the 18 British species of *Platycheirus* (including the two species previously placed in the genus *Pyrophaena*, now a synonym: Vockeroth, 1990) shows three main branches (Fig. 6). The first leads to *P. manicatus*, and contains species where males have disc-like fore-tarsi and a thick brush of hairs along the fore-femur: although defined on the basis of non-metric adult leg characters, the

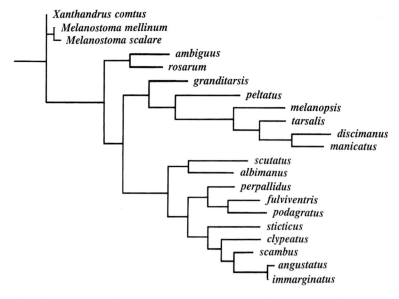

Figure 6 The phylogeny of the genera *Xanthandrus*, *Melanostoma* and *Platycheirus*, based upon hairs and cuticular modifications of the legs of adult males, and produced using the Hennig86 program.

more apomorphic species of this line are also larger, and have increasingly long tongues. A second group has two species with peculiar hair clumps at the base of the fore-femur. The final branch leads to *P. immarginatus*, and the males of these species show a generalised flattening of the end of the foreleg: species of this line are also all specialist feeders on grass and other anemophilous pollens as adults, and have very short tongues. *Melanostoma* species also have these characteristics.

With this phylogeny, we can now explore the evolution of quantitative characters of morphology and ecology much more accurately than in the previous generic-level analyses.

Morphological characters

Morphology is highly conserved during evolution: nearly all evolutionary change appears to represent size change with its associated allometry. We used data on head width, proboscis length and thorax volume: together these represent 'size' and 'proboscis shape', the two main determinants of foraging behaviour and activity (Gilbert, 1985a,b,d) and the major morphological differences between species (Gilbert, 1985c). All show high evolutionary heritabilities (thorax volume $h^2 = 0.35 \pm 0.13$; head width $h^2 = 0.64 \pm 0.13$; proboscis length $h^2 = 0.97 \pm 0.02$). It is interesting that the main food-niche determinant, proboscis length (Gilbert, 1981, 1985a), has the highest heritability.

Directional comparisons involve using the reconstructed ancestors at the nodes of the tree (Harvey & Pagel, 1991; Harvey & Purvis, 1991); such comparisons show a very tight covariance between evolution of size (thorax volume) and evolution in head width and proboscis length (Fig. 7). This in turn implies that the genetic links creating the allometry between size and shape have not changed during the evolution of the genus. Therefore the history of diversification in *Platycheirus* morphology is mainly one of size change.

Ecological characters

For the 18 species of *Platycheirus* and the two *Melanostoma* we extracted from the literature (and from our own unpublished material) data on various ecological and morphological features, and tested whether they were associated with phylogenetic relatedness. These data were on the number of generations per year, the number of recorded prey species, morphology (head width, thorax volume, proboscis length: see Gilbert, 1985c), and abundance in one well-studied community (Owen, 1991).

Because the number of generations per year (1, 2 or 3) is not even approximately a continuous character, we used the non-parametric Mantel test (taken from Manly, 1985) to check for similarity between ecological and phylogenetic distance matrices. To generate distance matrices, we calculated a Euclidean distance between pairs of species using either the number of generations, or the number of character changes occurring on the tree (character states for nodes were taken from optimizations produced by the Hennig86 program). There is a significantly low value for the test statistic ($G = -2.64$, $P < 0.01$), indicating that the ecological and phylogenetic distance matrices are indeed related. This implies that the number of generations per year evolved conservatively during the evolution of the genus.

We also used the number of prey species per species, as recorded in the

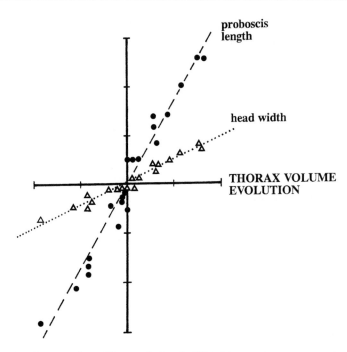

Figure 7 Directional comparison (Harvey & Pagel, 1991; Harvey & Purvis, 1991) of the evolution of morphological characters in the genera *Melanostoma* and *Platycheirus*.

literature. Although these data are better than the generic-level data discussed above, there are major problems with these numbers, since *Melanostoma* and *Platycheirus* larvae are rarely found and even more rarely identified in field research work. Nevertheless, Lynch's phylogenetic heritability is low but significant ($h^2 = 0.09 \pm 0.038$, $P < 0.05$, $N = 20$), and its additive genetic covariance with size (head width) is negative. Thus in evolving to larger body sizes, there is an associated evolutionary decrease in prey range. This result is all the more remarkable because, with some notable exceptions, there is a very strong emphasis in the literature on ecological rather than historical forces explaining prey range (but see Futuyma & Moreno, 1988; Berenbaum, 1990a,b). Prey range has almost always been viewed as a product of current-day forces, at an equilibrium with natural selection. However, this result shows a historical component: prey range in one species is constrained by the prey range of its ancestors. These species-level data also support our contention that the evolution of prey range is a history of increasing specialization.

Even abundance is significantly associated with the phylogeny. Abundance measures are taken from Jennifer Owen's long-term study of insect populations in a suburban garden (Owen, 1981, 1991) which are correlated with abundances elsewhere in Europe (Owen & Gilbert, 1989). We used the sum of 20 years of trap catches as our estimate of abundance, and find that this measure contains a significant phylogenetically heritable additive component ($h^2 = 0.09 \pm 0.038$). Since abundance is measured at only one site, we expect only a crude relationship if any. The direction of change is that the more apomorphic species are rarer, and again there is a negative additive covariance with size evolution. Nee *et al.* (1991) have already demonstrated that there are taxonomic differences in the relationship between body size and abundance. Here we have shown explicitly an effect of phylogenetic relatedness on relative abundance.

Behaviour and growth

Finally we focus on six species studied intensively in the laboratory under controlled conditions (Zafar, 1987). We are interested in the evolution of behavioural components of predation in the two *Melanostoma* species and four *Platycheirus*. The behavioural components of predation are exceptionally easily studied in syrphines (Rotheray, 1983, 1987, 1989), because the behavioural process of searching, attacking and feeding on aphids is so easily quantified. For each species under standardized conditions of age, starvation and temperature, we measured the average casting rate, capture rate, capture efficiency, handling time and the casting-rate response to a short period of starvation. In addition we fitted two-parameter growth curves (Koijman, 1986) of the form:

$$dW/dt = aW^{2/3} + 3bW \qquad (1)$$

We fitted this equation to data for each species using the statistical program Statgraphics™, and tested the two fitted parameters, *a* and *b*, for the influence of phylogeny.

If increasing specialization involves becoming more efficient as a predator (Futuyma & Moreno, 1988), we predict that specialization will entail increases in the casting rate, capture rate and growth rate, whilst also involving decreases in handling time. We have used casting rate (the rate of making searching movements: see Rotheray, 1983, 1987, 1989) as a reference for directional comparisons because of our doubts about the quality of the data on prey range. The additive-value regression of casting rate on prey range is negative (-0.35 ± 0.15), as predicted. Because we have data for only six species, significance tests are of dubious validity (see the example used by Lynch, 1992). We are, however, able to consider the sign of the slope of the directional comparisons, and in every case the slope of evolutionary change is as predicted (Fig. 8). The additive-value regressions are strongly influenced by the large amount of evolutionary change occurring between the genera *Melanostoma* and *Platycheirus*: however, only one additive-value regression slope (handling time) changes when this point is removed.

CONCLUSIONS

The new techniques now available constitute a significant advance for the comparative method. In conjunction with new multivariate methods, developed for relating environmental factors to species abundances (CANOCO, see ter Braak, 1986), we can now relate evolutionary change in a set of multivariate characters to environmental differences between the habitats in which they live. We have shown here that phylogenetic influence extends to all types of ecological, behavioural and morphological characters. This means that any sort of comparison between species must involve a phylogenetic component. The new techniques open up comparative analysis in an entirely new way. As Harvey & Pagel (1991: 203) discuss, the

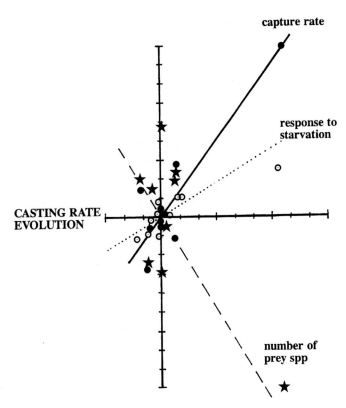

Figure 8 Directional comparisons of the evolution of ecological and behavioural characters of the foraging behaviour of *Melanostoma* and *Platycheirus* larvae. The outlier points document the evolutionary change from *Melanostoma* to *Platycheirus*.

voluminous and accurate comparative data we have will soon be treated with just as much respect as experimental results on individual variation.

REFERENCES

BERENBAUM, M., 1990a. Evolution of specialisation in insect-umbellifer *Annual Review of Entomology, 35*: 319–343.

BERENBAUM, M., 1990b. Coevolution between herbivorous insects and plants: tempo and orchestration. In F.S. Gilbert (Ed.), *Insect Life Cycles: Genetics, Coordination and Evolution*, 87–99. London: Springer-Verlag.

ter BRAAK, C.J.F., 1986. Canonical correspondance analysis: a new eigenvector technique for multivariate direct-gradient analysis. *Ecology, 67*: 1167–79.

BRISTOWE, C., 1988. What makes a predator specialize? *Trends in Ecology & Evolution, 3*: 1–2.

CHAMBERS, R.L., 1986. Preliminary experiments on the potential of hoverflies for the control of aphids under glass. *Entomophaga, 31*: 197–204.

CHEVERUD, J.M., DOW, M. & LEUTENEGGER, W., 1985. The quantitative assessment of phylogenetic constraints in comparative analyses: sexual dimorphism in body weight among primates. *Evolution, 36*: 914–933.

CREAGER, D.B. & SPRUJIT, F.J., 1935. The relation of certain fungi to larval development of

Eumerus tuberculatus Rond. *Annals of the Entomological Society of America, 28*: 425–437.

DAVIDSON, W.M., 1922. Notes on certain species of *Melanostoma* (Diptera, Syrphidae). *Transactions of the American Entomological Society, 48*: 35–47.

DISNEY, R.H.L., 1990. Some myths and the reality of scuttle-fly biology. *Antenna, 14*: 64–67.

DISNEY, R.H.L., 1991. The aquatic Phoridae (Diptera). *Entomologica Scandinavica, 22*: 171–191.

DONISTHORPE, H., 1927. *The Guests of British Ants.* London: Routledge & Sons.

EBERHARD, W.G., 1985. *Sexual Selection and the Evolution of Animal Genitalia.* Cambridge, MS: Harvard University Press.

FARRIS, J.S., 1988. Hennig86, version 1.5. Manual and MS-DOS program. Available from author, 41 Admiral St, Port Jefferson Station, New York 11776, U.S.A.

FUTUYMA, D. & MORENO, G., 1988. The evolution of ecological specialisation. *Annual Review of Ecology & Systematics, 19*: 207–233.

GARNETT, W.B., AKRE, R.D. & SEHLKE, G., 1985. Cocoon mimicry and predation by myrmecophilous Diptera. *Florida entomologist, 68*: 615–621.

GILBERT, F.S., 1981. Foraging behaviour of hoverflies (Diptera, Syrphidae): morphology of the mouthparts in relation to feeding on nectar and pollen in some common urban species. *Ecological Entomology, 6*: 245–262.

GILBERT, F.S., 1985a. Ecomorphological relations in hoverflies (Diptera, Syrphidae). *Proceedings of the Royal Society of London, Series B, 224*: 91–105.

GILBERT, F.S., 1985b. Size and shape in Syrphus ribesii (Diptera, Syrphidae). *Proceedings of the Royal Society of London, Series B, 224*: 107–114.

GILBERT, F.S., 1985c. Morphometric patterns in hoverflies (Diptera, Syrphidae). *Proceedings of the Royal Society of London, Series B, 224*: 79–90.

GILBERT, F.S., 1985d. Diurnal activity patterns in hoverflies (Diptera, Syrphidae). *Ecological Entomology, 10*: 385–392.

GILBERT, F.S., 1986. Hoverflies (1st edition). Naturalists' Handbooks (5). Cambridge: Cambridge University Press.

GILBERT, F.S., 1990. Size, phylogeny and life-history in the evolution of feeding specialization in insect predators. In F.S. Gilbert (Ed.), *Insect Life Cycles: Genetics, Evolution and Coordination*, 101–124. London: Springer-Verlag.

GILBERT, F.S., 1991. Feeding in adult hoverflies. *Hoverfly Newsletter, (13)*: 5–11.

GILBERT, F.S. & OWEN, J., 1990. Size, shape, competition and community structure in hoverflies (Diptera, Syrphidae). *Journal of Animal Ecology, 59*: 21–39.

GILBERT, F.S., HARDING, E.H., LINE, J.M. & PERRY, I., 1985. Morphological approaches to community structure in hoverflies (Diptera, Syrphidae). *Proceedings of the Royal Society of London, Series B, 224*: 115–130.

GITTLEMAN, J.L. & KOT, M., 1990. Adaptation: statistics and a null model for estimating phylogenetic effects. *Systematic Zoology, 39*: 227–241.

GITTLEMAN, J.L. & LUH, H.-K., 1992. On comparing comparative methods. *Annual Review of Ecology & Systematics, 23*: 383–404.

GITTLEMAN, J.L. & LUH, H.-K., 1994. Phylogeny, evolutionary models and comparative methods: a simulation study. In P. Eggleton & R.I. Vane-Wright (Eds), *Phylogenetics and Ecology*, 103–122. London: Academic Press.

GOELDLIN, P., 1974. Contribution a l'étude systématique et écologique des Syrphidae (Dipt.) de la Suisse orientale. [A contribution to the systematics and ecology of the Syrphidae (Dipt.) of eastern Switzerland]. *Mitteilungen der Schweizerischen Entomologischen Gesellschaft, 47*: 151–252.

HARVEY, P.H. & PAGEL, M., 1991. *The Comparative Method in Evolutionary Biology.* Oxford: Clarenden Press.

HARVEY, P.H. & PURVIS, A., 1991. Comparative methods for explaining adaptations. *Nature, 351*: 619–624.

HUTCHINSON, G.E., 1959. Homage to Santa Rosalia, or why are there so many kinds of animals? *American Naturalist, 95*: 145–159.

KARR, J.R. & JAMES, F.C., 1975. Ecomorphological configurations and convergent evolution of species and communities. In M.L. Cody & J.M. Diamond (Eds), *Ecology and Evolution of Communities*, 258–291. Cambridge, MS: Belknap Press.

KIKKAWA, J., 1977. Ecological paradoxes. *Australian Journal of Ecology, 2*: 121–136.

KOIJMAN, S.A., 1986. Energy budgets can explain body size relations. *Journal of Theoretical Biology, 121*: 269–282.

LOSOS, J.B., 1990a. Ecomorphology, performance capability, and scaling of West Indian *Anolis* lizards: an evolutionary analysis. *Ecological Monographs, 60*: 369–388.

LOSOS, J.B., 1990b. The evolution of form and function: morphology and locomotor performance in West Indian *Anolis* lizards. *Evolution, 44*: 1189–1203.

LYNCH, M., 1992. Methods for the analysis of comparative data in evolutionary biology. *Evolution, 45*: 1065–80.

McALPINE, J.F., 1989. Phylogeny and classification of the Muscomorpha. In J.F. McAlpine & D.M. Wood (Eds), *Manual of Nearctic Diptera, 3*, 174–232. Ottawa: Agriculture Canada Monographs.

MANLY, B.M., 1985. *The Statistics of Natural Selection*. London: Chapman & Hall.

MARTINS, E.P. & GARLAND, T., 1991. Phylogenetic analyses of the correlated evolution of continuous characters: a simulation study. *Evolution, 45*: 534–557.

MILES, D.B. & DUNHAM, A.E., 1992. Comparative analysis of phylogenetic effects in the life history patterns of iguanid reptiles. *American Naturalist, 139*: 848–869.

MILES, D.B. & RICKLEFS, R.E., 1984. The correlation between ecology and morphology in deciduous passerine birds. *Ecology, 65*: 1629–1640.

NEE, S., READ, A.F., GREENWOOD, J.D. & HARVEY, P.H., 1991. The relationship between abundance and body size in British Birds. *Nature, 351*: 312–314.

OWEN, J., 1981. Trophic variety and abundance of hoverflies (Diptera, Syrphidae) in an English suburban garden. *Holarctic Ecology, 4*: 221–228.

OWEN, J., 1991. *Ecology of a Garden*. Cambridge: Cambridge University Press

OWEN, J. & GILBERT, F.S., 1989. On the abundance of hoverflies. *Oikos, 55*: 183–193.

PIANKA, E.R., 1975. Competition and niche theory. In R.M. May (Ed.), *Theoretical Ecology: Principles and Applications*, 114–141. Oxford: Blackwell Scientific.

RICHMAN, A.D. & PRICE, T., 1992. Evolution of ecological differences in the Old World leaf warblers. *Nature, 355*: 817– 821.

ROTHERAY, G.E., 1983. Feeding behaviour of *Syrphus ribesii* and *Melanostoma scalare* on *Aphis fabae*. *Entomologia experimentalis et applicata, 34*: 148–154.

ROTHERAY, G.E., 1986. Colour, shape and defence in aphidophagous syrphid larvae. *Zoological Journal of the Linnean Society, 88*: 201–216.

ROTHERAY, G.E., 1987. Larval morphology and searching efficiency in aphidophagous syrphid larvae. *Entomologia Experimentalis at Applicata, 43*: 49–54.

ROTHERAY, G.E., 1988. Larval morphology and feeding patterns of four Cheilosia species associated with *Cirsium palustre* L. Scopoli in Scotland. *Journal of Natural History, 22*: 17–25.

ROTHERAY, G.E., 1989. *Aphid Predators*. Naturalists Handbooks (11). Slough, U.K.: Richmond Publishing Co.

ROTHERAY, G.E., 1990. The relationship between feeding mode and morphology in *Cheilosia* larvae (Diptera, Syrphidae). *Journal of Natural History, 24*: 7–19.

ROTHERAY, G.E. & GILBERT, F.S., 1989. Systematics and phylogeny of the European predacious Syrphidae (Diptera) from larval and pupal stages. *Zoological Journal of the Linnean Society, 95*: 27–70.

RUPP L., 1989. Die mitteleuropäischen Arten der Gattung *Volucella* (Diptera, Syrphidae) als Kommensale und Parasitoïde in den Nestern von Hummeln und Sozialen Wespen. Untersuchungen zur Wirtsfindung, Larval Biologie und Mimikry. [The mid-European species of the genus *Volucella* as commensals and parasitoids in the nests of bees and social wasps: studies in host finding, larval biology and mimicry]. Unpublished dissertation. Freiburg-im–Breisgau: Fakultät für Biologie, Albert-Ludwigs-Universität.

SCHOENER, T.W., 1974. Resource partitioning in ecological communities. *Science, 185*: 27–35.

SEDLAG, U., 1966. *Triglyphus primus* – eine weitgehend übersehene Syrphide. [*Triglyphus primus* – a much-overlooked hoverfly]. *Entomologische Berichte, Berlin, 1966*: 88–90.

STRONG, D.R., 1980. Null hypotheses in ecology. *Synthèse, 43*: 271–285.

THOMPSON, F.C., 1972. A new *Platycheirus* from New Zealand: first records of a melanostomine syrphid fly associated with ants. *New Zealand Journal of Science, 15*: 77–84.

VOCKEROTH, R., 1990. Revision of the Nearctic species of *Platycheirus* (Diptera, Syrphidae). *Canadian Entomologist, 122*: 659–766.

WADA, S., 1991. Morphologische Indizien für das unmittelbare Schwestergruppenverhältnis der Schizophora mit den Syrphoidea ('Aschiza') in der phylogenetischen Systematik der Cyclorrhapha (Diptera: Brachycera). [Morphological evidence for the immediate sister-group of the Schizophora being the Syrphoidea ('Aschiza') in the phylogenetic systematics of the Cyclorrhapha (Diptera: Brachycera)]. *Journal of Natural History, 25*: 1531–1570.

ZAFAR, R., 1987. Morphology and foraging behaviour in the Melanostomini (Diptera, Syrphidae). Unpublished M.Phil thesis, Nottingham University.

chapter 16

Some principles of phylogenetics and their implications for comparative biology

PAUL EGGLETON & R.I. VANE-WRIGHT

CONTENTS

Keywords: Phylogenetic trees – evolutionary polarity – tree construction – interpretation of trees and classifications.

Abstract

Phylogenetic trees form a basis for most studies in contemporary comparative biology. We discuss the nature of such trees and their terminology, how evolutionary character polarity is assessed and how phylogenetic trees are constructed from character matrices. Finally, we touch on some problems in the interpretation of phylogenetic trees. We argue that phylogenetic trees are usually not as robust as they appear and that they must be used with informed caution in comparative studies.

Phylogenetics and Ecology
ISBN 0-12-232990-2

INTRODUCTION

The aims and methods of the forms of comparative biology discussed in this book are varied, but one element is nearly always assumed *a priori*: the existence of a phylogenetic tree. In this chapter we outline how phylogenetic trees are obtained, discuss the problems of interpreting such trees and emphasize that a comparative study is only as sound as its systematic framework.

We define phylogenetics as the sub-discipline of systematics concerned with the estimation of phylogenetic relationships. As such it is clearly not only a part of systematics but also a part of evolutionary biology (de Quieroz & Donoghue, 1990). Despite the fundamental distinction between cladograms (which summarize patterns of character distribution amongst taxa) and trees (which summarize theories of ancestor–descendant relationships), we subsume all phylogenetic methods under the umbrella term 'cladistics', in that they use, broadly, the methods and terminology of Willi Hennig (1950, 1966, etc.). We thus use phylogenetics in a very general sense, and regard any systematic method that attempts to reconstruct evolutionary (i.e. genealogical) history as phylogenetic. Some methods, such as DNA hybridization and other broadly phenetic molecular clock methods, do not really fit under this cladistic umbrella. They are included here for completeness because, however open to methodological criticism they might be, they do produce tree diagrams that purport to estimate phylogenetic relationships. Every chapter in this book relies on the existence of phylogenetic trees – and these form the most direct link between systematics and ecology.

TERMS

The systematics literature is full of complex jargon that, moreover, is not always used consistently. In addition, *Cladistics*, the only journal concerned exclusively with phylogenetics (in our sense), is not yet found in many university libraries. This chapter is not a methodological or philosophical treatise (admirable and extensive texts already exist: Wiley, 1981; Sober, 1988; Brooks & McLennan, 1991; Forey *et al.*, 1992), but it is intended to introduce readers to the phylogenetics literature and to define some of the terms which make up the building blocks of phylogenetics (with the caveat that some systematists will not agree with all of them).

Most systematists now believe that the most satisfactory way to present their results is as a diagram showing postulated phylogenetic relationships. In order to present this data, they use **trees**, which are, in mathematical terms, "connected acyclic graphs" (Penny, Hendy & Steel, 1992). For our purposes we can think of them simply as branching diagrams in which each terminal element (species, taxon) is linked once, and only once, to one or more other taxa, to specify a particular hierarchy (Fig. 1a). Trees can be used to show the relationships between any objects which can be linked in hierarchical clusters without reticulations (that is in this context, the phylogenetic relationships between taxa but not the tokogenetic relationships between individuals produced by sexual reproduction). Indeed, classical phenetic procedures produce trees, as do any number of other statistical techniques. Trees can be **rooted** (that is, with an additional identified point at their base, Fig. 1a) or **unrooted** (Fig. 1b). When a tree is used to show only a **pattern**

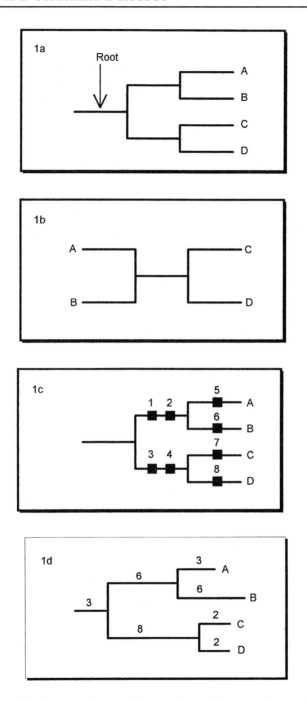

Figure 1 Some simple diagrams of trees. (a) rooted tree; (b) unrooted tree; (c) rooted cladogram with characters marked 1–9 on the tree; (d) rooted tree with branch lengths.

of synapomorphies (see below) it is generally known as a **cladogram** (Fig. 1c). A cladogram can also be rooted or unrooted, but either way, it has no necessary evolutionary significance (Platnick, 1979; Patterson, 1988).

A **phylogenetic** (or **evolutionary**) **tree**, as its name suggests, is a tree with an evolutionary interpretation. This phylogenetic tree is often derived from a clado-gram, but does not have to be (for example molecular trees are often not derived from cladograms). An evolutionary tree may also have branch lengths (see Fig. 1d), but one derived from a cladogram will not, strictly speaking, have them. Phylo-geneticists who use morphological data alone generally only accept evolutionary trees that are identical to cladograms. An evolutionary tree must have a root, as it is the root that indicates evolutionary direction (see section on polarity).

A phylogenetic tree should not be confused with the **phylogeny**. The phylogeny is the actual pattern of relationships between taxa, whereas a tree can only be an estimate of those relationships. Phylogeny results from a series of historical events. As such it is unrecoverable (unknown), in the same way that we can never know the exact sequence of events at Crecy or Agincourt. Unfortunately, the term phylogeny is often used loosely to mean 'a phylogenetic tree', and some unwary readers might be misled into believing that an actual evolutionary history was being presented (see, for example, Gilbert *et al.*, Chapter 15).

Although a long debate has raged about the validity of inferring evolutionary trees from cladograms (e.g. Patterson, 1982a), most systematists and ecologists routinely treat cladograms as if they were evolutionary trees, and that is a vital assumption to the majority of papers in this book.

All phylogenetic trees generally have the following elements (Penny, Hendy & Steel, 1992). **Nodes** (or internal branching **points**) represent putative ancestors. **Tips** (or **leaves**, or external points) are the ends of branches that lead to all taxa under investigation (the **terminal taxa**, that may be extant or fossil). **Branches** (or **edges**) of evolutionary trees are lines connecting nodes together, or lines connecting a node to a tip (hence branch or edge lengths/weights). In a rooted tree a point is identified to form the base of the tree (Fig. 2), or **root**. Branch lengths can be arbitrary (in the case in many morphological studies), or related to estimates of phenetic or **genetic distance**.

In cases where branch lengths are related to genetic distances, trees may be constrained to have all tips the same distance from the root (**ultrametric** trees, usually derived from analyses assuming an exact molecular clock, see below) or the tips may be at differing distances from the root (**metric** trees, that do not necessarily assume an exact molecular clock). Cladograms are often drawn as if they were ultrametric trees, but it is important to realize that branch lengths are then usually

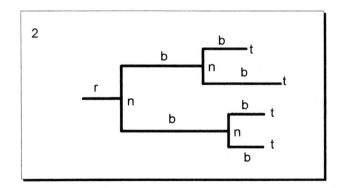

Figure 2   Tree with constituent parts marked. b = branch, n = node, r = root, t = tip.

arbitrary and do not imply a 'morphological' clock (see Williams, Gaston & Humphries, in press).

There is no general consensus amongst phylogeneticists about the meaning of the terms **character** and **character state**. Two main schools of thought exist. The first defines the two terms separately, such that characters are parts (or categories of parts) of a group of organisms studied in order to find differences that may be phylogenetically meaningful. The differences are then known as character states (or cladistic variables: Pimentel & Riggins, 1987). In this formulation 'bill' is a character, while its particular shape (tapered, hooked, blunt) is one of two or more mutually exclusive character states.

We will define characters here in the second, more general sense, as observable features of organisms (for example 'presence of bill' is a character of birds, while 'thick, downcurved bill' is a character of parrots). Characters (or character state gains or losses) are what are usually indicated on the branches of phylogenetic trees or cladograms. This implies that characters and homologies are synonymous, and we discuss this later.

Characters can be qualitative (presence/absence, number of digits etc.) or quantitative (measurements, distances etc.). However, there are great problems with using quantitative variables as cladistic data, especially within morphological studies (Pimentel & Riggins, 1987; Chappill, 1989).

BUILDING BLOCKS

Phylogenetic analysis requires only a very limited number of assumptions pertaining to patterns of organismic diversity. At the most basic level it is that "nature is ordered in a single specifiable pattern" (Platnick, 1979). This is an essentially evolution-free assumption, and phylogeneticists who take this position claim to be concerned only with the hierarchy of life rather than making unwonted assumptions about genealogical relationships (Platnick, 1979, 1985; Patterson, 1980, 1982a). It is important to realize that under this assumption such a hierarchy does not have to be expressed as a tree, and any set-theory type diagram will suffice (Fig. 3) – although this does not prevent such diagrams from being interpreted as phylogenetic trees. Supporters of this view of phylogenetics see the discovery of pattern as separate from the investigation of evolutionary process. They attempt to escape the potential circularity of using trees to test evolutionary theories by reference to a tree diagram constructed under the assumption of particular evolutionary processes, however reasonable those processes might appear (Patterson, 1988)

Most phylogeneticists, however, explicitly use phylogenetic trees, and then they would generally accept two low-level evolutionary assumptions as axiomatic (Farris, 1986):

1. "evolution as descent with modification",
2. "observed similarities [are] explicable by inheritance and common ancestry".

Although these assumptions are generally accepted and uncontroversial (and we accept them here), some phylogeneticists make more complex process assumptions in their analyses and interpretation of trees (see Friday, Chapter 9). Amongst molecular phylogeneticists this often involves developing probabilistic models of

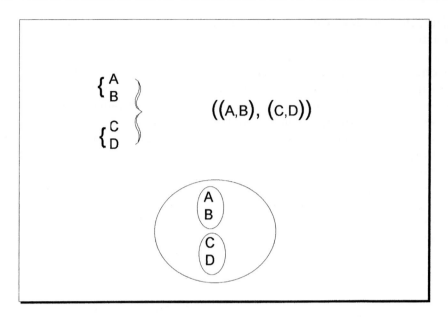

Figure 3 Set diagrams indicating non-tree methods of summarising cladistic data. All these diagrams are identical in topology to the trees in Fig. 1.

evolutionary processes at the molecular level (see Felsenstein, 1981, 1983a,b, 1988a; Bishop & Friday, 1985; and below).

All phylogeneticists search for phylogenetically meaningful similarities between taxa. They look for **homology** and attempt to reject **analogy**. Even in cases where apparently phenetic distance methods are employed (e.g. DNA hybridization), they are used because direct sampling of, say, DNA strands, has been thought to lead to meaningful sampling of a majority of homologous nucleotide sites (Gould, 1985, 1986).

Homology is perhaps the most important concept in phylogenetics, even if there is some disagreement about its definition. Essentially it refers to systematically meaningful similarity between taxa (i.e. synapomorphy, see below). Patterson (1982b) recognized three criteria for testing homology: *conjunction* (homologous characters cannot occur in the same organism); *similarity* in topology, ontogeny and composition; and *congruence* with other homologies (see tree building section below). These criteria apply even without the assumption of genealogical relationships between taxa, when considering hierarchical patterns of characters (see above). In the context of a particular phylogenetic tree, however, **homologous** structures are similarities between taxa that are due to shared ancestry, although they may not now have the same function (for example limbs are homologous across all terrestrial vertebrates, but have been modified as swimming structures in whales, wings in bats, etc.).

Analogous structures are similarities between taxa that are not due to shared ancestry (again with reference to a particular phylogenetic tree). They are usually recognized when they have the same function, although this is not always so (for example the wings of bats, insects and birds are all analogous; they look grossly similar but are not derived from a common winged ancestor, using generally accepted ideas about relationships).

Some workers (e.g., Saether, 1979) recognize the existence of parallelism (= homology, Hennig, 1966), distinct from convergence (= analogy, *sensu* Scotland, 1992), and stress that the sharing of parallel states between 'closely' related organisms (involving state changes in homologous structures) is qualitatively different from the sharing of convergent states by 'distantly' related organisms (state changes in non-homologous structures). However, Farris (1986) follows Hennig (1966) in treating convergence and parallelism as functionally identical (i.e. as analogy) – indeed, there appears to be no *a priori* way of distinguishing between them in a phylogenetic analysis.

In all cases decisions about homology and analogy are *a posteriori* decisions based on a particular obtained tree topology. It is important to realize that changes in tree topology may thus turn homologies recognized in one analysis into analogies in another, and vice-versa.

Homologies alone may not tell a phylogeneticist much about relationships. Two types of homology can be recognized: first, **synapomorphy** (= shared **apomorphy**, a **derived** homologous character state; sometimes known as 'special similarity'), where homologous structures are found in a limited subgroup of the study taxa and thus may indicate a phylogenetic relationship; and second, **symplesiomorphy** (= shared **plesiomorphy**, a primitive character state), where homologies are either found in all the taxa under scrutiny, or where they are found not only in certain of the study taxa but also outside the study taxa. In the second case, although the homologies suggest relationships at more inclusive phylogenetic levels, they tell us nothing about relationships at the level of the study taxa (that is they are irrelevant to the particular question at hand; Patterson, 1980).

Groups based on synapomorphies are **monophyletic** and are known as **clades** (or lineages). Groups based on **symplesiomorphies** are **paraphyletic** and are known as **grades**, while groups defined by **analogy** (see above) are **polyphyletic**. Polyphyletic groups do not share a common ancestor (within a particular study group – clearly most, if not all, groups have a common ancestor in some time frame); paraphyletic groups are based on characters that define a larger, more inclusive, group.

Some examples may make these points more clearly. Figure 4a shows a simplified phylogenetic tree for extant members of the clade Amniota (all terrestrial vertebrates; this tree, intended for discussion purposes only, does not represent a complete summation of present systematic thinking). According to this tree, mammals are monophyletic, and there are clear synapomorphies upon which this clade is based (e.g. possession of fur, mammary glands etc.). Other monophyletic groups recognized using this tree include birds + crocodiles and birds + crocodiles + lepidosaurians, and, of course, the amniotes as a whole.

In contrast, according to this tree the group 'reptiles' is not monophyletic, lacking uniquely shared characters (they are only linked by symplesiomorphies of amniotes as a whole). Thus Romer & Parsons (1977) describe reptiles as "amniotes, but without advanced avian or mammalian characters (i.e. no feathers or hair)". The reptiles form a classic example of a paraphyletic group – a group defined only by the lack of one or more apomorphies (Patterson, 1980).

Although birds and mammals have been grouped together (as the, so called, Haematothermia – see also below) such a grouping is not supported by Fig. 4a. According to a widely held view, the main character that birds + mammals share

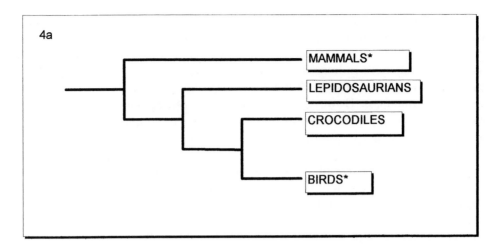

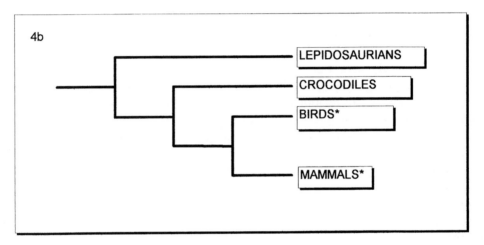

Figure 4 Two simplified phylogenetic tree diagrams for the terrestrial vertebrates (amniotes). A number of groups are excluded for clarity. * – the postulated group 'Haemothermia'. Diagram based on trees in Gauthier, Kluge & Rowe (1988).

uniquely (endothermy) appears to have arisen convergently in the two groups (and there is good evidence that certain extinct amniotes may have been endothermic). If we accept this as a group based on an analogous character, then the group birds + mammals is polyphyletic.

We must stress that Fig. 4a represent one possible summation of the available systematic data on the amniotes. Recent molecular analyses (reviewed in Patterson, Williams & Humphries, in press), have supplied evidence supporting the monophyly of mammals + birds (Fig. 4b). If this evidence is accepted, then endothermy is correctly to be interpreted not as a convergence between two clades but as a synapomorphy of a single clade (Gardiner, in press). No tree is infallibly true, and our conclusions about the status of groups and characters often depend upon which of a number of possible trees we choose to accept (see tree building section).

Two separate clades that share a node uniquely (**sister-groups**, see Figs 1 and 2) are of the same age (as they must share the same most-recent ancestor). This has major consequences for evolutionary work dealing with comparisons between taxa, where only sister-group comparisons are valid (see, for example, Mitter, Farrel & Wiegemann, 1988). Phylogenetics does not tell us anything about **ancestor– descendent** sequences, as all we can infer from a phylogenetic tree is that sister-groups have a shared common ancestor (which, itself, cannot be recognized as a 'real' species, as real species must be at the tips of trees). Patterns of homology cannot tell us if one taxon (whether extinct or extant) is the ancestor or descendant of another. Some authors even doubt the validity of assuming that the nodes of phylogenetic trees represent hypothetical ancestral species (Nelson, in press).

Synapomorphies that define a particular monophyletic group are known as **autapomorphies** (Hennig, 1966). In the context of phylogenetic analysis, autapo-morphies usually refer to unique synapomorphies of the study taxa (that is the tips of the tree, be they species, genera, families or whatever).

From an early stage, molecular systematists recognized the difficulties of identifying homology in data from molecules (particularly sequence data). Measures such as overall similarity (i.e. distance methods) suffer from the same problems as phenetic approaches to morphological systematics. Indeed, the general tendency for multiple copies of genes to exist within organisms led to a number of additional problems related to the difficulty of distinguishing truly homologous DNA sequences (**ortho- logous** sequences, reflecting the phylogeny of species) from apparently homolo-gous sequences, due to gene duplication (**paralogous** sequences, reflecting the phylogeny of genes – these may occur within the same organism). There is no room here to go into the details of these problems (see Patterson 1982b, 1988; Moritiz & Hillis 1990; Nelson, in press), but it is worth noting that molecular trees are prone to error in essentially the same ways as morphological trees (e.g. Smith, 1989).

Homology is used here as a synonym for synapomorphy (de Pinna's (1991) secondary or phylogenetic homology). A number of authors (see Roth, 1988) have suggested that we need to distinguish between the biological basis of homology, and the criteria by which we may recognize homologies. Our use of homology here thus relates to the latter, operational sense of the term – to mean a real pattern in nature that we are attempting to uncover. The search for potential homologies within the study taxa (typological homologies; that is, typological similarities between taxa that fulfil the first two of Patterson's (1982b) criteria) is an initial step prior to tree construction (see below).

POLARITY

A potential homology is not always the same as a potential apomorphy. As mentioned above, some homologies may be symplesiomorphic (shared primitive states) at the level at which phylogenetic relationships are of interest. In order to assess whether a character is plesiomorphic or apomorphic, it is necessary to estimate the direction of evolutionary change (i.e. the **polarity** of the character). Several ways have been proposed to achieve this (reviewed in Stevens, 1980), but all but two are generally considered inadequate. The two accepted methods are: a direct method (**ontogeny**), and an indirect method (**outgroup comparison**).

Nelson (1978) was the first to suggest, in conformation to von Baer's law, that within ontogenetic character state transformations, more general states are plesiomorphic and less general states are apomorphic. Although there are some criticisms of this approach (for example, Brooks & Wiley, 1985; Kluge, 1985; Mabee, 1989), it is a direct method of judging polarity from inspection of just the study taxa themselves (Weston, 1988). For some systematists it also represents a method that is without evolutionary implications, as it does not rely on phylogenetic assumptions about related taxa (Nelson, 1985), unlike the outgroup method. Critics have identified its main weaknesses as the paucity of information on the developmental stages of most organisms, and the difficulty of dealing with evolutionary developmental changes that are not simply terminal additions (Brooks & Wiley, 1985).

Outgroup comparison involves using taxa from outside the study group (the **ingroup**) to assess evolutionary direction. Generally it is considered safest to have at least two taxa in the outgroup, and it is helpful if the relationships within the outgroup are known (see Brooks & McLennan, 1991). Characters that occur unambiguously in the outgroup are plesiomorphic, those that occur only in the ingroup are apomorphic (Fig. 5). Where characters occur in some outgroup taxa but not all, simple algorithms exist for calculating character polarity (Maddison, Donoghue & Maddison, 1984). This method is by far the commonest used to assess polarities in cladistic studies.

One inadequate method for assessing polarity deserves further discussion in the context of comparative biology. The **commonality** criterion proposes that the commonest character state of any character in any study group will be plesio-

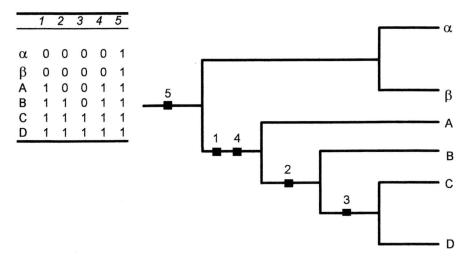

Figure 5 Diagram showing the basic principle of polarity determination by outgroup comparison. A–D are ingroup taxa, α and β are outgroup taxa, 1–5 are characters, and the matrix shows the distribution of the characters among the taxa. The outgroup taxa together form the sister-group of the ingroups (assumed). Characters 1 and 4 are synapomorphies of the whole ingroup clade, as they are present in all ingroup taxa but neither of the outgroup taxa. Character 2 is a synapomorphy of taxa B, C and D, as it only occurs in those taxa. By the same logic, character 3 is a synapomorphy of taxa C and D. Character 5 is invariant across ingroup and outgroup, and so is a symplesiomorphy of the ingroup (it is uninformative about relationships within the ingroup).

morphic, and the rarest apomorphic. This would make egg-laying in mammals, and flightlessness in apteryygote insects apomorphic. Trees that have assessed polarity by this criterion should be treated with caution (Watrous & Wheeler, 1981).

One other criterion is used in molecular studies, **midpoint rooting**. When the outgroup is not known the root is placed on the middle of the longest branch of the unrooted tree. This assumes that the two most divergent taxa in the analysis have the same rate of evolutionary change, and therefore is open to the same criticisms as with other procedures that assume a molecular clock (Farris, 1981; see below)

There is no room here to discuss the ordering (polarity) of characters with more than two character states (**multistate characters** that are often split into a **transformation series**), but it is a considerable problem. Interested readers should consult Mikevich (1982), Mikevich & Weller (1990), Wilkinson (1992), Hauser (1992) and Lipscomb (1992). Strictly speaking, character states should not be ordered, as this represents an *a priori* constraint on tree topology. The use of characters, rather than character states, removes this problem, as each character is treated independently (as either present or absent) and thus there is no *a priori* ordering of characters states.

TREE CONSTRUCTION

A cladistic analysis starts with a set of potentially homologous characters (see above). The chosen tree construction algorithm then attempts to filter out those proposed homologies that are either completely false (analogous), or homologies at too high a level (plesiomorphous), leaving true (i.e. informative) homologies (within the study taxa). Although this sounds straightforward, in practice there are many ways to attempt this and no general agreement on the best way.

Tree construction is problematical because data sets rarely consist entirely of postulated homologies that all agree with each other (i.e. that are congruent). An example is given in Fig. 6. Usually there are a number of contradictory characters because some of the proposed homologies will, in fact, be analogies. These add to the uncertainty (incongruity) in the data set and this uncertainty due to analogy is known as **homoplasy**. Levels of overall homoplasy can be calculated for a data set (see tree statistics section), but it is never easy to tell which characters are actually analogous and which are homologous. Each character is being compared with all the others that go into making up the tree, and so there are no objective criteria by which to judge each individual character separately.

Some cladists get around this problem by constructing trees by hand, making subjective decisions about the resolution of homoplasy. Although this may produce trees that summarize the data well, there is often no way of testing whether this is true or repeating exactly the decision processes made to produce a particular tree. This approach is essentially the one advocated by Hennig (1966), who suggested that incongruence in the data set should be dealt with by reassessing character states so as to split analogy into a number of separate homologies and continue by coding each new homology as a separate character.

The usual approach of phylogeneticists using morphological characters (and other broadly qualitative, non-continuous characters) is to employ **global parsimony** to resolve homoplasy. They use algorithms (usually using computer programs) that

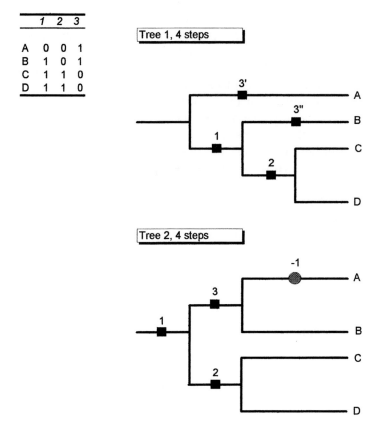

	1	2	3
A	0	0	1
B	1	0	1
C	1	1	0
D	1	1	0

Figure 6 Simple data matrix that generates two equally parsimonious (see text) trees. Character 1 (grouping taxa B, C and D) conflicts with character 3 (grouping A and B). The two trees that result require the same number of postulated independent character changes ("steps") (4). Extra characters would be needed to resolve this tree (that is to choose a single tree from these two trees).

search for the tree(s) that minimize(s) the number of characters that have to be assumed to be analogous (i.e. really two or more characters) rather than homologous (i.e. remaining as single characters) (example, Fig. 7). In character state terms, this minimizes the number of state changes when building a tree. If there is a lot of homoplasy in the data set, this approach often produces a large number of equally parsimonious trees (e.g. Naylor, 1992).

There is a sharp and continuing debate about the applicability of global parsimony to tree construction. Much of this has focused on situations where parsimony will give the wrong result, given assumed rates of change within different parts of trees (Felsenstein, 1978, 1983a; Hendy & Penny, 1989, Friday, Chapter 9). These studies suggest that when there are very unequal rates of change along different branches of a tree, global parsimony will give an incorrect result. How realistic these conditions are remains unknown.

The common occurrence of highly homoplasious data sets has led many systematists to seek ways of **weighting** characters so that the most homoplasious characters have a smaller impact on the overall tree structure. This weighting can be made before the analysis (*a priori*) or after the analysis (*a posteriori*). Weighting made *a priori* cannot be based on any properties of the combined data set, and is

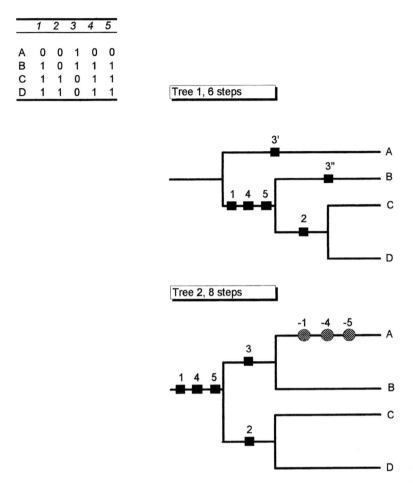

	1	2	3	4	5
A	0	0	1	0	0
B	1	0	1	1	1
C	1	1	0	1	1
D	1	1	0	1	1

Figure 7   A simple data matrix showing the resolution of trees by the method of maximal global parsimony. Tree 1 requires 6 independently postulated character changes, while tree 2 requires 8. Tree 1 is therefore the most parsimonious. This figure can be considered as Fig. 6 with additional characters added, leading to a resolution of the data into one most parsimonious tree. In this case characters 4 and 5 corroborate the grouping [B, C and D], rather than [A and B].

therefore subjective. Weighting made *a posteriori* can take into account information about character congruence within the whole data set, and is plausibly more objective.

The most commonly used *a posteriori* method is **successive character weighting** (see Carpenter, 1988). An initial analysis is performed from which consistency indices (ci, or rescaled consistency indices – both of these are measures of homoplasy amongst characters, see below) are calculated. These indices are then used to weight characters according to their congruence within the data set. The set is then re-analysed with the calculated weights and the analysis is iterated until the weights do not change. This seems to be similar to clique analysis, as characters may end up with zero weights, and so become uninformative (Siebert, 1992; see below).

The most recent development in parsimonious tree building has been the **three-**

taxon statement method of Nelson & Platnick (1991). Essentially the method consists of splitting each character up into a series of statements about all combinations of taxa taken three at a time, and these are then used in a parsimony analysis. The method is still controversial (see Harvey, 1992; Nelson, 1992, 1993; Kluge, 1993, Platnick, 1993) but may begin to be used more widely over the next few years.

Parsimony is only one criterion that has been used to construct trees from homoplasious data. A number of statistical procedures exist to construct trees using maximum likelihood techniques. These differ from parsimony techniques in that they require an *a priori* probabilistic model of character evolution which is used to test the fit of characters to a tree under that particular model (Bishop & Friday, 1985; Williams, 1992; Friday, Chapter 9). How far such models (and indeed parsimony) are true representations of evolutionary processes is hotly debated (for a summary of some of the current issues in this area, see Friday, Chapter 9).

Clique analysis (or compatibility) is an alternative, but generally less often used approach to the problem of multiple trees in morphological analyses. The largest number of characters are sought that form a **clique** of compatible, non-homoplasious characters (see Meacham & Estabrook, 1985). Various algorithms exist to do this (e.g. Gauld & Underwood, 1986; Sharkey, 1989). Arguments against this approach have been strong and influential, mostly centred around the amount of information that is lost when characters are excluded from analyses because they do not form part of a clique (see, for example, Farris, 1979).

MOLECULAR CHARACTERS – SOME SPECIAL PROBLEMS

Although there are many methods of tree construction relevant to this topic, ones involving the **alignment** of DNA base sequences are becoming most widely used. Sequences from different taxa are aligned to minimize mismatch and identical positions form the basis for the proposal of character homology. Sequence data can thus be analysed using unweighted parsimony techniques (or can be converted into distance matrix data, see below). Each base is used as a character state, polarity is decided by outgroup comparison and a tree that minimizes the number of base pair changes is sought (Felsenstein, 1988a,b). However, it is now well established that not all character state transformations between bases are equally likely (e.g. Wheeler & Honeycutt, 1988), and so some weighting of transformations is often considered necessary (Wheeler, 1990), even if there is disagreement about how exactly this is best achieved (Albert & Mishler, 1992; Rodrigo, 1992; Wheeler, 1992).

Distance matrix data (e.g., from DNA hybridization or allozyme data) is usually treated in a rather different way (although it can be analysed using parsimony: Penny, Hendy & Steel, 1992). Such data can be used to measure overall similarity by estimating the total number of base positions that differ between taxa (Williams, 1992). These distances can then be used to generate trees directly, but only by assuming additivity (that the sum of individual similarity measures is equal to the total branch length of the entire tree: Swofford & Olsen, 1990). These procedures lead to trees that may have unequal branch lengths (i.e. are metric). The assumption of additivity is strictly only supported when there is no homoplasy in the data set (Fitch, 1981, 1984; Farris, 1985). However, there is increasing evidence of molecular processes that may lead to homoplasy. The differing thermodynamic stability of DNA G:C and A:T bonds and the effect that this may have on molecular phylogenetic reconstructions of homothermic animals is one good example (see Friday, Chapter 9).

Other analyses assume both additivity and the existence of an exact molecular clock (basically analyses assuming neutral theories of molecular evolution: Kimura, 1983). Trees derived using such methods are known as ultrametric (see building blocks section above): sister-taxa are constrained to have identical branch lengths and all taxa are the same distance from the root. Probably the most important phylogenetic reconstruction so far attempted using this method is that employing DNA hybridisation for birds (Sibley & Ahlquist, 1990). This analysis has been strongly criticized on theoretical and methodological grounds (see, for example, Mindel, 1992). Most importantly, perhaps, there is a growing body of evidence contradicting the whole idea of the molecular clock, as differing taxonomic groups and traits appear to have different rates of molecular evolution (Britten, 1986; Li & Tanimura, 1987).

There are severe computation problems associated with tree construction using any underlying criterion (Penny, Hendy & Steel, 1992). One major difficulty is that most exact tree construction algorithms belong to a class of mathematical problems known as NP-complete, meaning that although exact algorithms exist to solve them, the solutions are not computable in a short enough time to make them practically usable (Felsenstein, 1982; Day, 1983; Day & Sankoff, 1988). In practice this means that most tree construction algorithms represent computational compromises giving locally rather than globally optimal solutions (see Penny, Hendy & Steel, 1992; Friday, Chapter 9).

Results from molecular and morphological analyses do not habitually coincide (but see Brooks & McLennan, Chapter 1). An excellent review is given by Patterson, Williams & Humphries (in press; see also Patterson, 1987). However, combined molecular and morphological data sets can produce extremely well corroborated phylogenetic trees (Wheeler, Cartwright & Hayashi, 1993), emphasizing the importance of using all available evidence in phylogenetic analyses (Kluge, 1989).

Another, less fashionable area where molecular data can be assessed, compared or combined with morphological (or other types of data), involves use of the great variety of chemical compounds that occur within organisms. Morris & Cobabe (1991) review this field, and suggest that problems over homology may be less severe with this type of data (although there is no doubt that identical small molecules often occur independently in quite unrelated organisms). In practice, systems such as insect cuticular volatiles can yield large amounts of data suitable for cladistic analysis (e.g. Schulz, Boppre & Vane-Wright, 1993).

HOW MUCH FAITH CAN WE HAVE IN A TREE OR CLASSIFICATION?

Systematic data is, by its very nature, presented in a reassuringly orderly way. Both trees and classifications are esoteric constructions, in the sense that it is not usually possible to tell by inspecting them how good they are (however bad the data that they were obtained from). In this section we discuss the problem of estimating how good a representation of true evolutionary relationships is a tree or classification. In addition, we discuss some hidden difficulties with trees that may influence both their topologies and how comparative biologists use them.

Tree statistics

It is all too easy to use a tree without realizing that it is only *one* evolutionary chronicle (hypothesis), not the *true* evolutionary chronicle (phylogeny) (O'Hara, 1988). There are some pointers, though, to the strength of a chronicle implied by a phylogenetic tree, and we discuss them here.

The commonest indices seen quoted in morphological systematics are those recording the fit of characters to a tree. They include the consistency index (ci), the rescaled consistency index (rc) and the retention index (ri) (Farris, 1989). The ci is used as a measure of homoplasy (within a data set and referring to a particular tree), while the rc is simply the ci rescaled from 0 (all homoplasy) to 1 (pure synapomorphy). The ri is a measure of the "fraction of the apparent synapomorphy in the character that is retained as synapomorphy on the tree" (Farris, 1989), and also runs from 0 (worst) to 1 (best). They are often used together, especially in trees derived by the Hennig86 program. The homoplasy excess ratio (HER) is similar to the ri (Archie, 1989). All such tree statistics can either refer to individual characters, or be calculated across all characters on a tree. Thus trees in the literature will often have overall ri and ci (or rc) scores.

Other tree statistics calculate statistical confidence limits for phylogenetic trees using randomization null models. These are generally found less commonly, for published phylogenetic trees, than ci and ri, so are not discussed in detail here. Such approaches are mathematically extremely difficult (Siebert, 1992), but authors have suggested using: data decisiveness (Goloboff, 1991); skewness (Huelsenbeck, 1991); bootstrapping (Felsenstein, 1985; Sanderson, 1989); permutational tail probability (PTP) (Faith & Cranston, 1990; Faith, 1991, 1992; used by and discussed by Crisp, Chapter 13).

Some phylogeneticists object on theoretical grounds to the general use of randomization tree statistics; see, for example, Carpenter's (1992; see also Wenzel & Carpenter, Chapter 4) critique of the PTP. Trueman (1993) argues, however, that the only meaningful randomization approach is the PTP, as the others have no bearing upon the question as to whether a "given most parsimonious tree can be considered a good current estimate of phylogenetic relationships among a set of terminal taxa". It seems likely that the PTP will be quoted more extensively in the phylogenetics literature in the future.

Missing data

Perhaps the most cryptic problem with interpreting a given tree is missing data (that is the data not used in the original data set but which, if included, could change the tree's topology). No phylogenetic analysis can use all characters (there are a near infinite number of ways to code the characters of a group of organisms) or taxa (some taxa are extinct and unknown). These two classes of missing data make the use of trees in comparative biology problematical.

In the case of missing character data, the problem often lies with computer algorithms. In parsimony analyses missing characters are always assumed to be present in their most parsimonious distribution (Kitching, 1992). Clearly in the case where there is much missing data the resulting trees may be strongly biased towards stable, low homoplasy, phylogeny reconstructions.

The effects of extinct taxa on the structure of phylogenetic trees is an important area for comparative biology and the comparative method. Studies are beginning to show that phylogenetic trees obtained from combined fossil and extant taxa can have different topologies from those derived from extant taxa alone (Gauthier, Kluge & Rowe, 1988, for amniotes; Doyle & Donoghue, 1992, for seed plants). In both cases apparent synapomorphies linking extant groups become analogous when fossil groups are added (i.e. [birds + crocodiles] + mammals amongst amniotes; conifers + anthophytes amongst seed plants). As nearly all phylogenetic trees are derived from extant taxa, this may be a serious problem for the interpretation of phylogenetic trees.

Consensus trees

Consensus trees are often quoted in phylogenetic analyses, especially when a highly homoplasious data set is employed. A consensus tree is one that summarizes topologies from a number of different trees as a single tree. Several methods exist to construct consensus trees, including: strict consensus, combinable consensus, Nelson consensus and Adams consensus (see Siebert, 1992). The exact details of these methods are not relevant here, but it is important for comparative biologists to realize that a consensus tree is always a less resolved tree than the trees from which it was derived (Miyamoto, 1985; Carpenter, 1988). They also often give topologies that are not represented in any of the original trees, and summarize less of the raw data. For most purposes any one of the initial trees is preferable to the consensus tree. In our opinion consensus trees should be avoided in comparative studies.

Trees and classifications

In the absence of phylogenetic trees, comparative biologists often fall back on traditional classifications to provide hierarchical patterns for their studies (see discussion in Gittleman & Luh, Chapter 5). Traditional classifications have a number of pitfalls for comparative biologists really concerned with the effects of phylogeny. The first of these is that such classifications are usually non-phylogenetic. Most traditional classifications accept paraphyletic groups ('grade' groups; e.g. the classification in Romer & Parsons, 1977, discussed above). Obviously, inspection of a non-phylogenetic classification will give little information about the possible phylogenetic status of the included taxa. There has been little work dealing with the inclusion of non-monophyletic groups within comparative studies, but where evidence exists in a wider context (e.g., Patterson & Smith, 1989), the consequence appear to be serious errors.

Coddington (Chapter 3) and Wenzel & Carpenter (Chapter 4) touch on difficulties of predicting how close a taxonomic classification will be to a phylogenetic reconstruction. For our part we see a major problem in the acceptance of paraphyletic groups in classifications. Such groups are always more inclusive than monophyletic ones in unpredictable ways (i.e. are artifactually more, or less inclusive). Thus if they are treated as evolutionary units for the purposes of comparative studies a large number of paraphyletic groups will artificially reduce or inflate the number of degrees of freedom, thus leading to possible Type I or Type II errors.

CONCLUSIONS

Phylogenetic trees are generally not as robust as they appear. This is a fact clearly appreciated by many comparative biologists (see Mooers, Nee & Harvey, Chapter 11), but perhaps not by all. It is all too easy for trees to be interpreted as phylogenies. However, we can only work with the trees that are available to us and hope that with time there will be more and better resolved trees, enabling comparative biology to proceed on a sounder footing. Until then it is important that phylogeneticists and other evolutionary biologists should not overstate the comprehensiveness of their findings.

ACKNOWLEDGEMENTS

We thank Robert Belshaw, Paul Williams and Dan Faith who commented on various drafts of this manuscript.

REFERENCES

ALBERT, V.A. & MISHLER, B.D., 1992. On the rationale and utility of weighting nucleotide sequence data. *Cladistics, 8*: 73–83.

ARCHIE, J.W., 1989. Homoplasy excess ratios: new indices for measuring levels of homoplasy in phylogenetic systematics and a critique of the consistency index. *Systematic Zoology, 38*: 253–269.

BISHOP, M.J. & FRIDAY, A.E., 1985. Evolutionary trees from nucleic acid and protein sequences. *Proceedings of the Royal Society London, Series B, 226*: 271–302.

BRITTEN, R.J. 1986. Rates of sequence evolution differ between taxonomic groups. *Science, 231*: 1393–1398.

BROOKS, D.R. & McLENNAN, D.A., 1991. *Phylogeny, Ecology, and Behavior: a Research Program in Comparative Biology.* Chicago: University of Chicago Press.

BROOKS, D.R. & McLENNAN, D.A., 1994. Historical ecology as a research programme: scope, limitations and future. In P. Eggleton & R.I. Vane-Wright (Eds), *Phylogenetics and Ecology,* 1–27. London: Academic Press.

BROOKS, D.R. & WILEY, E.O., 1985. Theories and methods in different approaches to phylogenetic systematics. *Cladistics, 1*: 1–11.

CARPENTER, J.M. 1988. Choosing amongst multiple equally parsimonious cladograms. *Cladistics, 4*: 291–296.

CARPENTER, J.M., 1992. Random cladistics. *Cladistics, 8*: 147–153.

CHAPPILL, J.A., 1989. Quantitative characters in phylogenetic analysis. *Cladistics, 5*: 217–234.

CRISP, M., 1994. Evolution of bird-pollination in some Australian legumes (Fabaceae). In P. Eggleton & R.I. Vane-Wright (Eds), *Phylogenetics and Ecology,* 281–309. London: Academic Press.

DAY, W.H.E., 1983. Computationally difficult parsimony problems in phylogenetic systematics. *Journal of Theoretical Biology, 103*: 429–438.

DAY, W.H.E. & SANKOFF, D., 1988. Computational difficulties of inferring phylogenies by compatibility. *Systematic Zoology, 35*: 224–229.

DE PINNA, M. 1993. Concepts and tests of homology in the cladistic paradigm. *Cladistics, 7*: 367–394.

DE QUEIROZ, K. & DONOGHUE, M.J., 1990. Phylogenetic Systematics or Nelson's version of cladistics? *Cladistics, 6*: 61–75.

DOYLE, J.A & DONOGHUE, M.J., 1992. Fossil seed plant phylogeny and reanalyzed. *Brittonia, 44*: 89–106.

FAITH, D.P., 1991. Cladistic permutation for monophyly and non-monophyly. *Cladistics, 7*: 366–375

FAITH, D.P., 1992. Probability, parsimony and Popper. *Systematic Biology, 41*: 252–257.

FAITH, D.P. & CRANSTON, P.S., 1990. Could a cladogram this short have arisen by chance alone? *Cladistics, 6*: 1–28.

FARRIS, J.S., 1979. The information content of the phylogenetic system. *Systematic Zoology, 28*: 483–519.

FARRIS, J.S., 1981. Distance data in phylogenetic analysis. *Advances in Cladistics, 1*: 3–23.

FARRIS, J.S., 1985. Distance data revisited. *Cladistics, 1*: 67–85.

FARRIS, J.S., 1986. The boundaries of phylogenetics. *Cladistics, 2*: 14–27.

FARRIS, J.S., 1989. The retention index and the rescaled consistency index. *Cladistics, 5*: 417–419.

FELSENSTEIN, J., 1978. Cases in which parsimony or compatibility methods will be positively misleading. *Systematic Zoology, 28*: 49–62.

FELSENSTEIN, J., 1981. Evolutionary trees from DNA sequences: a maximum likelihood approach. *Journal of Molecular Evolution, 17*: 368–376.

FELSENSTEIN, J., 1982. Numerical methods for inferring evolutionary trees. *Quarterly Review of Biology, 57*: 381–404.

FELSENSTEIN, J., 1983a. Parsimony in systematics: biological and statistical issues. *Annual Review of Ecology and Systematics, 14*: 313–333.

FELSENSTEIN, J., 1983b. Statistical inferences of phylogenies. *Journal of the Royal Statistical Society, Series A, 146*: 246–272.

FELSENSTEIN, J., 1985. Confidence limits on phylogenies: an approach using the bootstrap. *Evolution, 39*: 783–791.

FELSENSTEIN, J., 1988a. Phylogenies from molecular sequences: inference and reliability. *Annual Review of Genetics, 22*: 521–565.

FELSENSTEIN, J., 1988b. Phylogenies and quantitative characters. *Annual Review of Ecology and Systematics, 19*: 445–471.

FITCH, W.M., 1981. A non-sequential method for constructing trees and hierarchical classifications. *Journal of Molecular Evolution, 18*: 30–37.

FITCH, W.M., 1984. Cladistics and other methods: problems, pitfalls and potentials. In T. Duncan & T.F. Stuessy (Eds), *Cladistics: Perspectives on the Reconstruction of Evolutionary History*, 221–252. New York: Columbia University Press.

FOREY, P.L., HUMPHRIES, C.J., KITCHING, I.J., SCOTLAND, R.W., SIEBERT, D.J. & WILLIAMS, D.M., 1992. *Cladistics: a Practical Course in Systematics*. Oxford: Oxford University Press.

FRIDAY, A.E. 1994. Adaptation and phylogenetic inference. In P. Eggleton & R.I. Vane-Wright (Eds), *Phylogenetics and Ecology*, 207–217. London: Academic Press.

GARDINER, B.G., in press. Haematothermia: warm-blooded vertebrates. *Cladistics.*

GAUTHIER, J., KLUGE, A.G. & ROWE, T., 1988. Amniote phylogeny and the importance of fossils. *Cladistics, 4*: 105–210.

GAULD, I.D. & UNDERWOOD, G., 1986. Some applications of the LeQuesne compatibility test. *Biological Journal of the Linnean Society 29*: 191–222.

GILBERT, F., ROTHERAY, G., EMERSON, P. & ZAFAR, R., 1994. The evolution of feeding strategies. In P. Eggleton & R.I. Vane-Wright (Eds), *Phylogenetics and Ecology*, 323–343. London: Academic Press.

GITTELMAN J.L. & LUH H.-K., 1994. Phylogeny, evolutionary models, and comparative methods: a simulation study. In P. Eggleton & R.I. Vane-Wright (Eds), *Phylogenetics and Ecology*, 103–122. London: Academic Press.

GOLOBOFF, P.A., 1991. Homoplasy and the choice among cladograms. *Cladistics, 7*: 215–232.

GOULD, S.J., 1985. A clock of evolution: we finally have a method for sorting out homologies from 'subtle as subtle be' analogies. *Natural History, 94*: 12–25.

GOULD, S.J., 1986. Evolution and the triumph of homology, or why history matters. *American Scientist, 74*: 60–69.

HARVEY, A.W., 1992. Three-taxon statements: more precisely, an abuse of parsimony? *Cladistics, 8*: 345–354.

HAUSER, D.L., 1992. Similarity, falsification and character state order – a reply to Wilkinson. *Cladistics, 8*: 339–344.

HENDY, M.D. & PENNY, D., 1989. A framework for the quantitative study of evolutionary trees. *Systematic Zoology, 38*: 297–309

HENNIG, W., 1950. *Grundzuge einer Theorie der Phylogenetischen Systematik.* Berlin: Deutscher Verlag.

HENNIG, W., 1966. *Phylogenetic Systematics.* Urbana: University of Illinois Press.

HUELSENBECK, J.P., 1991. Tree length distribution skewness: an indicator of phylogenetic information. *Systematic Zoology, 40*: 257–270.

KIMURA, M., 1983. *The Neutral Theory of Molecular Evolution.* Cambridge: Cambridge University Press.

KITCHING, I.J., 1992. Tree-building techniques. In P. Forey *et al., Cladistics: a Practical Course in Systematics*, 44–71. Oxford: Oxford University Press

KLUGE, A.G., 1985. Ontogeny and phylogenetic systematics. *Cladistics, 1*: 13–28.

KLUGE, A.G., 1989. A concern for evidence and a phylogenetic hypothesis for relationships among *Epicrates* (Boidae, Serpentes). *Systematic Zoology, 38*: 1–25.

KLUGE, A.G., 1993. Three-taxon transformation in phylogenetic inference: ambiguity and distortion as regards explanatory power. *Cladistics, 9*: 246–259.

LI, W.H. & TANIMURA, M., 1987. The molecular clock runs more slowly in man than apes and monkeys. *Nature, 326*: 93–96.

LIPSCOMB, D.L., 1992. Parsimony, homology and the analysis of multistate characters. *Cladistics, 8*: 45–65.

MABEE, P.M., 1989. An empirical rejection of the ontogenetic polarity criterion. *Cladistics, 5*: 409–416.

MADDISON, W.P., DONOGHUE, M.J. & MADDISON, D.R., 1984. Outgroup analysis and parsimony. *Systematic Zoology, 33*: 83–103.

MEACHAM, C.A. & ESTABROOK, G.F., 1985. Compatibility methods in systematics. *Annual Review of Ecology and Systematics, 16*: 431–446.

MIKEVICH, M.F., 1982. Transformation series analysis. *Systematic Zoology, 31*: 461–478.

MIKEVICH, M.F. & WELLER, S., 1990. Evolutionary character analysis: tracing character evolution on a cladogram. *Cladistics, 6*: 137–170.

MINDEL, D.P., 1992. DNA-DNA hybridization and avian phylogeny. *Systematic Biology, 41*: 126–134.

MITTER, C., FARREL, B. & WIEGEMANN, J., 1988. The phylogenetic study of adaptive zones: has phytophagy promoted insect diversification? *American Naturalist, 132*: 107–128.

MIYAMOTO, M.M., 1985. Consensus cladograms and general classifications. *Cladistics, 1*: 186–189.

MOOERS, A.Ø., NEE. S. & HARVEY, P.H., 1994. Biological and algorithmic correlates of phenetic tree pattern. In P. Eggleton & R.I. Vane-Wright (Eds), *Phylogenetics and Ecology*, 233–251. London: Academic Press.

MORITZ, C. & HILLIS, D.M., 1990. Molecular systematics: context and controversies. In D.M. Hillis & C. Moritz (Eds), *Molecular systematics*, 1–10. Sunderland, MA: Sinauer.

MORRIS, P. & COBABE, E., 1991. Cuvier meets Watson and Crick: the utility of molecules as classical homologies. *Biological Journal of the Linnean Society, 44*: 307–324.

NAYLOR, G.J.P., 1992. The phylogenetic relationships among requiem and hammerhead sharks: inferring phylogeny when thousands of equally most parsimonious trees result. *Cladistics, 8*: 295–318.

NELSON, G., 1978. Ontogeny, phylogeny, paleontology, and the biogenetic law. *Systematic Zoology, 27*: 324–345.

NELSON, G., 1985. Outgroups and ontogeny. *Cladistics, 1*: 29–46.

NELSON, G., 1992. Reply to Harvey. *Cladistics, 8*: 355–360.

NELSON, G., 1993. Reply to Kluge. *Cladistics, 9*: 261–265.

NELSON, G., in press. Homology and systematics. In B.K. Hall (Ed.) *Homology: the Hierarchical Basis of Comparative Biology*. San Diego: Academic Press.

NELSON, G. & PLATNICK, N.I., 1991. Three-taxon statements: a more precise use of parsimony? *Cladistics, 7*: 351–366.

O'HARA, R.J., 1988. Homage to Clio, or, toward an historical philosophy for evolutionary biology. *Systematic Zoology, 37*: 142–155.

PATTERSON, C., 1980. Cladistics. *The Biologist, 27*: 234–239.

PATTERSON, C., 1982a. Classes and cladists or individuals and evolution. *Systematic Zoology, 31*: 284–386.

PATTERSON, C., 1982b. Morphological characters and homology. In K.A. Joysey & A.E. Friday (Eds), *Problems in Phylogenetic Reconstruction*, 21–74. London: Academic Press.

PATTERSON, C. (Ed.), 1987. *Molecules and Morphology in Evolution: Conflict or Compromise?* Cambridge: Cambridge University Press.

PATTERSON, C., 1988. The impact of evolutionary theories on systematics. In D.L. Hawksworth (Ed.), *Prospects in Systematics*, 59–91. Oxford: Clarendon Press.

PATTERSON, C. & SMITH, A.B., 1989. Periodicity in extinction: the role of systematics. *Ecology, 70*: 802–811.

PATTERSON, C., WILLIAMS, D.M. & HUMPHRIES, C.J., in press. Congruence between molecular and morphological phylogenies. *Annual Review of Ecology and Systematics*.

PENNY, D., HENDY, M.D. & STEEL, M.A., 1992. Progress with methods for constructing evolutionary trees. *Trends in Ecology and Evolution, 7*: 73–79.

PIMENTEL, R.A. & RIGGINS, R., 1987. The nature of cladistic data. *Cladistics, 3*: 201–209.

PLATNICK, N.I., 1979. Philosophy and the transformation of cladism. *Systematic Zoology, 28*: 537–546.

PLATNICK, N.I., 1985. Philosophy and the transformation of cladistics revisited. *Cladistics, 1*: 87–94.

PLATNICK, N.I., 1993. Character optimization and weighting: differences between the standard and three-taxon approaches to phylogenetic inference. *Cladistics, 9*: 267–272.

RODRIGO, A.G., 1992. A modification to Wheeler's combinatorial weights calculations. *Cladistics, 8*: 165–170.

ROMER, A.S. & PARSONS, T.S., 1977. *The Vertebrate Body* (5th edition). Philadelphia: Saunders.

ROTH, V.L., 1988. The biological basis of homology. In C.J. Humphries (Ed.), *Ontogeny and Systematics*, 1–26. New York: Columbia University Press.

SAETHER, O., 1979. Underlying synapomorphies and anagenetic analysis. *Zoologica Scripta, 8*: 305–312.

SANDERSON, M.J., 1989. Confidence limits on phylogenies: the bootstrap revisited. *Cladistics, 5*: 113–129.

SCHULZ, S., BOPPRE, M. & VANE-WRIGHT, R.I., 1993. Specific mixtures of secretions from male scent organs of Kenyan milkweed butterflies (Danainae). *Philosophical Transactions of the Royal Society of London, Series B, 342*: 161–181.

SCOTLAND, R.W., 1992. Cladistic theory. In P. Forey *et al.*, *Cladistics: a Practical Course in Systematics*, 14–21. Oxford: Oxford University Press.

SHARKEY, M.J., 1989. A hypothesis-independent method of character weighting for cladistic analysis. *Cladistics, 5*: 63–86.

SIBLEY, C.G. & AHLQUIST, J.E., 1990. *Phylogeny and Classification of Birds: a Study in Molecular Evolution.* Connecticut: Yale University Press.

SIEBERT, D.J., 1992. Tree statistics; trees and 'confidence'; consensus trees; alternatives to parsimony; character weighting; character conflict and its resolution. In P. Forey *et al.*, *Cladistics: a Practical Course in Systematics*, 72–88. Oxford: Oxford University Press.

SMITH, A.B., 1989. RNA sequence data in phylogenetic reconstruction: testing the limits of its resolution. *Cladistics, 5*: 321–344.

SOBER, E., 1988. *Reconstructing the Past: Parsimony, Evolution, and Inference.* Cambridge, MA: MIT.

STEVENS, P.F., 1980. Evolutionary polarity of character states. *Annual Review of Ecology and Systematics, 11*: 333–358.

SWOFFORD, D.L. & OLSEN, G.J., 1990. Phylogeny reconstruction. In D.M. Hillis & C. Moritz (Eds), *Molecular Systematics*, 411–501. Sunderland, MA: Sinauer.

TRUEMAN, J.W.H., 1993. Randomization confounded: a response to Carpenter. *Cladistics, 9*: 101–109.

WATROUS, L.E. & WHEELER, Q.D., 1981. The outgroup comparison method of character analysis. *Systematic Zoology, 30*: 1–11.

WENZEL, J.W. & CARPENTER, J.M., 1994. Comparing methods: adaptive traits and tests of adaptation. In P. Eggleton & R.I. Vane-Wright (Eds), *Phylogenetics and Ecology*, 79–101. London: Academic Press.

WESTON, P.H., 1988. Indirect and direct methods in systematics. In C.J. Humphries (Ed.), *Ontogeny and Systematics*, 27–56. New York: Columbia University Press.

WHEELER, W.C., 1990. Combinatorial weights in phylogenetic analysis: a statistical parsimony procedure. *Cladistics, 6*: 269–275.

WHEELER, W.C., 1992. Quo vadis? *Cladistics, 8*: 85–86.

WHEELER, W.C. & HONEYCUTT, R.L., 1988. Paired sequence differences in ribosomal RNAs: evolutionary and phylogenetic implications. *Molecular Biology and Evolution, 5*: 90–96.

WHEELER, W.C., CARTWRIGHT, P. & HAYASHI, C.Y., 1993. Arthropod phylogeny: a combined approach. *Cladistics, 9*: 1–39.

WILEY, E.O., 1981. *Phylogenetics: the Theory and Practice of Phylogenetic Systematics.* New York: Wiley.

WILKINSON, M., 1992. Ordered versus unordered characters. *Cladistics, 8*: 375–385.

WILLIAMS, D.M., 1992. DNA analysis: theory; DNA analysis: methods. In P. Forey *et al.*, *Cladistics: a Practical Course in Systematics*, 89–101, 102–123. Oxford: Oxford University Press.

WILLIAMS, P.H., GASTON, K.J. & HUMPHRIES, C.J. In press. Do conservationists and molecular biologists value differences between organisms in the same way? *Biodiversity Letters.*

Index

Linnean Society Symposium Series

Number 1. **The Evolutionary Significance of the Exine**, edited by I.K. Ferguson and J. Miller, 1976, xii + 592pp., 0.12.253650.9.

Number 2. **Tropical Trees – Variation, Breeding and Conservation**, edited by J. Burley and B.T. Styles, 1976, xvi + 244pp., 0.12.145150.X

Number 3. **Morphology and Biology of Reptiles**, edited by A. d'A. Bellairs and C. Barry Cox, 1976, xvi + 290pp., 0.12.085850.9

Number 4. **Problems in Vertebrate Evolution**, edited by S. Mahala Andrews, R.S. Miles and A.D. Walker, 1977, xii + 412pp. 0.12.059950.3

Number 5. **Ecological Effects of Pesticides**, edited by F.H. Perring and K. Mellanby, 1977, xii + 194pp., 0.12.551350.X

Number 6. *Botanical Society of the British Isles Conference Report No. 16,* **The Pollination of Flowers by Insects**, edited by A.J. Richards, 1977, xii + 214pp., 0.12.587460.X

Number 7. **The Biology and Taxonomy of the Solanaceae**, edited by J.G. Hawkes, R.N. Lester and A.D. Skelding, 1979, xviii + 738pp. 0.12.333150.1

Number 8. **Petaloid Monocotyledons – Horticultural and Botanical Research**, edited by C.D. Brickell, D.F. Cutler and Mary Gregory, 1980, xii + 222pp., 0.12.133950.5

Number 9. **The Skin of Vertebrates**, edited by R.I.C. Spearman and P.A. Riley, 1980, xiv + 322pp., 0.12.656950.9

Number 10. **The Plant Cuticle**, edited by D.F. Cutler, K.L. Alvin and C.E. Price, 1982, x + 462pp., 0.12.199920.3

Number 11. **Ecology and Genetics of Host–Parasite Interactions**, edited by D. Rollinson and R.M. Anderson, 1985, xii + 266pp., 0.12.593690.7

Number 12. **Pollen and Spores: Form and Function**, edited by S. Blackmore and I.K. Ferguson, 1986, xvi + 443pp., 0.12.103460.7

Number 13. **Desertified Grasslands: Their Biology and Management**, edited by G.P. Chapman, 1992, xii + 360pp., 0.12.168570.5

Number 14. **Evolutionary Patterns and Processes**, edited by D.R. Lees and D. Edwards, 1993, xii + 326pp., 0.12.440895.8

Number 15. **The Biology of Lemmings**, edited by N.C. Stenseth and R.A. Ims, 1994, xvi + 683pp., 0.12.666020.4

Number 16. **Shape and Form in Plants and Fungi**, edited by D.S. Ingram and A. Hudson, 1994, x + 381pp., 0.12.371035.9

Number 17. **Phylogenetics and Ecology**, edited by P. Eggleton and R.I. Vane-Wright, 1994, xi + 378pp., 0.12.232990.2